The Nucleus of Cells: Emerging Ideas

The Nucleus of Cells: Emerging Ideas

Edited by:

Shivsanjeevi Sripathi

Delve
PUBLISHING

www.delvepublishing.com

The Nucleus of Cells: Emerging Ideas

Shivsanjeevi Sripathi

Delve Publishing
224 Shoreacres Road
Burlington, ON L7L 2H2
Canada
www.delvepublishing.com
Email: orders@arclereducation.com

ISBN: 978-1-77469-046-8 (Hardcover)

DECLARATION

ABOUT THE EDITOR

ShivSanjeevi Sripathi completed his Masters in Biotechnology from Mumbai University in 2008. He was awarded for academic excellence in both his Bachelors and Masters for securing second rank in Mumbai University in 2006 and first rank in his college: Kishinchand chellaram College. For his Masters he secured first rank in his college KET's V.G.Vaze College. He qualified CSIR and NET and and TOEFL in September 2008. He then worked on a stem cell project at the Specialized Centre for Cell Based Therapy (SCCT),KEM Hospital at Mumbai on a project entitled, : Isolation & detection of stem cells from Human Umbilical cord/ amniotic membrane" following which he worked at Junior Research Fellow at Microbiology & Cell Biology Department, Indian Institute of Science, Bangalore on cloning of cell wall genes and transcription factors in E.coli & M.smegmatis. As a writer, he has authored and co-authored 35 books on various aspects of biology such as bionics, molecular wires, cloning, hypertension, the epidemics of the 21st century, handling depression, camouflage, hygiene, immunology and many more with international publishers. He loves to read and share on interesting aspects of life sciences in books. In his free time he loves to travel and explore and give talks on spirituality, ancient customs and traditions.

TABLE OF CONTENTS

LIST OF CONTRIBUTORS

Shivsanjeevi Sripthi

Thomas Cavalier-Smith
Department of Zoology, University of Oxford, South Parks Road, Oxford, OX1 3PS, UK

Srujana Neelam
Utilization and Life Sciences, NASA John F. Kennedy Space Center, Kennedy Space Center, Merritt Island, FL 32899, USA
Department of Botany, University of Wisconsin-Madison, Madison, WI 53706, USA
Universities Space Research Association, Columbia, MD 21046, USA

Brian Richardson
Department of Population and Quantitative Health Sciences, Case Western Reserve University, Cleveland, OH 44106, USA

Richard Barker
Department of Botany, University of Wisconsin-Madison, Madison, WI 53706, USA

Ceasar Udave
Department of Biomedical Engineering, Arizona State University, Tempe, AZ 85281, USA

Simon Gilroy
Department of Botany, University of Wisconsin-Madison, Madison, WI 53706, USA

Mark J. Cameron
Department of Population and Quantitative Health Sciences, Case Western Reserve University, Cleveland, OH 44106, USA

Howard G. Levine
Utilization and Life Sciences, NASA John F. Kennedy Space Center, Kennedy Space Center, Merritt Island, FL 32899, USA

Ye Zhang
Utilization and Life Sciences, NASA John F. Kennedy Space Center, Kennedy Space Center, Merritt Island, FL 32899, USA

Sergio Sarnataro
Institut de Ge´ne´tique et de Biologie Mole´culaire et Cellulaire, 67404 Illkirch, France

Andrea M. Chiariello
Dipartimento di Fisica, Universita´ di Napoli Federico II, and INFN Napoli, CNR-SPIN, Complesso Universitario di Monte Sant'Angelo, 80126 Naples, Italy

Andrea Esposito
Dipartimento di Fisica, Universita´ di Napoli Federico II, and INFN Napoli, CNR-SPIN, Complesso Universitario di Monte Sant'Angelo, 80126 Naples, Italy
Max Delbru¨ck Center for Molecular Medicine in the Helmholtz Association (MDC), Berlin, Germany

Antonella Prisco
CNR-IGB, via Pietro Castellino 111, Naples, Italy

Mario Nicodemi
Dipartimento di Fisica, Universita´ di Napoli Federico II, and INFN Napoli, CNR-SPIN, Complesso Universitario di Monte Sant'Angelo, 80126 Naples, Italy

Raffaele Fabrini
Department of Chemical Sciences and Technologies, University of Rome Tor Vergata, Rome, Italy

Alessio Bocedi
Department of Internal Medicine, University of Rome Tor Vergata, Rome, Italy

Valentina Pallottini
Department of Biology, University of Roma Tre, Rome, Italy

Lorena Canuti
Department of Biology, University of Rome Tor Vergata, Rome, Italy

Michele De Canio
Department of Internal Medicine, University of Rome Tor Vergata, Rome, Italy

Andrea Urbani
Department of Internal Medicine, University of Rome Tor Vergata, Rome, Italy
S. Lucia Research Institute IRCCS, Rome, Italy

Valeria Marzano
Department of Internal Medicine, University of Rome Tor Vergata, Rome, Italy
S. Lucia Research Institute IRCCS, Rome, Italy

Tommaso Cornetta
Department of Biology, University of Roma Tre, Rome, Italy

Pasquale Stano
Department of Biology, University of Roma Tre, Rome, Italy

Anna Giovanetti
Institute of Radiation Protection, ENEA-CR Casaccia, Rome, Italy

Lorenzo Stella
Department of Chemical Sciences and Technologies, University of Rome Tor Vergata, Rome, Italy

Antonella Canini
Department of Biology, University of Rome Tor Vergata, Rome, Italy

Giorgio Federici
Department of Internal Medicine, University of Rome Tor Vergata, Rome, Italy

Giorgio Ricci
Department of Chemical Sciences and Technologies, University of Rome Tor Vergata, Rome, Italy

Jaroslav Albert
BioModeling, BioInformatics & BioProcesses, Université Libre de Bruxelles, Brussels, Belgium
Applied Physics Research Group, Vrije Universiteit Brussel, Brussels, Belgium

Marianne Rooman
BioModeling, BioInformatics & BioProcesses, Université Libre de Bruxelles, Brussels, Belgium

Jón Pol Gales
Institut de Biologie Moléculaire des Plantes, The French National Center for Scientific Research (CNRS) UPR2357, Université de Strasbourg, F-67084 Strasbourg, France

Julie Kubina
Institut de Biologie Moléculaire des Plantes, The French National Center for Scientific Research (CNRS) UPR2357, Université de Strasbourg, F-67084 Strasbourg, France
SVQV UMR-A 1131, INRAE, Université de Strasbourg, F-68000 Colmar, France

Angèle Geldreich
Institut de Biologie Moléculaire des Plantes, The French National Center for Scientific Research (CNRS) UPR2357, Université de Strasbourg, F-67084 Strasbourg, France

Maria Dimitrova
Institut de Biologie Moléculaire des Plantes, The French National Center for Scientific Research (CNRS) UPR2357, Université de Strasbourg, F-67084 Strasbourg, France

Florian Bo Otto
Department of Cellular Biochemistry, University Medical Center Göttingen, Humboldtallee 23, D-37073 Göttingen, Germany
Michael Thumm
Department of Cellular Biochemistry, University Medical Center Göttingen, Humboldtallee 23, D-37073 Göttingen, Germany

LIST OF ABBREVIATIONS

AIDS	Acquired Immunodeficiency Syndrome
CRM1	Chromosome region maintenance 1
CL	Clathrin vesicles
CPSF	Cleavage and polyadenylation specificity factor complex
CFFM	Coherent feed-forward motif
CTEs	Constitutive transport elements
CK	Creatine kinase
DR	Direct repeat
EAGs	Early activation genes
ER	Endoplasmic reticulum
ESCRT	Endosomal sorting complexes required for transport
EJC	Exon junction complex
ECM	Extracellular matrix
FCs	Fibrillar centers
FWHM	Full width half maximum
GSVA	Gene Set Variation Analysis
GANP	Germinal-centre-associated nuclear protein
HA	Haemagglutinin
HBSP	HBV splice-generated protein
HBV	Hepatitis B virus
HUVECs	Human Umbilical Vein Endothelial Cells
IGS	Intergenic spacer sequences
IPP	Isopentenyl diphosphate
JSRV	Jaagsiekte Sheep Retrovirus 1
LADs	Lamin-associated chromatin domains
LBR	Lamin B receptor
LMB	Leptomycin B
LRPPRC	Leucine-rich PPR motif-containing

LINC	Linker of Nucleoskeleton and Cytoskeleton
LDH	L-lactate dehydrogenase
MPMV	Mason–Pfizer Monkey virus
mRNPs	messenger ribonucleoprotein complexes
MIPA	Micropexophagic membrane apparatus
MTOCs	Microtubule Organization Centers
MSigDB	Molecular Signatures Database
MMTV	Mouse Mammary Tumor Virus
NPP	NASA Postdoctoral Program
NRE	Negative regulatory element
NEXT	Nuclear exosome targeting complex
ONM	Outer nuclear membrane
OSBPs	Oxysterol-binding proteins
PAR	Photosynthetically active radiations
PMN	Piecemeal nucleophagy
PTB	Polypyrimidine tract-binding
RHD	Reticulon homology domain
RdRp	RNA-dependent RNA polymerase
SRV1	Simian Retrovirus type 1
SMG	Simulated microgravity
STRE	Stress-response element
TPR	Translocator promoter region protein
UV	Ultraviolet
vRNPs	Viral ribonucleoproteins

PREFACE

The high school biology textbook describes the nucleus as the "master controller of the cell". This organelle of the cell is the basis of classifying organisms into prokaryotes and eukaryotes. It contains the genetic material and the nucleolus that is associated with ribosomes. The nucleus is emerging to be a "dynamic" organelle as opposed to the traditional concept of it being "stiff".

The following quote summarizes the current scenario regarding the nucleus as of the latest research.

"Currently, most studies on nuclear structure and function focus on a very small fraction of species, perhaps because of their perceived medical relevance or their proven suitability as model systems. Yet the diversity in nuclear shape, size, and even number throughout all living organisms is immense. What is certain, however, is that the nucleus is a dynamic structure, both in health and disease."-Research by Orna Cohen-Fix and Ulrike Kutay. The cell nucleus: Plastic, elastic and fantastic. *Curr Opin Cell Biol.* 2016 June; 40: iv–v. doi:10.1016/j.ceb.2016.05.007.

This makes this book come at an apt time to present the emerging ideas of the nucleus in one concise volume. While the organelle is large and contains the genetic material, the saying "handle with care", seems apt for the nucleus. While it is indeed true, research has uncovered that this organelle can undergo deformation in migrating cells and can handle the stress. It has also been suggested to sense shear, handle mechanical stress, and generate the propelling force to allow cells to migrate across narrow spaces.

The book first presents the various models of how the nucleus was first formed and its structure with respect to the nucleolus. The association of chromatin and nuclear shape is also presented along with the various nuclear shapes across different organisms. The next chapter discussed how the nucleus plays a role in aging and the functioning of nuclear proteins in various diseases. This is followed by interesting and latest research as to how the nucleus functions to detect shear in endothelial cells and how light dictates the growth of the nucleus in plant embryos.

To summarize, the idea that the nucleus is a "bag containing chromatin" is being replaced by the concept of a very dynamic organelle. Let us recollect what Charles Darwin said,

"A cell is a complex structure, with its investing membrane, nucleus, and nucleolus". The use of the latest technologies shall pave the way for further exciting research on the nucleus.

Editors

CHAPTER 1

THE STRUCTURE AND DYNAMICS OF THE NUCLEUS

Shivsanjeevi Sripathi

CONTENTS

1.1 THE PAST

Let us start with the following quotes that described the nucleus many years ago but still resonate with what current research is revealing.

"A cell is a complex structure, with its investing membrane, nucleus, and nucleolus"- Charles Darwin.

"A nutritive centre, anatomically considered, is merely a cell, the nucleus of which is the permanent source of successive broods of young cells, which from time to time fill the cavity of their parent, and carrying with them the cell wall of the parent, pass off in certain directions, and under various forms, according to the texture or organ of which their parent forms a part"

— John Goodsir, Anatomical and Pathological Observations (1845), 2.

The earliest record of how the nucleus was observed is not clear. Antony van Leeuwenhoek saw it in amphibian and avian erythrocytes while Felice Fontana observed it in skin cells of eels. Franz Bauer showed the nucleus in orchid sketches and the term "vesicula germanitiva" was given to it by Jan Purkyne. The term nucleus can be ascribed to Robert Brown (Pederson, 2011)

Given that eukaryotes are defined on the basis of the nucleus, there are several theories describing the origin of the nucleus. One is the *endosymbiotic theory* for the origin of the nucleus that was put forth by Mereschkowsky. This states that the origin of the nucleus is from a mycoplasma (prokaryote) that was engulfed by an ameboid cell (similar to the cytoplasm). This is shown in Figure 1.1 (a) below.

Another theory describes that invaginations of the plasma membrane of a prokaryotic cell account for the origin of the nuclear and ER membranes. This is shown in Figure 1.1 (b) below. It was hypothesized that the prokaryote could phagocytose food particles due to a loss of its cell wall. The ribosomes from the plasma membrane were internalized to form the rough ER and out of it the nuclear envelope.

Figure 1.1 below summarizes the various models accounting for the origin of the nucleus (Martin et al, 2015; *Published by the Royal Society under the terms of the Creative Commons Attribution License http://creativecommons.org/licenses/by/4.0/, which permits unrestricted use, provided the original author and source are credited*):

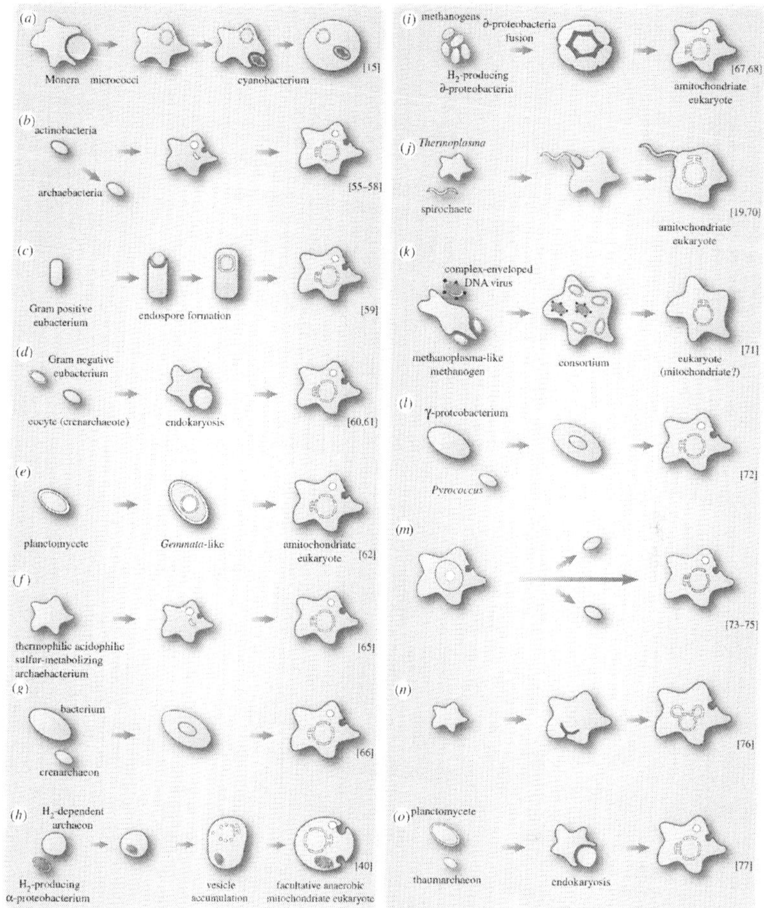

Figure 1.1

Another model describes that the nucleus can be traced back to the endospores of Gram-positive bacteria. Gould & Dring (1979) describe that the engulfment of a part of the cytoplasm of a cell by itself during endospore formation followed by the formation of a double membrane to form the nucleus. Figure 1.1 (c) describes this model.

The late nineties saw the development of endokaryotic theories where the nucleus arose from the engulfment of archaebacterial endosymbiont by a eubacterial host to later form the nucleus. Figure 1.1 (d) describes this model.

Figure 1.1 (e) shows a structure that was described as on the lines of the nucleus in *Gemmata obscuriglobus (a budding eubacterium* of the *Planctomyces-Pirella* group) by Fuerst & Webb.

The formation of eukaryotic cells was also described as having an origin from archaebacteria that metabolize sulfur with a flexible cell wall and a well-developed cytoskeleton. This model put forth by Searcy & Hixon shown in Figure 1.1 (f) accounts for eukaryotic cell development from archaebacteria.

Endosymbiosis where a crenarchaeon was engulfed by a bacterium has also been put forth by Lake & Rivera, shown in Figure 1.1 (g).

In another theory, a mitochondrial endosymbiont accounts for the origin of the nucleus as represented in Figure 1.1 (h). This opens up the formation of the endomembrane system and nucleus–cytosol compartmentation.

The origin of the nucleus according to Moreira & López-García is the combination of δ-proteobacteria that trapped a methanogenic archaebacterium via the fusion of plasma membranes (Figure 1.1 (i)). While such plasma membrane fusion has not yet been reported for cells-it has been seen in archaea.

In another model, the nucleus and the flagella of eukaryotes have been attributed to the symbiosis between a spirochaete and an archaebacterium lacking a cell wall (Figure 1.1 (j)).

There is a model that attributes the origin of the nucleus to viruses. To quote the author Bell, "the eukaryotic nucleus evolved from a complex DNA virus. It is proposed that the virus established a persistent presence in the cytoplasm of a methanogenic mycoplasma and evolved into the eukaryotic nucleus by acquiring a set of essential genes from the host genome and eventually usurping its role." (Bell, 2001). This has been shown in Figure 1.1 (k).

According to Horiike, a γ-proteobacterium engulfed an archaeal endosymbiont that developed as the nucleus (Figure 1.1(l)). The process of Forterre's thermoreduction hypothesis or the formation of prokaryotes from eukaryotes due to high temperature accounts for the formation of eukaryotes first as an ancestor. This is in Figure 1.1 (m).

Figure 1.1(n) shows that mutations in an archezoon (a phagocytosing amitochondriate eukaryote) can account for changes thus, ruling out the theory of symbiosis, as shown in Figure 1.1 (n).

According to Forterre, who put forth the PTV (for PVC–thaumarchaeon–virus) fusion hypothesis, the host (Planctomycetes, Verrucomicrobia, Chlamydiae-PVC) engulfed of a thaumarchaeon as the nucleus, that was followed by a retrovirus and nucleo-cytoplasmic large DNA viruses (NCLDV) invasions. This has been shown in Figure 1.1 (o).

(Martin et al, 2015)

It remains to be established as to how the nucleus developed from the primitive cells. The possible development of a ribozyme RNA replicase could have facilitated the development of this organelle based on free energy. As far as DNA is concerned, an endosymbiotic relationship has been proposed between a prokaryote and an "outsider" cell. While hypotheses are plenty, the exact mechanism is yet to be ascertained.

(Pederson, 2011)

1.2 WHAT WE KNOW OF THE STRUCTURE

The nucleus is known to be the largest organelle in the cell and is the control center of our cell. It is the site where the genetic material is present and also the location where the processes of replication and transcription occur. While the nucleus is considered to have rigidity ten-fold that of the cytoplasm, the idea that the nucleus is static is changing. The organelle is capable of undergoing changes in its shape and also propagates force for cell migration. An amalgamation of forces to and fro among the nucleus and the cytoskeleton allows for the determination of the morphology and location of the nucleus. The type and functions of the cell are the determining factors for the exact nature of the forces. (More is revealed in this aspect in Chapter 3).

The nuclear envelope encases the nucleus at the boundary to serve as a boundary between the nuclear contents and the cytoplasm. This nuclear membrane is actively involved in the biomechanics of the nucleus and is quite rigid to stabilize the structure. The nuclear lamina, transmembrane proteins, and a bilipid membrane constitute the nuclear envelope. Proteins belonging to the SUN and Nesprin families form the linker of nucleoskeleton and cytoskeleton (LINC) complex that span the nuclear double membrane. The complex allows for connections with the cytoskeleton (such as microtubules or actin or motor proteins), chromatin, nuclear lamins, and plectin. It facilitates the connection between the nucleoplasm and the cytoplasm.

original author(s) and the source, provide a link to the Creative Commons license, and indicate if changes were made):

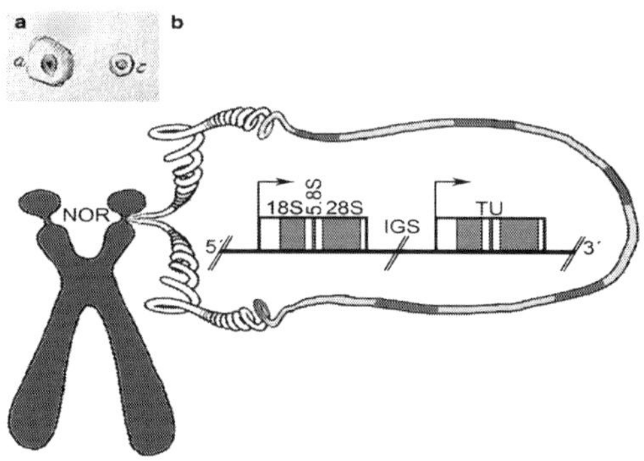

Figure 1.3

Subsequent research showed that specific chromosomal sites interact with nucleoli (nucleolus organizer region [NOR]) and the presence of ribosomal RNA and DNA in the nucleolus. While many aspects of the nucleolus are being unveiled, much remains to be elucidated. The use of molecular tools in conjunction with microscopy is allowing further research in detail.

The nucleolus is composed of ribosomal DNA (rDNA) that is present as tandem arrays in nucleolus organizer regions (NORs) in humans. These are on the p arms of acrocentric NOR chromosomes (10 in number). The rDNA arranged as repetitive element clusters are composed of transcription units or rRNA genes with intergenic spacer sequences (IGS) in between. The 47S nascent transcripts are synthesized from these genes. The 18S, 5.8S, and 28S rRNAs along with external transcribed spacer sequences (at the 5' and 3') and internal transcribed spacers (flank 5.8S rRNA) constitute the 47S nascent transcript. Maturation of these 3 rRNAs along with the 5S rRNA (imported from outside the nucleolus) make up the two ribosomal subunits. The IGS codes for RNA molecules that control the rDNA array organization and enhancers.

(Schöfer and Weipoltshammer, 2018)

Figure 1.4 below shows a scheme of nucleolar organizer regions (NOR) containing the rDNA array, TU transcription unit (rRNA gene), IGS intergenic spacer sequence; DNA loop: red: TU, gray: IGS (Schöfer and Weipoltshammer, 2018; *Open Access This article is distributed under the terms of the Creative Commons Attribution 4.0 International*

License (http://creativeco mmons.org/licenses/by/4.0/), which permits unrestricted use, distribution, and reproduction in any medium, provided you give appropriate credit to the original author(s) and the source, provide a link to the Creative Commons license, and indicate if changes were made):

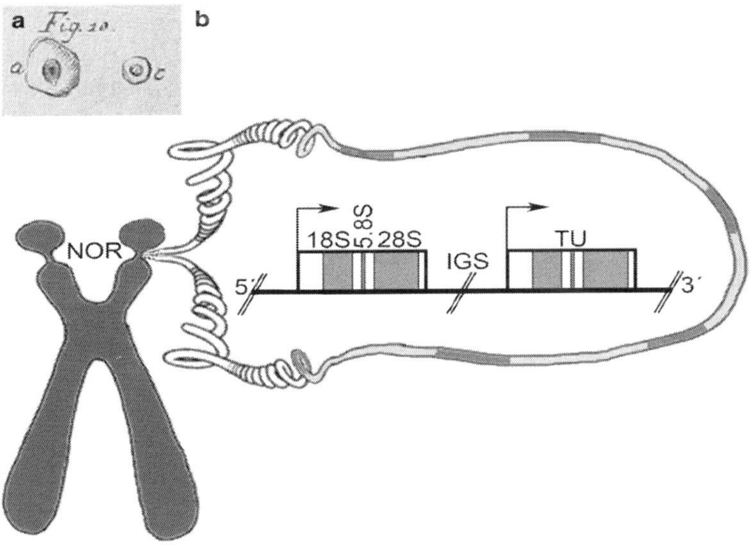

Figure 1.4

The ultrastructure of nucleoli is suggestive of a tri-partite organization. The overall shape is dictated by the granular component (GC) that contains 2 types of fibrillar structures: dense fibrillar component (DFC) that surrounds lighter fibrillar centers (FCs). The periphery of the nucleolus shows patches of heterochromatin and euchromatin of which few condensed chromatin strands form the intranucleolar chromatin by entering the nucleolus.

Interestingly, spermatogenesis in mice has shown the association of the sex vesicle with the nucleolus. Many interphase human cells have demonstrated the nucleoli and nuclear lamina to be associated with the sex chromosomes. Sequencing has suggested that the silencing of the inactive X-chromosome (Xi) in cats is due to its location close to the nucleolus.

Figure 1.5 below shows the a) a compact nucleolus in a rat hepatoma cell with copious amounts of perinucleolar heterochromatin b) rat fibroblast showing copious levels of intranucleolar chromatin c) The sex chromosomes and nucleoli: rDNA in red (FISH analysis) with the sex vesicle (X plus Y chromosomes); FC fibrillar center, DFC dense fibrillar component, GC granular component; (Schöfer and Weipoltshammer, 2018;

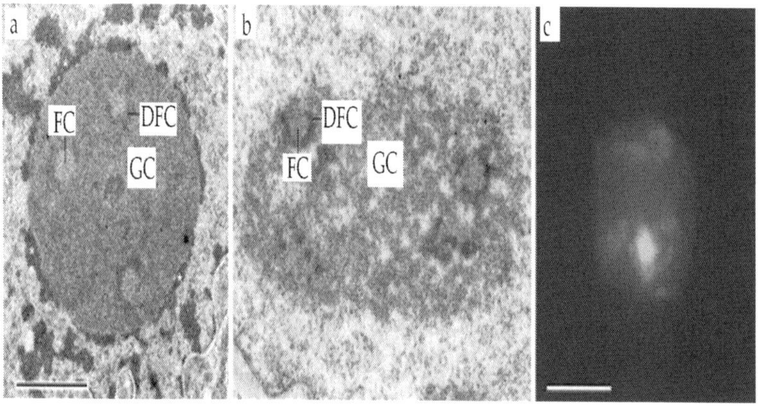

Figure 1.5

Based on the metabolic needs of the cell, the nucleolus monitors the biogenesis of ribosomes. Research in mutant yeast cells that had limited rDNA highlighted a higher response of protein synthesis with more numbers of PolI enzymes actively transcribing rRNA genes that were not observed in normal yeast cells. During the cell cycle and constant metabolic conditions, the gene number remains constant in mammals. Based on changes, these genes are actively recruited or silenced.

The location and the actions of rDNA genes are yet to be assessed in greater detail. From what has been found out, there are hypotheses that describe how the rDNA chromatin changes based on the transcriptional status of a cell.

In Figure 1.6 below, **(a)** represents when the demand for ribosomes is low in a cell; the nucleolar surface shows the presence of silent NORs (shown in black) with no transcription. (A) shows the protrusion of intranucleolar chromatin as silent rRNA genes (black) that form the majority plus rRNA genes that are actively transcribed (in red) and IGS (in blue).

(B) it remains to be ascertained as to how the poised genes (in orange) are placed: they are either with silent rDNA in NORs that are transcriptionally silent (C) the intranucleolar chromatin stretches (C, D).

(b) represents the scenario proposed when there is a demand for ribosomes; there is the formation of a fibrillar complex made of the NOR-rDNA. (A) shows an increase in PolI occupied by rRNA in yeast

(B) Poised genes are activated by activation of NORs near FC/DFC domains

(C) Reactivation of genes previously active or new FC/DFC complexes to activate poised genes within intranucleolar chromatin

(D) "Yellow—fibrillar center, blue—dense fibrillar component, pale brown—granular component; green dots— Pol I molecules, and red-orange fibers—nascent rRNA transcripts" (Schöfer and Weipoltshammer, 2018: *Open Access This article is distributed under the terms of the Creative Commons Attribution 4.0 International License (http://creativeco mmons. org/licenses/by/4.0/), which permits unrestricted use, distribution, and reproduction in any medium, provided you give appropriate credit to the original author(s) and the source, provide a link to the Creative Commons license, and indicate if changes were made*):

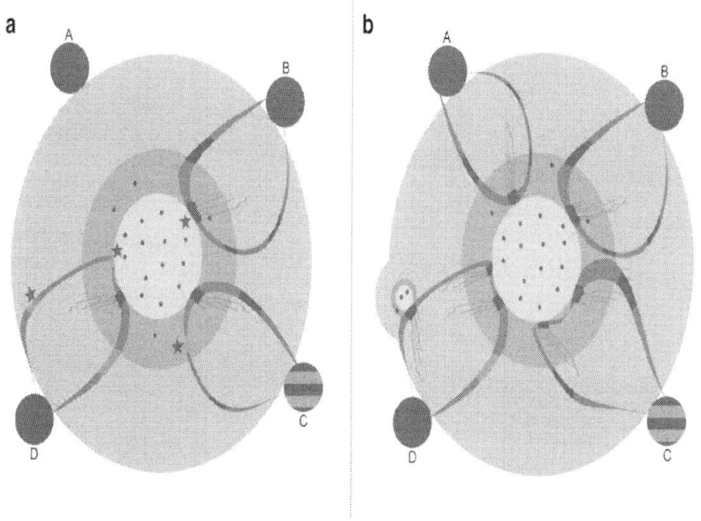

Figure 1.6

The rDNA array displays variability and epiallelic composition accounting for heterogeneity. This becomes relevant for oncology given the recent development of drugs targeting PolI and ribosomopathies or diseases that cause ribosome dysfunctions.

(Schöfer and Weipoltshammer, 2018)

1.4 THE SHAPE

The early notion that the nucleus serves as a bag for the chromatin is being replaced by research that shows the dynamic and complicated aspects of this organelle. In school, we would draw the nucleus as a spherical object that indeed is true for many cells. However, there are variations with research highlighting the influence of the function and surroundings on the nucleus shape and the effect of the shape of the nucleus on gene expression.

In cells that have ovoid or spherical nuclei, the chromatin may be present in various arrangements based on the functions of the cell and its status. The ratio of nuclear volume to cytoplasmic volume plays a role in influencing the size of the nucleus. Such a connection between the cell and nucleus size allows for reducing the genome size for organisms such as bats and birds that need rapid metabolism.

Figure 1.7 below shows different shapes of nuclei **a** Nuclei that are spherical and ovoid. **b** The lobed granulocytes. **c** The lobed monocyte **d** Other shapes are shown such as the polyploid megakaryocyte, fusiform fibrocyte and smooth muscle nuclei, and the condensed nucleus of a sperm (Skinner and Johnson, 2017; *This article is distributed under the terms of the Creative Commons Attribution 4.0 International License (http:// creativecommons. org/licenses/by/4.0/), which permits unrestricted use, distribution, and reproduction in any medium, provided you give appropriate credit to the original author(s) and the source, provide a link to the Creative Commons license, and indicate if changes were made*):

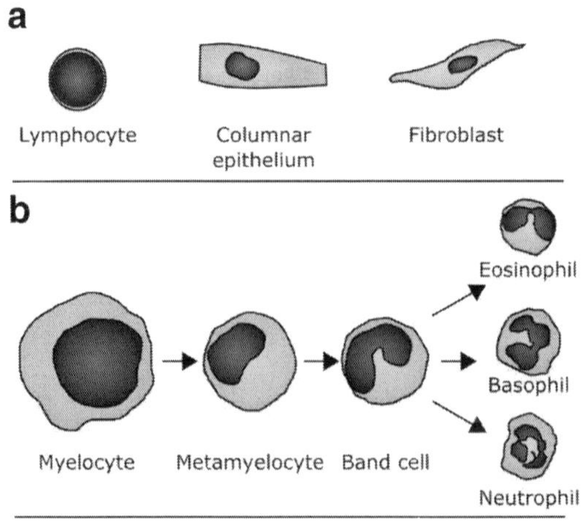

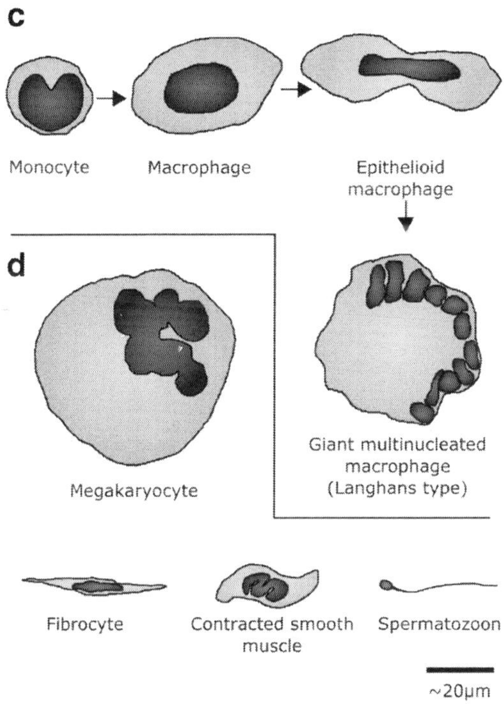

Figure 1.7

1.4.1 Lobed Nuclei

The granulocytes of the immune system (called so given the appearance upon staining and examination) show the presence of nuclei with many lobes. The many lobes are connected by nucleoplasm. What is interesting is that the nuclei are used for identifying the type of cell. For example, eosinophils show a biolobed nucleus described as "a sunburned face wearing dark sunglasses".

Neutrophils (heterophils in reptiles and birds) have multi-lobed nuclei (2-5) containing hardly any chromatin and are separated by nucleoplasm. The arrangement of the chromatin is gene dense regions towards the inside and the chromatin with few genes on the outside. The X chromosome that is inactivated in women is present near a lobe at either extremity.

The presence of lobules allows for the neutrophils to bend easily and migrate across narrow spaces (this has been described in Chapter 3). The formation of these lobes involves many factors. Deficiencies in iron, folic acid and vitamin B12 and megaloblastic anemias show the presence of increased lobes (more than 6).

In the case of monocytes, the nucleus is bilobed appearing as a horseshoe or a "U" in microscopy. When these cells are recruited into tissues to form macrophages, the chromatin and nucleus show changes with increased rounding. In monocytes, the chromatin as aggregates and large spaces in between facilitate the reshaping of the nucleus for easy migration.

(Skinner and Johnson, 2017)

1.4.2 Spindle-shaped Nuclei

The smooth muscle cells or mesenchyme demonstrate the presence of spindle-shaped nuclei. It has been demonstrated that the shape and the position of the nucleus are maintained via the attachment of actin fibers. The contraction of the muscles allows the spindles to be squeezed to form a corkscrew shape showing that the nuclei themselves are too subject to the contraction.

Fibroblasts that are hematopoietic cells associated with the inflammatory response also show the presence of cells and nuclei that are spindle-shaped.

In the case of flowering plants, spindle-shaped nuclei are seen in the endosperm of *Coronopus didymus* and *Arabidopsis thaliana*. The shape becomes rounded with development.

1.4.3 Asymmetric Nucleus

The female gamete of metazoans is usually immotile with a spherical nucleus while the motile male gamete is called a spermatozoan that shows an asymmetric nucleus. There is a vast array of shapes of the sperm head with beluga whales showing a square shape; ensiform heads in some bats; rodents with falciform heads and the ovate shape of pigs and humans. *Tubiluchus troglodyte*, a psudocoelomate worm shows a double-helical nucleus. *Noctilio leporinus*, the greater bulldog bat has a ridged and spatulate sperm. Interestingly, there is a wide array of shapes in the ejaculate of the naked mole-rat sperm, a eusocial animal. The sperm are compressed or lobed or miniature or even double-headed.

This shape is associated with pathology as seen in the case of teratozoospermia where abnormal sperm account for infertility. Further research can highlight the factors involved and the importance of nuclear shape and sperm shape in biology.

Figure 1.8 below shows sperm morphologies: a) The typical ovate b) Rodent sperm c) Odd shapes d) Eurasian bullfinch (*Pyrrulah pyrrulah*), the rounded sperm head of *Sparus aurata* or the sea bream, the double spiral nucleus of *Tubiluchus troglodytes*- The dashed outline shows nuclei cross-sections while grey regions show the acrosomal regions (Skinner and Johnson, 2017; *This article is distributed under the terms of the Creative Commons Attribution 4.0 International License (http:// creativecommons. org/licenses/by/4.0/), which permits unrestricted use, distribution, and reproduction in any medium, provided you give appropriate credit to the original author(s) and the source, provide a link to the Creative Commons license, and indicate if changes were made*):

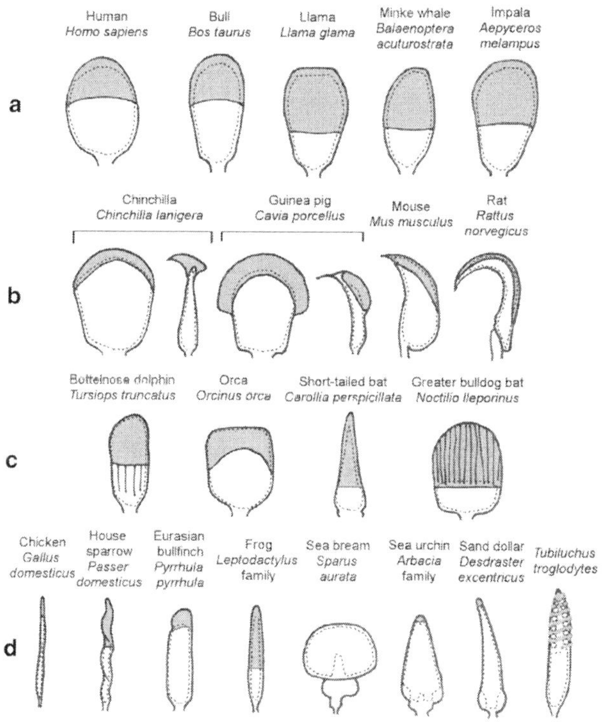

Figure 1.8

(Skinner and Johnson, 2017)

1.4.4 Other Shapes

Early research in 2007 showed that the plasticity and pliability of the nucleus in a stem cell was more than that of nuclei in differentiated cells. The use of micromanipulation showed that there is a stiffening of the nucleus by a factor of 6 during differentiation. The knocking out of Lamin A/C in human epithelial cells resulted in the deformability of the nucleus on the lines of adult hematopoietic stem cells. It was suggested that the nucleoplasm and chromatin decide the rheological character of the nucleus while the lamina controls the deformation (Pajerowski et al, 2007).

In the case of embryonic stem cells, lamins A/C are not expressed making them flexible. It has been suggested that these nuclei later become "hardened" to handle mechanical stress.

The nucleus of a T-lymphocyte, a cell that can migrate across endothelial tissues is *spherical*. The cell does not deform its nucleus during migration but the nucleus is forced through the gaps via actin and myosin.

There are certain human embryonic stem cells and few adenocarcinomas that display nuclei with a *bell shape*. The cells have been indicated to undergo syncytial division and also show very unique chromosomal pairing with condensation of an arm at the base, at the mouth, and a decondensed stretch between.

In the case of skeletal muscle, the nucleus is ovoid and is present just below the cell membrane at the periphery of the cells. There are invaginations composed of cytoskeletal proteins to allow mRNA trafficking to the cells.

(Skinner and Johnson, 2017)

The shape of the nucleus displays alterations in disease. An increased alteration of nuclear shapes called pleomorphy has been observed by researchers in carcinomas. In fact, the presence of large nuclei in urothelial carcinomas allows the diagnosis of this disease. Defects in lamins result in Hutchinson–Gilford progeria syndrome. The lamina is thickened to make the cells more susceptible to mechanical stress.

(Skinner and Johnson, 2017)

Figure 1.9 below shows skin fibroblasts from a healthy person (on the left) vs. a Hutchinson–Gilford progeria syndrome patient subject to immunostaining (lamin B1 in green and Lamin A/C in red) (Jean-Ha Baek, Tomás McKenna and Maria Eriksson (January 9th, 2013). Hutchinson-Gilford Progeria Syndrome, Genetic Disorders, Maria Puiu, IntechOpen, DOI: 10.5772/53794. Available from: https://www.intechopen.com/books/genetic-disorders/hutchinson-gilford-progeria-syndrome. *Licensee IntechOpen. This chapter is distributed under the terms of the Creative*

Commons Attribution 3.0 License, which permits unrestricted use, distribution, and reproduction in any medium, provided the original work is properly cited):

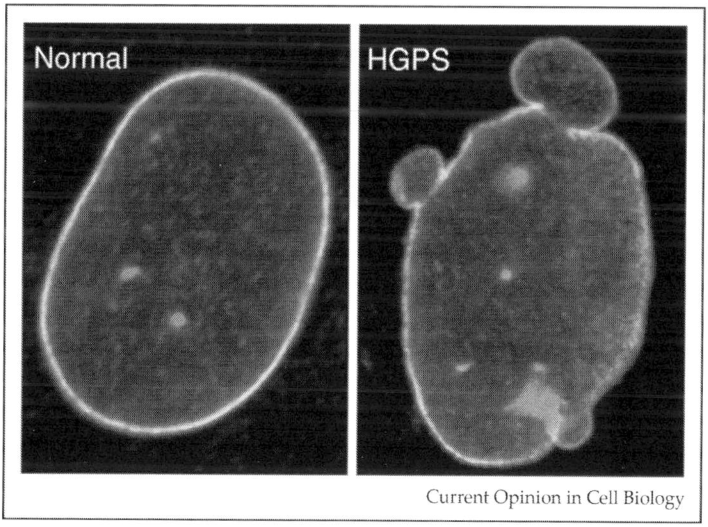

Figure 1.9

Abnormal nuclear architecture is the outcome of repeats in *C9orf72* seen in frontotemporal dementia and Amyotrophic lateral sclerosis. Even pathogens can affect the morphology of the nucleus. For example, specific nucleoporins (Nups) present in the nuclear pore complex (NPC) involved in nuclear permeability are targeted by Human immunodeficiency virus type 1 (HIV-1) to show the involvement of the architecture of the nucleus in the life cycle of the pathogen (Mukherjee et al, 2016).

The expression of nuclear lamins is different in cells infected by the herpes simplex virus 1 (HSV-1). Given the large size of these viruses, the transport of viral particles from the nucleus involves changes in the lamina. The net content of lamins shows a dip in infected cells with higher lamin A solubility and diffusion of the lamin B receptor in the inner nuclear membrane. The presence of lamin A/C is vital for localizing the viral protein called pUL34 involved in viral envelope formation (Cibulka et al, 2012).

Figure 1.10 below shows the involvement of lamin B (red) and lamin A/C (green) to form a nuclear egress complex upon infection with herpes simplex virus 1 (HSV-1) and human cytomegalovirus (HCMV). (Cibulka et al, 2012; *This article is an open-access article distributed under the*

the Creative Commons Attribution License, which permits unrestricted use, distribution, and reproduction in any medium, provided the original author and source are credited):

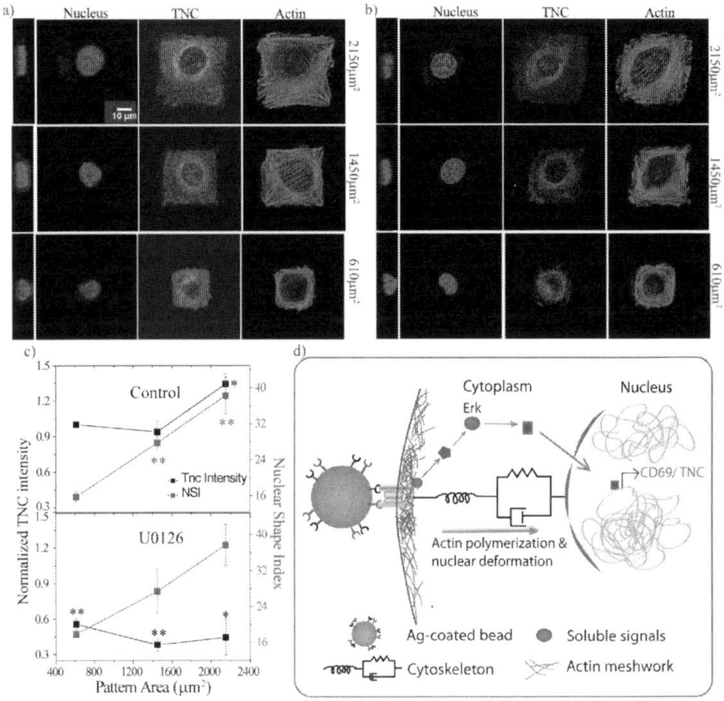

Figure 1.12

The shape of the nucleus is controlled by several factors. The cytoskeleton of the cell interacts with the nuclear envelope and the lamina beneath. The LINC complex (Linker of Nucleoskeleton and Cytoskeleton) is a major mediator of the mechanistic associations between the nucleus and the cytoplasm. The proteins with KASH domains and SUN domains are major players of the LINC. The former enters the perinuclear space from the cytoplasm to interact with the latter that enters the nuclear lamina and the nucleoskeleton.

Several molecules of the cytoskeleton and KASH proteins can interact with each other such as nesprin-3α with intermediate filaments; nesprin-1G with actin, and nesprin-1, nesprin-2 with microtubules. Lamin A is bound to by the SUN proteins. Nesprins (nuclear envelope associated spectrin-repeat proteins) are KASH proteins of the LINC that link the cytoskeleton and the proteins of the nuclear envelope. These proteins have also been implicated in the development of the nuclear shape of sperm in rat falciform sperm and also facilitate the nuclei of smooth muscles to be flexible.

The shape of the nucleus is also affected by Brahma-related gene 1 or BRG1. This is an ATPase subunit of the SWI/SNF chromatin remodeling enzyme complex. The number of lobules in the nucleus was elevated in experiments where this protein was knocked out showing its role in nuclear shape and gene expression.

(Skinner and Johnson, 2017)

The role of chromatin in the shape of the nucleus has also been shown by researchers Schreiner and team in 2015. The authors showed that the association of the inner nuclear membrane (INM) with the chromatin influenced the stiffness of the nucleus. A model was developed that shows that the stiffness of the nucleus was increased when the chromatin is tethered to the periphery of the nucleus. This tethering also limits the flow of chromatin to allow it to handle stress. The untethering resulted in the flowing of chromatin that caused the deformation of nuclei.

Figure 1.13 below shows a model showing how an (a) wild-type yeast nucleus (b) a *heh2Δ* nucleus (a mutant of the Heh2- an ortholog of the LEM domain family (LAP2, emerin, Man1) of inner nuclear membrane proteins) responds. In the former, the stiffness of the nucleus is facilitated by the tethering of the chromatin to avoid large fluctuations. The mutant shows the flow of chromatin due to microtubule forces to have long-lasting changes in the shape of the nucleus on account of lowered chromatin tethering (Schreiner et al, 2015; *This work is licensed under a Creative Commons Attribution 4.0 International License. The images or other third party material in this article are included in the article's Creative Commons license, unless indicated otherwise in the credit line; if the material is not included under the Creative Commons license, users will need to obtain permission from the license holder to reproduce the material. To view a copy of this license, visit http://creativecommons.org/licenses/by/4.0/*):

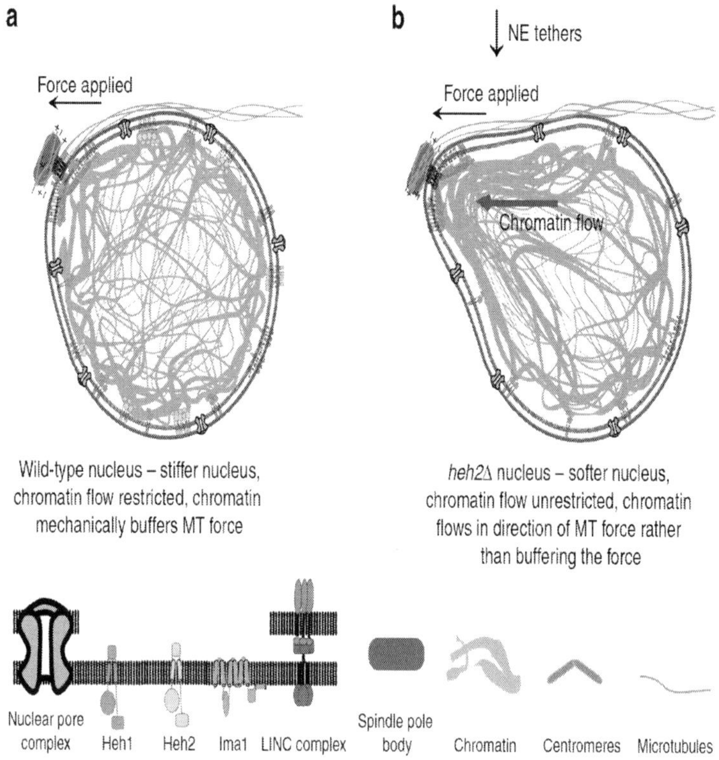

Figure 1.13

Thus, modulating the attachment of the chromatin to the nucleus is seen to influence the shape of the nucleus.

(Schreiner et al, 2015)

Apart from internal factors, external stress can also control the nuclear shape and gene expression. Lamin phosphorylation is influenced by cytoskeleton stress that alters the status of lamins to make the nucleus more rounded. The change in gene expression in response to such mechanical stress has been shown to allow certain pathways to be triggered such as the formation of mesenchyme from epithelium in mammary glands.

(Skinner and Johnson, 2017)

Figure 1.14 shows the different factors that control the shape of the nucleus. This involves the repositioning of the nucleus by actin/microtubules or the change in gene expression to mechanical stress (Skinner and Johnson, 2017; *This article is distributed under the terms of the Creative Commons Attribution 4.0 International License (http:// creativecommons.org/licenses/*

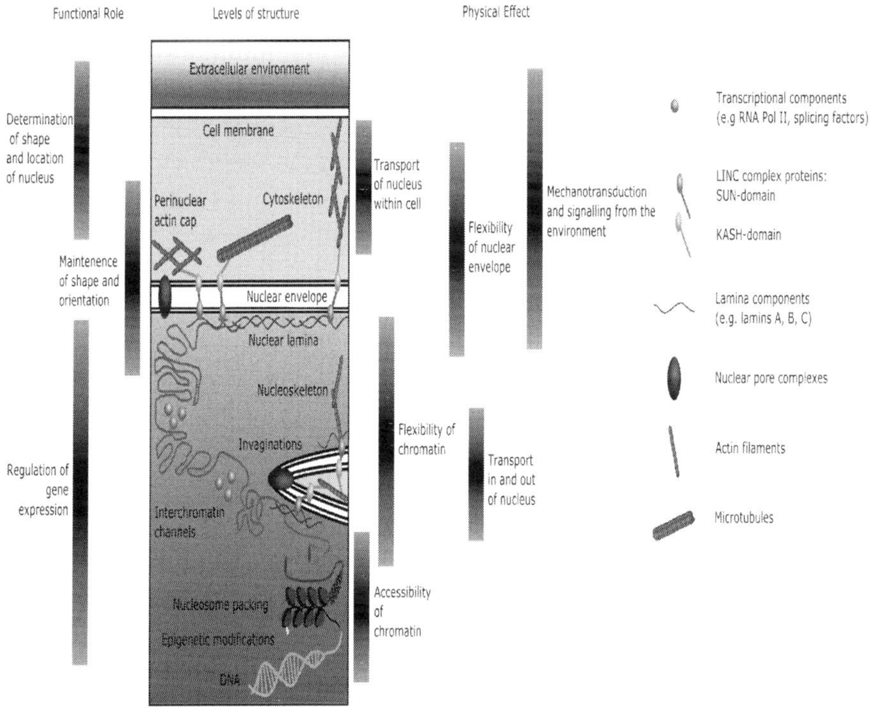

Figure 1.14

1.6 THE NUCLEUS AND CHROMATIN

A cell is subject to dynamic changes in their surroundings that dictate changes in the biochemistry and physical structures in both short-term (seconds/minutes) and long-term (hours/days). Research is now casting light on the ability of the nucleus to respond to such changes in the environment. On the lines of the plasma membrane, the nuclear envelope on account of possessing the LINC plus additional proteins (an example is emerin) can serve as a barrier to force. The nuclear lamina can serve as a buffer to mechanical signals from outside and inside of the nucleus.

The process of post-translational modification allows the shift between the loosely packed and active euchromatin and the tightly packed inactive chromatin. This process becomes vital for epigenetics. Research has shown that the nuclear envelope is connected to heterochromatin via lamin-associated chromatin domains (LADs). The heterochromatin is capable of changes in topology in chromosomes to serve as a stimulus to the nucleus. These arrangements albeit over short distances can circumvent genetic material damage by cushioning the mechanical stimuli.

The nucleus can deform physically on account of phosphorylation or mechanical changes in the lamins A/C. This facilitates optimal nuclear mechanics: allow the nucleus to be softened for migration of cells across small pores or hardening the nucleus in stem cells to facilitate subsequent differentiation. In either case, maintaining genetic integrity is of vital importance. This modification of the rheology can allow for handling mechanical stress. Interaction between the genetic material and lamin A/C also allows alterations in the genome organization to facilitate the handling of mechanical stress. The integrity of the nucleus is also maintained by the involvement of nuclear actin. Studies in *Xenopus laevis* oocytes revealed the ability of nuclear actin to form a mesh to prevent damage to the nuclear content due to gravity. Interestingly, research has also shown chromatin to directly receive mechanical strain via nonmuscle myosin IIA (NMIIA) and emerin with nuclear actin to show the direct association of chromatin and this protein.

Studies are now showing that the nucleus is an adaptive organelle though the stiffness of the nuclear lamina is known. Multiple studies have shown that lowering lamin A/C and the presence of perinuclear F-actin in the nuclear lamina allowed for a more elastic nucleus in migratory blood cells to prevent DNA damage. However, these cells displayed lower survival indicative of the role of lamin A/C in DNA repair. Further, the deformation of the nucleus and damage to the DNA in migratory cells has been linked to cancer heterogeneity. Thus, there is a delicate and intricate system of lamins and perinuclear actin that allows for a dynamic nucleus in a cell.

(Miroshnikova et al, 2017)

Figure 1.15 below shows the dynamics of lamins, perinuclear actins, and chromatin as a scheme. A mesh of lamin A/C is associated with stiffness to protect the genetic material (shown on the left). The right side shows how mechanical stress is dissipated via the LINC and perinuclear actin. Research is indicative of the association of constitutive heterochromatin in the nuclear periphery regions rich in lamin A/C while the chromatin when loosely packed dissociates from the lamina when associated with perinuclear actin-rich nuclear domains (Miroshnikova et al, 2017; *This is an Open Access article distributed under the terms of the Creative Commons Attribution License (http://creativecommons.org/licenses/by/3.0), which permits unrestricted use, distribution and reproduction in any medium provided that the original work is properly attributed*):

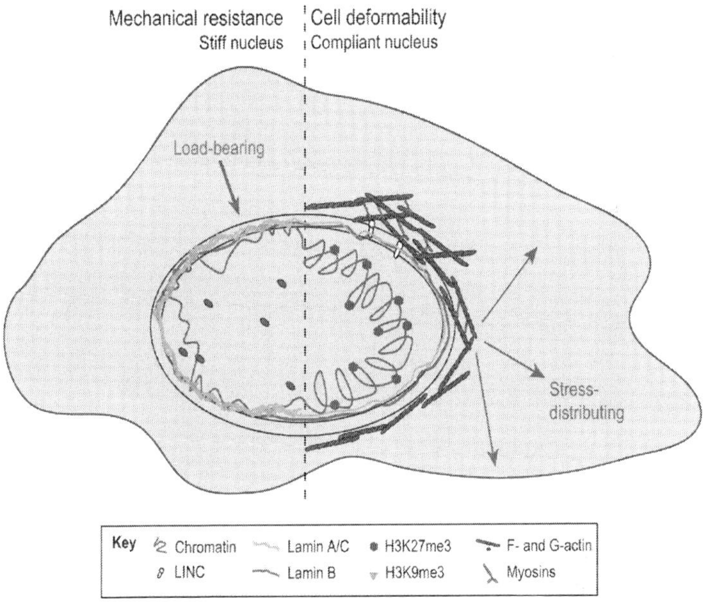

Figure 1.15

The chromatin itself serves as a mechanical element of the nucleus. This was highlighted in research back in 2007 when the separation of kinetochores and the spindle length in mitosis were extended by modifying the chromatin. That the chromatin functioned like a spring was shown by proportional forces exerted by kinesins and pericentric chromatin stretching in another research. The use of micropipette aspiration studies has highlighted that when a cell is migrating through a narrow pore, the integrity of chromatin is maintained despite the stretching of the nucleus.

Thus, a combination of changes in nuclear actin and lamin A/C along with the chromatin architecture accounts for the redistribution of stress in the nucleus. Research using molecules to condense and decondense the chromatin showed that the process of heterochromatin decondensation is associated with an increase in viscosity of the nucleus. Further, it was also observed that there is a reversible shift from H3K9me3-modified heterochromatin to H3K27me3-modified heterochromatin that shows more loose packing when there is mechanical stress. This shows the ability of chromatin to function as a stress absorber in the nucleus.

The process of differentiation is associated with an increase in lamin A in stem cells. This transforms the nucleus from a flexible conformation to a more rigid one due to elevated amounts of stiff heterochromatin and lamin A/C incorporation in the nuclear membrane.

The condensation of chromatin allows for control over transcription with access only to few genes for expression. It has also been demonstrated that mechanical perturbation of undifferentiated cells is characterized by elongation of both the cell and the nucleus while only the former is seen in differentiated cells.

(Miroshnikova et al, 2017)

Figure 1.16 below shows a proposed model of the involvement of chromatin in nuclear shape. The left shows a short -term response while the right shows a long-term one. The handling of mechanical stress involves a switch from H3K9me3-heterochromatin to H3K27me3-heterochromatin and a decrease in lamin A/C association to allow softening of the chromatin. Further, specific stretches of DNA are repositioned to handle shocks (Left). For the right, the viscosity of the nucleoskeleton shows an increase to allow high-stress to be handled with an open configuration of the chromatin along with silencing of promoters (Miroshnikova et al, 2017) (Miroshnikova et al, 2017; *This is an Open Access article distributed under the terms of the Creative Commons Attribution License (http://creativecommons.org/licenses/by/3.0), which permits unrestricted use, distribution and reproduction in any medium provided that the original work is properly attributed*):

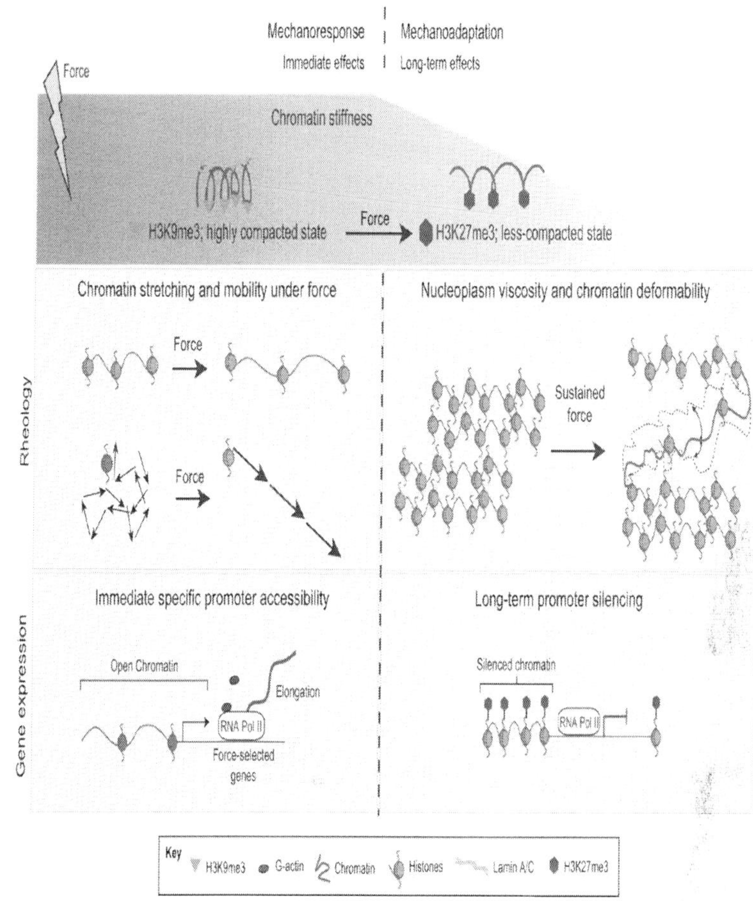

Figure 1.16

This becomes exciting that apart from nuclear proteins, the chromatin itself serves as a tool to adapt to different conditions.

Research published by scientists Ulianov and team (2019) showed the role of the nuclear lamina in maintaining the architecture of chromatin. Studies have already shown the presence of weakly expressed or silent genes in the lamin-associated chromatin domains (LADs) in mammals and *Drosophila*. The use of a high throughput chromosome conformation capture (Hi-C) technique has shown that chromatin in mammals is either open: DNase-sensitive or closed: DNase-resistant. The latter is associated with nuclear lamina contacts. It was observed that the packaging density of peripheral chromatin in *Drosophila* S2 cells (mutant for lamins) was increased with a decrease in total chromatin volume, as shown by FISH assays. In the case of active chromatin, the density of the topologically-associating domains (TADs) is increased with higher contact density.

A model has been put forth to show the involvement of nuclear lamins in allowing chromatin expansion and condensation. The tethering of chromatin to the lamins is associated with shrinkage of chromatin when the tethering is released. This is accounted for by the crowding of macromolecules and interactions among nucleosomes. This makes lamins to function as a surface to anchor interphase chromosomes. There can be further condensation of inactive chromatin by the action of histone deacetylases (HDACs) associated with nuclear lamins.

Similar work by another team in mouse embryonic stem cells shows that while the overall profile of TAD is similar between mouse and *Drosophila* cells, there are variations among individual domains. The condensation of some chromatin is lowered in both with a loss of nuclear lamins. The expression of few genes in the periphery and interior of the nucleus is altered in both the mouse and *Drosophila* cells.

Figure 1.17 below shows the role of nuclear lamins in chromatin condensation. a) shows a simulation of the conformation of the model polymer showing interaction among non-acetylated nucleosomes in chromatin (blue). There is no such interaction shown by the active chromatin (orange) b) Blue blocks represent LADs c) The impact of volume on the number of contacts with the surfaced) 3D structures e) The proposed model (Ulianov et al, 2019; *Open Access*- This article is licensed under a Creative Commons Attribution 4.0 International License, which permits use, sharing, adaptation, distribution and reproduction in any medium or format, as long as you give appropriate credit to the original author(s) and the source, provide a link to the Creative Commons license, and indicate if changes were made. The images or other third party material in this article are included in the article's *Creative Commons license, unless indicated otherwise in a credit line to the material. If material is not included in the article's Creative Commons license and your intended use is not permitted by statutory regulation or exceeds the permitted use, you will need to obtain permission directly from the copyright holder. To view a copy of this license, visit http://creativecommons.org/licenses/by/4.0/*):

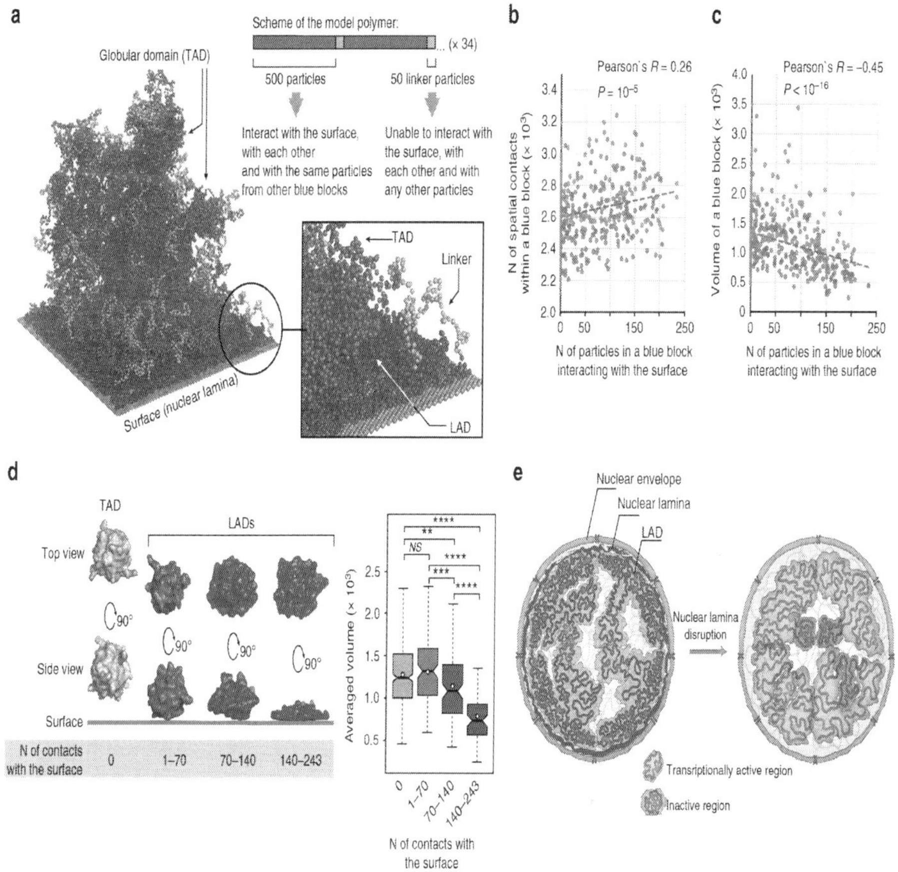

Figure 1.17

This shows the involvement of the nuclear lamins in making the associated chromatin more compact and also influences the interactions and expression of the chromatin.

(Ulianov et al, 2019)

1.7 FINAL POINTS

Described as the largest organelle in the cell, the idea that a nucleus is a rigid object is being replaced with a dynamic organelle that is capable of deformation. Further, various insights into the role of various nuclear proteins and the chromatin itself to handle mechanical stress on cells have also been revealed by research.

Future studies can only reveal more exciting findings of the nucleus to change our understanding of "the bag holding the chromatin". Science has already seen the use of the nucleus in somatic nuclear transfer that is the basis of cloning. In the case of the nucleus, the following quote resonates well, "Now, my suspicion is that the universe is not only queerer than we suppose but queerer than we can suppose."- J.B.S. Haldane.

1.8 REFERENCES

1. Thoru Pederson. The Nucleus Introduced. Cold Spring Harb Perspect Biol 2011;3:a000521.
2. Martin, W. F., Garg, S., & Zimorski, V. (2015). Endosymbiotic theories for eukaryote origin. *Philosophical transactions of the Royal Society of London. Series B, Biological sciences*, *370*(1678), 20140330. https://doi.org/10.1098/rstb.2014.0330.
3. Bell PJ. Viral eukaryogenesis: was the ancestor of the nucleus a complex DNA virus? J Mol Evol. 2001 Sep;53(3):251-6. doi: 10.1007/s002390010215. PMID: 11523012.
4. Manley, H. R., Keightley, M. C., & Lieschke, G. J. (2018). The Neutrophil Nucleus: An Important Influence on Neutrophil Migration and Function. *Frontiers in immunology*, *9*, 2867. https://doi.org/10.3389/fimmu.2018.02867.
5. Christian Schöfer and Klara Weipoltshammer. Nucleolus and chromatin. Histochemistry and Cell Biology (2018) 150:209–225.
6. Benjamin M. Skinner & Emma E. P. Johnson. Nuclear morphologies: their diversity and functional relevance. Chromosoma (2017) 126:195–212.
7. J. David Pajerowski, Kris Noel Dahl, Franklin L. Zhong, Paul J. Sammak, Dennis E. Discher. Physical plasticity of the nucleus in stem cell differentiation. Proceedings of the National Academy of Sciences Oct 2007, 104 (40) 15619-15624; DOI: 10.1073/pnas.0702576104.
8. Richik N. Mukherjee, Pan Chen & Daniel L. Levy (2016) Recent advances in understanding nuclear size and shape, Nucleus, 7:2, 167-186, DOI: 10.1080/19491034.2016.1162933
9. Cibulka, J., Fraiberk, M., & Forstova, J. (2012). Nuclear actin and lamins in viral infections. *Viruses*, *4*(3), 325–347. https://doi.org/10.3390/v4030325.
10. Gupta S, Marcel N, Sarin A, Shivashankar GV (2012) Role of Actin Dependent Nuclear Deformation in Regulating Early Gene Expression. PLoS ONE 7(12): e53031. https://doi.org/10.1371/journal.pone.0053031.
11. Schreiner, S., Koo, P., Zhao, Y. *et al.* The tethering of chromatin to the nuclear envelope supports nuclear mechanics. *Nat Commun* 6, 7159 (2015). https://doi.org/10.1038/ncomms8159.

12. Miroshnikova YA, Nava MM, Wickström SA. Emerging roles of mechanical forces in chromatin regulation. J Cell Sci. 2017 Jul 15;130(14):2243-2250. doi: 10.1242/jcs.202192. Epub 2017 Jun 23. PMID: 28646093.
13. Ulianov, S.V., Doronin, S.A., Khrameeva, E.E. *et al.* Nuclear lamina integrity is required for proper spatial organization of chromatin in *Drosophila*. *Nat Commun* 10, 1176 (2019). https://doi.org/10.1038/s41467-019-09185-y.

CHAPTER 2

THE NUCLEUS, DISEASE AND HEALTH

Shivsanjeevi Sripathi

CONTENTS

2.1 THE QUEST FOR YOUTH

The search for a magical potion to prevent aging and maintain youth can be dated back to the pages of history to the current slew of products available to "reverse aging".

The phenomenon of autophagy for cleaning up damaged organelles in the cell is associated with longevity. The word "auto" means self and "phagy" is eating-the process alluded to in physiology for recycling proteins or organelles that are damaged. There are several types of this system: autophagy where an autophagosome or double-membrane vesicle is formed. The enzymes in the lysosome digest the contents. There is microautophagy where the lysosome and the portion of the organelle to be degraded interact directly. There is also the involvement of chaperones present in the cytoplasm that target proteins to LAMP-2A-a protein associated with the lysosome. This process is known as chaperone-mediated autophagy.

The process occurs at basal levels as part of the normal physiology of cells. In the event of stress (DNA damage or starvation or oxidative stress), autophagy is stepped up. Many factors involved are recruited by omegasomes that serve as the sites of nucleation of autophagy. The association of class III phosphatidyl-inositol-3-kinase, Unc-51-like kinase 1 (ULK1) complex, and other factors leads to the formation of the phagophore. Further maturation is facilitated by protein1 light chain 3 (LC3) and Atg 5/Atg7/Atg 12: autophagy- related (Atg) proteins.

(Papandreou and Tavernarakis, 2019)

Figure 2.1 below shows the scheme of the maintenance process of autophagy: The autophagosome is formed with the initial activation of Unc-51-like kinase 1 and the phosphatidyl-inositol-3- kinase complex along with the Vps15/Vps34/Beclin 1 complex. The formation and maturation of the autophagosome also involve the ubiquitin conjugation systems: ATG5/ATG7/ATG12 and LC3.

(Papandreou and Tavernarakis, 2019; *Open Access*- This article is licensed under a Creative Commons Attribution 4.0 International License, which permits use, sharing, adaptation, distribution and reproduction in any medium or format, as long as you give appropriate credit to the original author(s) and the source, provide a link to the Creative Commons license, and indicate if changes were made. The images or other third party material

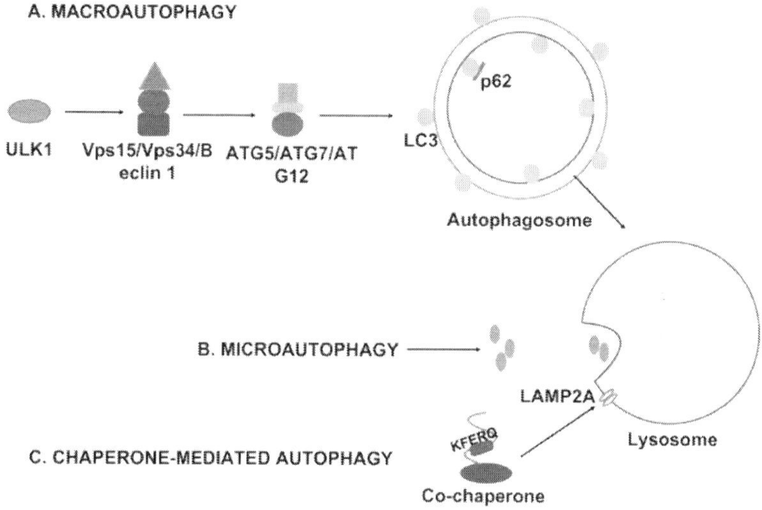

Figure 2.1

The process of autophagy is associated with lifespan. *Caenorhabditis elegans* mutants for genes involved in autophagy such as bec-1 or unc-51 are associated with a shorter lifespan than wild type. In the case of *Drosophila melanogaster*, neurons of aged flies show a low profile of autophagy gene expression. When the genes *Atg1* and *Atg8* are overexpressed, the lifespan is augmented. The process of autophagy and hence, lifespan is augmented in mice by overexpressing Atg5.

Geriatric human brains show lowered transcription of Atg5, Atg7, and Beclin-1. Interestingly, autophagy is induced by a compound present in vegetables and fruits called resveratrol that is shown to boost longevity. The induction of autophagy in mouse models of neurodegenerative disease results in the improvement of disease symptoms.

The maintenance of homeostasis involves selective autophagy. Given that the nucleus is the repository of the genetic material, the nucleolus, and vital proteins, the intricacies of the nuclear dynamics are slowly being studied. In the case of *Saccharomyces cerevisiae*, nuclear components are digested by microautophagy. This can be due to treatment with rapamycin or nitrogen starvation or a shortage of nutrients. These conditions set off a chain where mTOR (mammalian target of rapamycin pathway) is inactivated followed by the activation of the nuclear envelope morphology protein/sporulation 7/phosphatidic acid phosphohydrolase axis. There is an initial direct contact between the lytic vacuole and the nucleus. The vacuole then surrounds vesicles formed by protrusions of the outer nuclear membrane and the endoplasmic reticulum (ER) of the nucleus. This is followed by pinching off these vesicles from the nucleus. The process of piecemeal nucleophagy (PMN) involving vacuolar protein 8 membrane and nucleus to vacuole protein 1 (Nvj1p) follows that targets the nuclear envelope and the nucleolus.

(Papandreou and Tavernarakis, 2019)

Figure 2.2 below shows the scheme of piecemeal nucleophagy in yeast that is seen under conditions of starvation and also when nutrients are present. (a) Tight junctions (in blue) allow direct interactions between the lytic vacuole (green) and the nucleus (in purple). (b) formation of ER bulges with the targets (c) pinching off and (d) degradation in the lytic vacuole (Papandreou and Tavernarakis, 2019; *Open Access*- This article is licensed under a Creative Commons Attribution 4.0 International License, which permits use, sharing, adaptation, distribution and reproduction in any medium or format, as long as you give appropriate credit to the original author(s) and the source, provide a link to the Creative Commons license, and indicate if changes were made. The images or other third party material in this article are included in the article's *Creative Commons license, unless indicated otherwise in a credit line to the material. If material is not included in the article's Creative Commons license and your intended use is not permitted by statutory regulation or exceeds the permitted use, you will need to obtain permission directly from the copyright holder. To view a copy of this license, visit http://creativecommons.org/licenses/by/4.0/*):

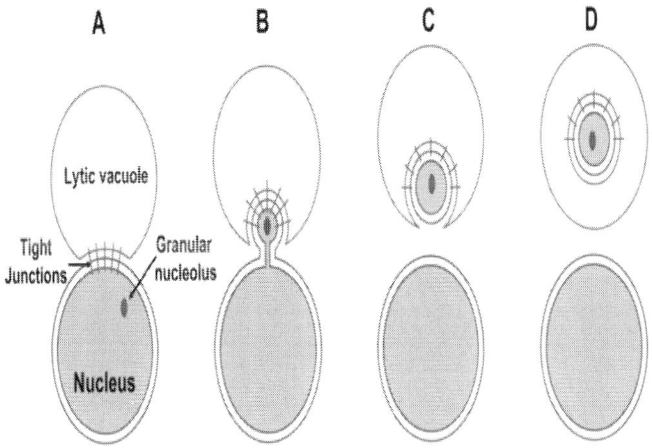

Figure 2.2

Interestingly, research has shown that *Aspergillus oryzae shows the degradation of entire organelles.* Filamentous fungi are made of multinucleate cells called hyphae. The basal cells in these hyphae in the interior were shown to display macroautophagy of entire nuclei! The process of autophagy in yeast involves only partial degradation as shown above. The autophagy of organelles: mitochondria, peroxisomes, and whole nuclei was observed using enhanced green fluorescent protein (eGFP)-labels. The whole nuclei are encircled by autophagosomes in the form of a ring that develops from crescents of autophagosome precursors. Given that nutrient starvation when autophagy was inhibited resulted in suppressed growth, the degradation of whole nuclei (with other organelles) allows for *Aspergillus oryzae* to tide over starvation (Shoji et al, 2010)

Figure 2.3 below shows (A) the maturation of an autophagosomal precursor crescent (green, arrow) into an autophagosome ring that encircles the nucleus. (B) The disappearance of the ring (green, arrow) around the nucleus (red). Arrowheads are vacuoles. Bars represent 5 μm (Shoji et al, 2010; *This is an open-access article distributed under the terms of the Creative Commons Attribution License, which permits unrestricted use, distribution, and reproduction in any medium, provided the original author and source are credited*):

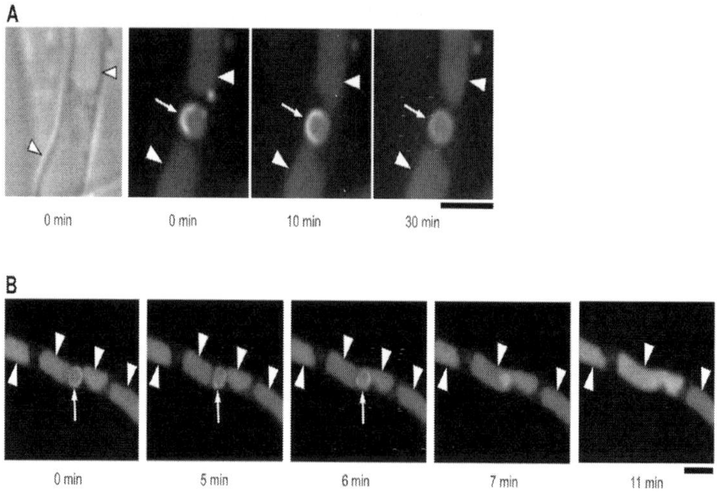

Figure 2.3

The integrity of the nuclear envelope (NE)-the bridge between the cytoplasm and the nucleus is vital for health. This has been reported in several studies that are suggestive of disturbed transport of molecules between the cytoplasm and the nucleus and even rupture of the nucleus when the NE is compromised. Several proteins of the nuclear envelope associated with aging and associated changes are summarized by Table 2.1 (Adapted from Martins et al, 2020):

Table 2.1

Protein of the nuclear envelope	**Changes**	**Aging model used**
Lamin A/C	Lowered levels of in the nucleoplasm Accumulated prelaminA at the periphery of the nucleus alters the lamina/C functions to affect the nuclear shape	Passaged human fibroblasts Aged induced pluripotent stem cells
	Progerin accumulates at the nuclear periphery that lowers lamin A/C levels	Premature aging models: Hutchinson–Gilford progeria (HGPS) fibroblasts and coronary arteries

Lamin B1	p53 and Rb activation in senescence lower the levels of lamin B1 This alters the selectivity of the nuclear envelope Lamin B1 reduction is seen as nuclear-to-cytoplasm chromatin blebbing Reduction in the protein is also implicated in SADs or senescence-associated distension of satellites The shape of the nuclei are misshapen when lamin B1 is silenced	Passaged human fibroblasts Fibroblasts from aged donors Human progeria fibroblasts Hutchinson–Gilford progeria (HGPS) fibroblasts Senescent human fibroblasts Aged induced pluripotent stem cells
LAP2	Aging, senescence, and progeria are associated with lowered levels of LAP2α and LAP2β The interaction between heterochromatin and lamin A/C is altered due to low levels of LAP2α	Skin samples from aged donors Hutchinson–Gilford progeria (HGPS) fibroblasts Senescent human fibroblasts
SUN1	During aging, senescence, and progeria, SUN1 is recruit ed to the nuclear envelope. However, an overaccumulation is associated with HGPS abnormalities.	Human centenarian fibroblasts Hutchinson–Gilford progeria (HGPS) fibroblasts
Nesprin-2	Aged induced pluripotent stem cells show elevated expression HGPS cells show a dip in Nesprin-2 at the nuclear periphery	Aged induced pluripotent stem cells Hutchinson–Gilford progeria (HGPS) fibroblasts

Nup153	Mislocalization is seen that adversely affects nuclear pore complexes (NPCs) and hence, cargo transport	Old vascular smooth muscle cells Hutchinson–Gilford progeria (HGPS) fibroblasts
Nup93 and Nup62	Lowered levels of both NPCs that affect the permeability barrier of the NPC	Aged rat neurons and *Caenorhabditis elegans*
Emerin	Increased emerin in the stem cells and mislocalized protein in the progeria fibroblasts	• Hutchinson–Gilford progeria (HGPS) fibroblasts • Aged induced pluripotent stem cells
Tpr	Mislocalization of Tpr due to compromised Nup153 import that leads to loss of organization of chromatin	Hutchinson–Gilford progeria (HGPS) fibroblasts

Thus, aging (either physiological or premature) is associated with loss of functioning of the nuclear envelope due to altered nuclear transport and chromatin organization.

Figure 2.4 below shows the changes in the nuclear envelope in gaining: *HP1, heterochromatin protein 1; LADs, lamina-associated domains; SAHF, senescence-associated heterochromatin foci*; *CCF, cytoplasmic chromatin fragment; NADS, nucleolar-associated domains; NPC, nuclear pore complexes* (Martin et al, 2020: *This is an open-access article under the terms of the http://creativecommons.org/licenses/by/4.0/* License, which permits use, distribution and reproduction in any medium, provided the original work is properly cited):

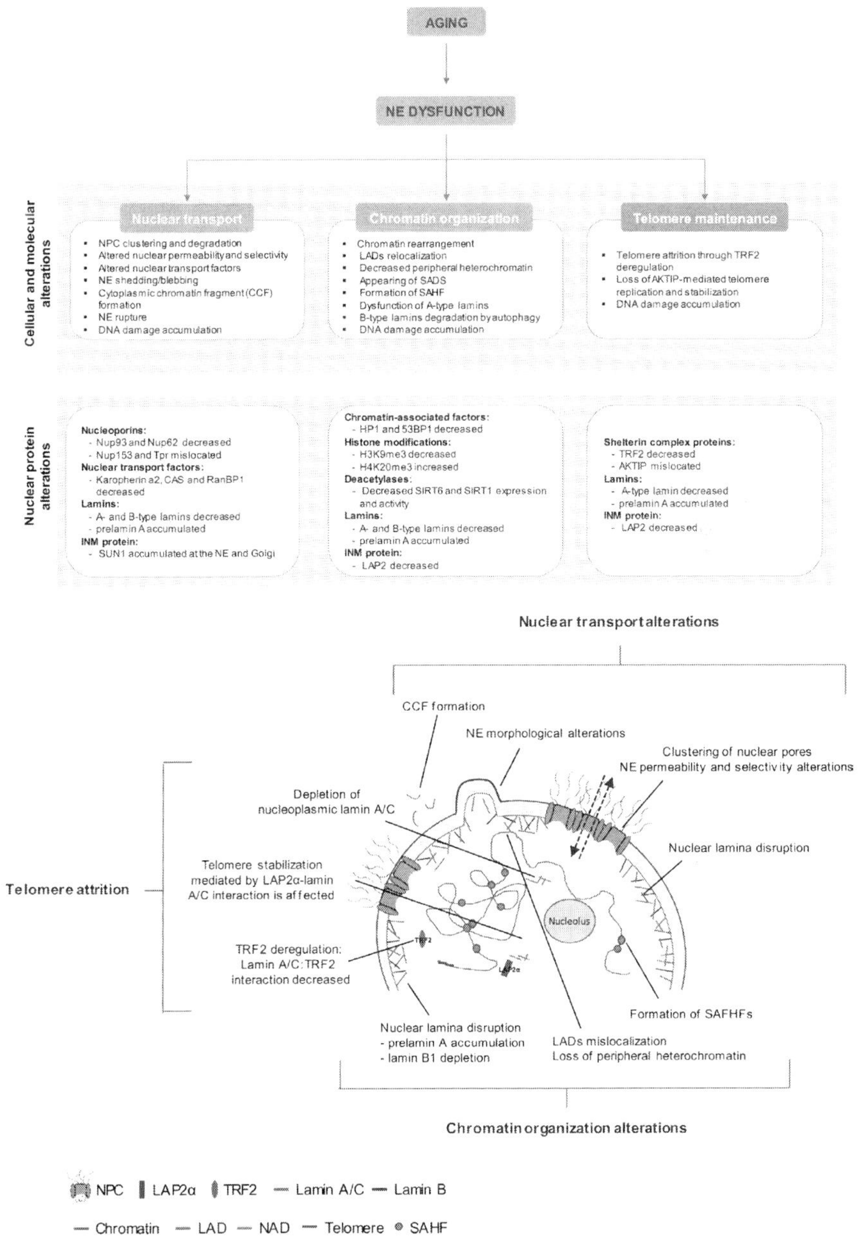
AGING
NE DYSFUNCTION
Cellular and molecular alterations
Nuclear transport
NPC clustering and degradation
Altered nuclear permeability and selectivity
Altered nuclear transport factors
NE shedding/blebbing
Cytoplasmic chromatin fragment (CCF) formation
NE rupture
DNA damage accumulation
Chromatin organization
Chromatin rearrangement
LADs relocalization
Decreased peripheral heterochromatin
Appearing of SADS
Formation of SAHF
Dysfunction of A-type lamins
B-type lamins degradation by autophagy
DNA damage accumulation
Telomere maintenance
Telomere attrition through TRF2 deregulation
Loss of AKTIP-mediated telomere replication and stabilization
DNA damage accumulation
Nuclear protein alterations
Nucleoporins:
- Nup93 and Nup62 decreased
- Nup153 and Tpr mislocated
Nuclear transport factors:
- Karopherin a2, CAS and RanBP1 decreased
Lamins:
- A- and B-type lamins decreased
- prelamin A accumulated
INM protein:
- SUN1 accumulated at the NE and Golgi
Chromatin-associated factors:
- HP1 and 53BP1 decreased
Histone modifications:
- H3K9me3 decreased
- H4K20me3 increased
Deacetylases:
- Decreased SIRT6 and SIRT1 expression and activity
Lamins:
- A- and B-type lamins decreased
- prelamin A accumulated
INM protein:
- LAP2 decreased
Shelterin complex proteins:
- TRF2 decreased
- AKTIP mislocated
Lamins:
- A-type lamin decreased
- prelamin A accumulated
INM protein:
- LAP2 decreased
Nuclear transport alterations
CCF formation
NE morphological alterations
Clustering of nuclear pores
NE permeability and selectivity alterations
Depletion of nucleoplasmic lamin A/C
Nuclear lamina disruption
Telomere attrition
Telomere stabilization mediated by LAP2α-lamin A/C interaction is affected
Nucleolus
TRF2 deregulation: Lamin A/C:TRF2 interaction decreased
TRF2
LAP2α
Formation of SAFHFs
Nuclear lamina disruption
- prelamin A accumulation
- lamin B1 depletion
LADs mislocalization
Loss of peripheral heterochromatin
Chromatin organization alterations
NPC
LAP2α
TRF2
Lamin A/C
Lamin B
Chromatin
LAD
NAD
Telomere
SAHF

Figure 2.4

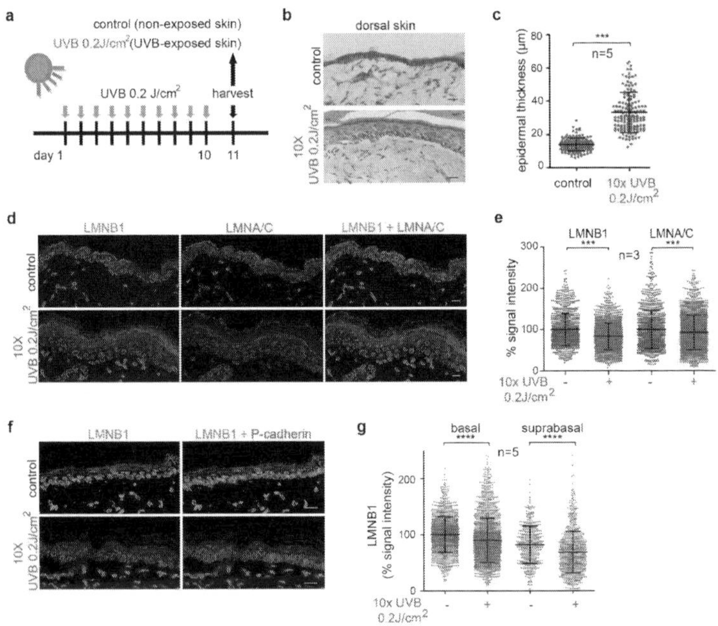

Figure 2.6

Thus, it was concluded by the research team that senescence can be detected by lamin B1 that can allow identifying age-related pathologies, look for aged cells in lesions, and evaluate aging of the skin in response to the diet and pollution.

(Wang et al, 2017)

Nucleophagy refers to autophagy of damaged nuclear DNA that can cause cancer and inflammation. The development of CAG repeats in polyQ stretches is associated with a neurological disease called Polyglutamine (PolyQ) disease. Degradation of lamin B along with compromised autophagy has been observed in Dentatorubral-pallidoluysian atrophy. This results in the collapse of the nucleus and damage to the neurons.

(Papandreou and Tavernarakis, 2019)

Figure 2.7 below shows diseases where molecules targeted in autophagy are implicated. DNA (black) when not degraded, nucleolar elements (in purple), and lamins (grey) have been implicated in cancer. In the case of neurodegeneration, the formation of aggregates or nucleophagy has been observed (Papandreou and Tavernarakis, 2019; *Open Access*- This article is licensed under a Creative Commons Attribution 4.0 International License, which permits use, sharing, adaptation, distribution and reproduction in any medium or format, as long as you give appropriate credit to the original author(s) and the source, provide a link to the Creative Commons license, and indicate if changes were made. The images or other third party material in this article are included in the article's *Creative Commons license, unless indicated otherwise in a credit line to the material. If material is not included in the article's Creative Commons license and your intended use is not permitted by statutory regulation or exceeds the permitted use, you will need to obtain permission directly from the copyright holder. To view a copy of this license, visit http://creativecommons.org/licenses/by/4.0/*):

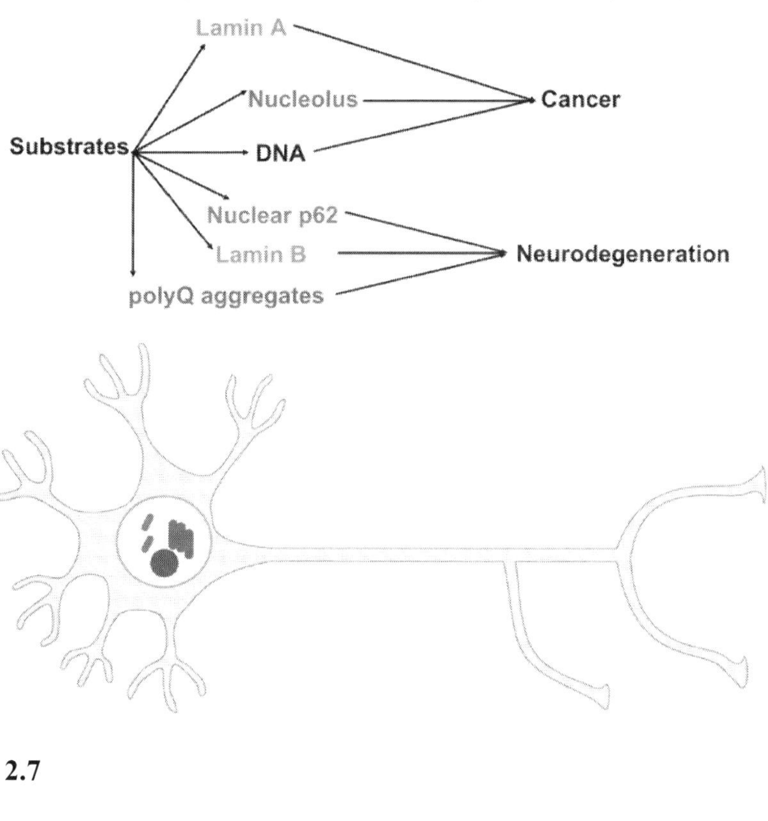

Figure 2.7

Mutations in lamin A/C of the nucleus are associated with progeria and muscular dystrophy, known as laminopathies, several cancer cells too such as that of the ovary, lung, prostate, and breast show lowered lamin A/C expression.

In the case of Hutchinson–Gilford progeria syndrome or HGPS, lamin A is altered to form progerin, which lacks a protease cleavage site to cause improper processing. This is attributed to mutations in the *LMNA* gene. The accumulation of this altered protein causes deformation of the nuclear envelope, cell cycle abnormalities, and early aging.

Lowered expression of lamin A/C is observed in epithelial cells of the human ovary and ovarian cancer cells that is associated with abnormalities in mitosis, mainly failure of mitosis. Research in cells with inhibited lamin A/C has shown tetraploid nuclei due to mitotic failure and also tripolar division.

In the case of ovarian cancer, extreme genetic variability can be linked to the formation of micronuclei in these cells. These micronuclei are often severed off from the nucleus to result in loss of chromosome(s). Also, the migration of cancer cells involves pressure on the nuclear envelope that causes the rupture of the nucleus. Further, lamin A/C inhibition is associated with nuclear budding. Ovarian cells with inhibited lamin A/C show breaking of nuclear material protrusions that lead to the formation of micronuclei to in turn cause aneuploidy. This has led to a model that associates the aneuploidy seen in ovarian cancers to aberrant expression of the nuclear lamin proteins.

(Smith et al, 2018)

Figure 2.8 below shows how the aneuploidy seen in cancer cells can be accounted for by defects in the nuclear envelope. The loss of proteins such as lamin A/C results in misshapen nuclei and higher occurrences of abnormal mitosis (Smith et al, 2018; *This is an open-access article distributed under the terms of the* ***Creative Commons Attribution License (CC BY)****. The use, distribution or reproduction in other forums is permitted, provided the original author(s) and the copyright owner(s) are credited and that the original publication in this journal is cited, in accordance with accepted academic practice. No use, distribution or reproduction is permitted which does not comply with these terms*):

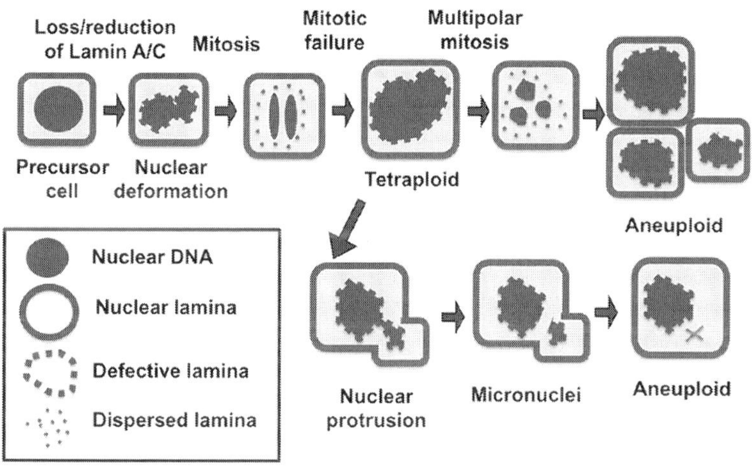

Figure 2.8

2.3 FINAL POINTS

Research is uncovering new attributes of the nucleus with alterations in its composition and shape influencing the health of a cell. The changes in nuclear proteins or shape can affect the cytoskeleton to influence the polarity of the cell while forces from the external surroundings and the cytoskeleton too have an effect on the nuclear envelope and the inner contents. Such changes become significant in the context of the migration of cells across narrow spaces. Researchers are slowly unveiling the role of the nuclear envelope, nuclear proteins, and the chromatin organization in health, gaining, and disease. Future studies on nuclear shape and dynamics can help to comprehend human disease and address the quest for new approaches to target diseases.

(Zwerger et al, 2011)

2.4 REFERENCES

1. Papandreou, M., Tavernarakis, N. Nucleophagy: from homeostasis to disease. *Cell Death Differ* **26,** 630–639 (2019). https://doi.org/10.1038/s41418-018-0266-5.
2. Shoji J-y, Kikuma T, Arioka M, Kitamoto K (2010) Macroautophagy-Mediated Degradation of Whole Nuclei in the Filamentous Fungus *Aspergillus oryzae*. PLoS ONE 5(12): e15650. https://doi.org/10.1371/journal.pone.0015650.
3. Martins F, Sousa J, Pereira CD, da Cruz E Silva OAB, Rebelo S. Nuclear envelope dysfunction and its contribution to the aging process. Aging Cell. 2020 May;19(5):e13143. doi: 10.1111/acel.13143. Epub 2020 Apr 15. PMID: 32291910; PMCID: PMC7253059.
4. Sadia Minhas, Muhammad Kashif, A. H. Nagi, "Evaluation of Various Nuclear Cytological Changes in Normal Buccal Mucosa and Peritumoural Area in Patients with Oral Squamous Cell Carcinoma Receiving Concomitant Chemoradiotherapy", Pathology Research International, vol. 2016, Article ID 6293795, 8 pages, 2016. https://doi.org/10.1155/2016/6293795.
5. Wang, A.S., Ong, P.F., Chojnowski, A. *et al.* Loss of lamin B1 is a biomarker to quantify cellular senescence in photoaged skin. *Sci Rep* 7, 15678 (2017). https://doi.org/10.1038/s41598-017-15901-9.
6. Smith ER, Capo-chichi CD and Xu X-X (2018) Defective Nuclear Lamina in Aneuploidy and Carcinogenesis. *Front. Oncol.* 8:529. doi: 10.3389/fonc.2018.00529.
7. Zwerger M, Ho CY, Lammerding J. Nuclear mechanics in disease. Annu Rev Biomed Eng. 2011 Aug 15;13:397-428. doi: 10.1146/annurev-bioeng-071910-124736. PMID: 21756143; PMCID: PMC4600467.

CHAPTER 3

THE NUCLEUS: INTERESTING RESEARCH

Shivsanjeevi Sripathi

CONTENTS

This chapter discusses several interesting and latest snippets of research on the nucleus of various cells.

3.1 THE VISCOELASTIC PROPERTIES OF THE NUCLEUS

The cells in our body on account of their active behavior and deformation due to mechanical stress from the environment can be described as exposed to complex and diverse surroundings. The extent of stress to which the cells of the cartilage experience can reach 10 MPa. Our cells have evolved to survive such mechanical stress arising due to fluids or strain by altering the metabolism and gene expression. In this context, researchers have been on the quest to explore how "flexible" the nucleus is.

Studies have pointed out at the involvement of the cytoskeletal proteins to monitor and control the shape of the nucleus. Based on the compression on the extracellular matrix, cells of the cartilage or chondrocytes have been shown to alter the volume and shape of the nucleus. Guilak et al (2000), isolated the nuclei of primary chondrocytes to assess the viscoelastic properties of cells and nuclei to assess if the nuclei could behave as a viscoelastic solid.

The isolated nuclei by both a mechanical approach (microdissection using micropipette aspiration) and a chemical one (detergent), as well as the cells, "displayed creep behavior typical of a viscoelastic solid" (in the words of the authors). This was described as "an initial jump in displacement with the application of step pressure followed by asymptotic creep displacement until equilibrium". It was thus concluded that the apparent mechanical properties of the cell can involve a major role by the nucleus given the viscoelastic properties of the cell nucleus demonstrated by the scientists.

(Guilak et al, 2000)

3.2 CELL MIGRATION AND THE NUCLEUS

The nucleus is described as the largest organelle in eukaryotic cells with its dynamics influencing the various aspects of a cell. In various activities of the cell including proliferation, differentiation or migration, the nucleus is specifically positioned at a place, based on the process. The functions of many specific cells such as immune cells, neurons, epithelial cells, and myofibers are dependent on the precise nuclear position and architecture.

A loss of regulation in these nuclear characteristics manifests as dysfunctional cells to cause disease. Research has shown the vital involvement of nuclear structure and positioning for cell migration. It is a two-sided coin: tissue development and homeostasis require cell migration while inflammation and metastasis in cancer too involve cell migration. The role of the cytoskeleton, cell adhesions, and nuclear positioning has been studied using 2D substrates while three dimensional (3D) substrates have uncovered novel factors functioning in cell migration.

So let us see how the nucleus is positioned during cell migration given that most cells have to pass through pores that are small. While the plasma membrane, smaller organelles, and cytoplasm can pass through pores, the large size of the nucleus makes it a challenge.

Cell migration necessitates a change in the architecture of cells. The cell gets polarity on the basis of the rearrangement of the cytoskeleton and organelles. In the case of mesenchymal cells, neurons, cancer cells, and polarized fibroblasts, the placement of the nucleus is at the rear of the cell in the direction of migration. Myosin and Cdc42 mediate the actin retrograde flow to cause this rearward nuclear movement. This involves the vital functioning of the LINC complex made of SUN proteins and nesprin to tether the nuclear envelope (NE) and the cytoskeleton. The tethering of the actin to the nucleus involves the formation of the TAN (Transmembrane Actin-associated Nuclear) lines: made of actin filaments, SUN2, and Nesprin-2. These lines facilitate the nucleus to be moved due to the actin retrograde flow.

The formation and dynamics of the TAN lines involve several proteins that can bring in several levels of regulation based on the process in a cell. The assembly of the TAN lines and the retrograde flow of dorsal perinuclear actin is mediated by AAA+ TorsinA and its activator LAP1. The LINC complex is stabilized at the TAN lines via SUN2 while the complex is anchored to the nuclear lamina by Samp1. Other proteins that connect the actin filaments to the LINC complex include regulators of actin dynamics such as Fascin and the formin FHOD1.

(Calero-Cuenca et al, 2018)

Figure 3.1 below shows the mechanisms unveiled so far for the positioning of the nucleus during cell migration. The actin cytoskeleton is connected to the nucleus via TAN lines with the main link as the LINC complex. The TAN lines are formed by the interaction of SUN2 and Nesprin-2G that binds to the actin filaments on the top of the nucleus. The nucleus is placed properly by the actin retrograde flow. The process involves several proteins such as TorsinA and LAP1 that assemble TAN lines; Lamin A/C and Samp1 facilitate the stabilization of the LINC complex at the nuclear envelope; the levels of regulation are increased by linking of Nesprin-2G and actin via Fascin and FHOD1

(Calero-Cuenca et al, 2018; *This is an open-access article under the CC BY license (http://creativecommons. org/licenses/by/4.0/): You are free to share — copy and redistribute the material in any medium or format; Adapt — remix, transform, and build upon the material for any purpose, even commercially*):

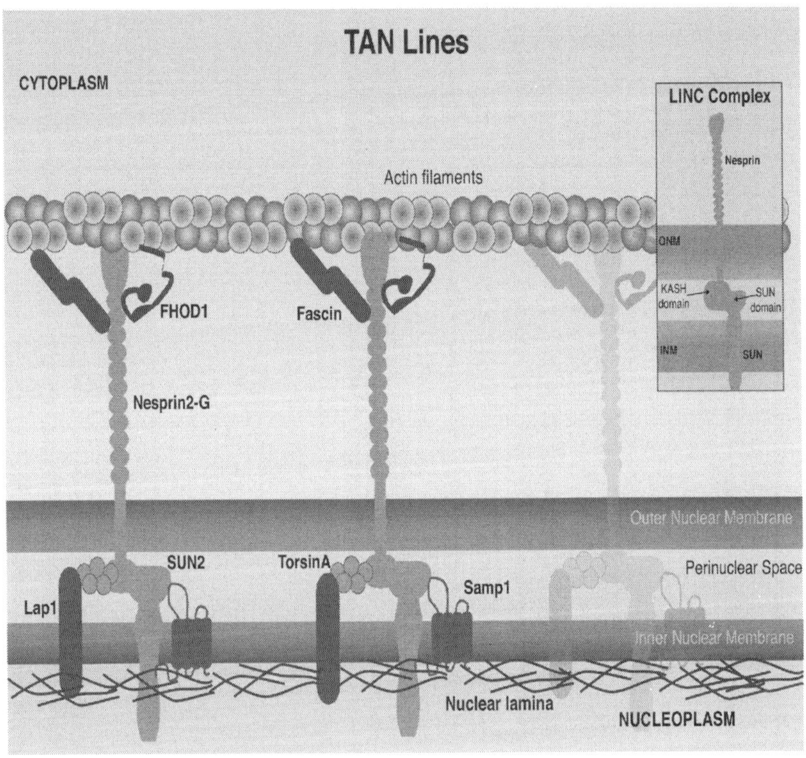

Figure 3.1

The translocation of the nucleus and the cell occurs in a myosin II-dependent mechanism with the involvement of the perinuclear actin network that confers protection to the nucleus. Studies have also shown the role of the formin FMN2 to offer protection from DNA damage and also influence nuclear positioning in melanoma cells. The nucleation of actin is augmented around the nucleus by the Arp2/3 complex in dendritic cells that perturbs Lamin A/ C to deform the nucleus and allow the cell to migrate through narrow pores. In the case of fibroblasts, the rotation of the nucleus involves the role of dynein.

Though the nucleus is considered stiff and large, live imaging of migrating cells has shown that the organelle is subject to deformation, pulling, and pushing to allow movement. Several possibilities have been proposed such as modulating the levels of euchromatin and heterochromatin to influence the stiffness of the nucleus. The functioning of the nucleus as a piston to boost the hydraulic pressure in the front of the cell has also been suggested. In the case of cells with higher migratory abilities, the expression of lamin A/C that controls the rigidity of the nucleus is lower.

It has also been observed that the nuclei of cancer cells and immune cells go through extreme pressure while passing through narrow spaces. This results in the deformation of the nucleus to cause a breakdown of the NE that is later addressed by the ESCRT III machinery. In the case of the development of muscle fibers, there is no such NE break down as the myonuclei squeeze to the periphery of myofibers. Alterations in lamin A/C are accounted for this mechanism.

Research on how immune cells are able to pass through the endothelial cells termed as the trans-endothelial migration (TEM) has revealed interesting insights. These cell barriers could be transposed across by both three-lobular neutrophils and uni-lobular T cells. The drilling through endothelial cells involves the vital role of the nucleus where small nuclear lobes enter lamellipodia to form a gap or pore to facilitate the migration.

(Calero-Cuenca et al, 2018)

Further research can highlight interesting mechanisms.

The following figure 3.2 below shows the positioning of the nucleus during cell migration. In **(a)**, TAN lines are formed by actin filaments and Nesprin-2 and SUN2 in the nuclear envelope to led to actin retrograde flow. In **(b)**, the pushing and pulling of the rigid nucleus results in deformation and breakdown of the nuclear envelope that is protected by the perinuclear actin network or repaired by the ESCRT-III system. In **(c) immune cells can place their nucleus in the front to induce bending of the actin filaments**

of the endothelial cells to facilitate transendothelial migration (Calero-Cuenca et al, 2018; *This is an open-access article under the CC BY license (http://creativecommons. org/licenses/by/4.0/): You are free to share — copy and redistribute the material in any medium or format; Adapt — remix, transform, and build upon the material for any purpose, even commercially*):

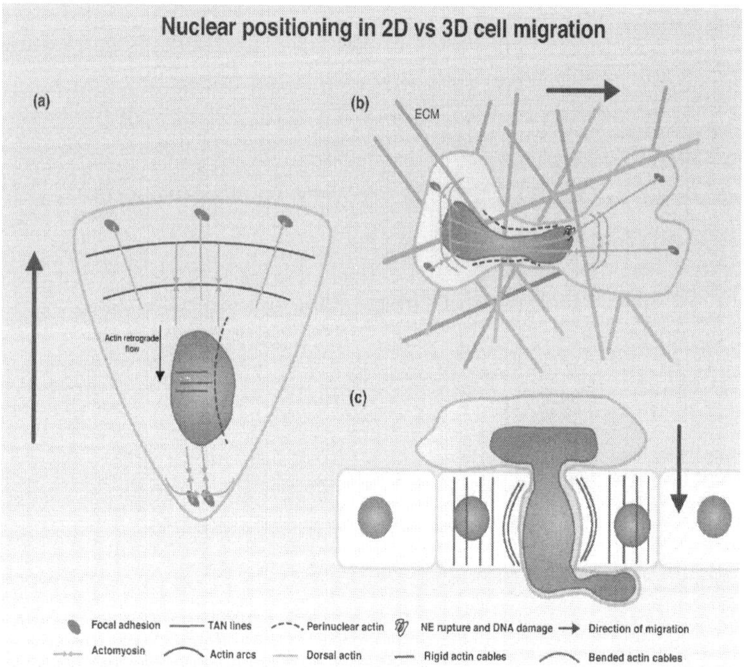

Figure 3.2

While it has been thought that the nucleus being the largest organelle in eukaryotic cells cannot be moved, research is showing that cell migration involves significant pushing or pulling or deformation of the nucleus. Interestingly, the ability of immune cells to use the nucleus to "drill" across endothelial cells has also been shown.

(Calero-Cuenca et al, 2018)

Figure 3.3 below shows the dynamics of the nucleus of an immune cell called a neutrophil. The nuclei of the cells are extensively deformed during transmigration across the endothelial cells. The proposed mechanism involves the generation of forces and contractility at the back of the cell to allow the nucleus to be pushed forward while the front end shows actin dynamics (Manley et al, 2018; *This is an open-access article distributed under the terms of the Creative Commons Attribution License (CC BY). The use, distribution, or reproduction in other forums is permitted, provided the original author(s) and the copyright owner(s) are credited and that the original publication in this journal is cited, in accordance with accepted academic practice. No use, distribution or reproduction is permitted which does not comply with these terms*):

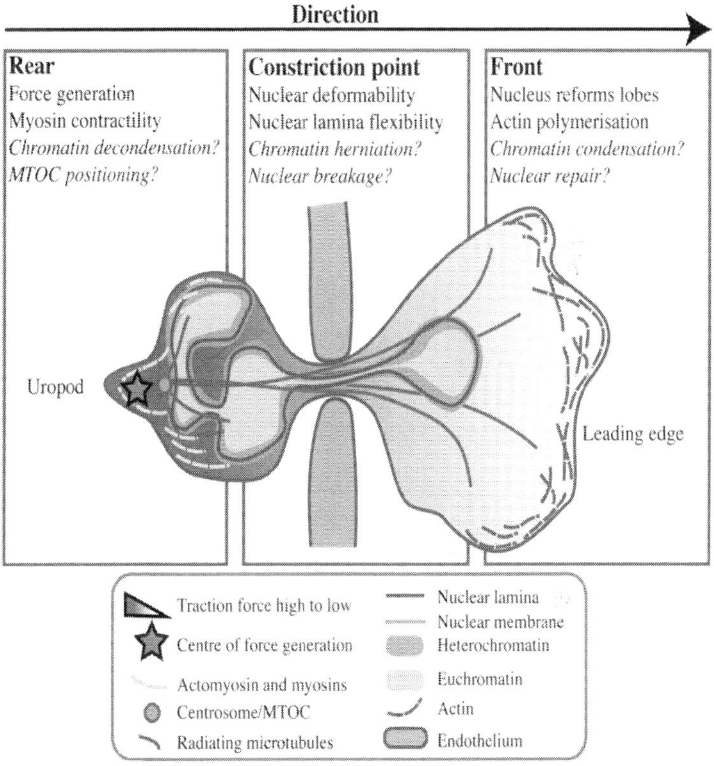

Figure 3.3

Further research can highlight interesting mechanisms regarding the dynamics of the nucleus in cell migration.

3.3 THE NUCLEUS AS A SENSOR

The phenomenon of planar cell polarity or precise coordination between opposite sides of individual cells that are described as having unique features is seen in the majority of cells. The optimum development and functioning of organs involve a vital functioning of the planar cell polarity. In the case of endothelial cells, shear flow has been shown to induce planar cell polarity.

Research has uncovered the role of several proteins/factors for endothelial cell polarization such as lowered GSK-3b activity or the tilting of the primary cilia in response to the shear stress induced by the flow. These mechanisms further require elucidation to address certain gaps. Interesting research from Tkachenko et al (2013) showed the ability of the nucleus of endothelial cells to sense the strength and direction of the blood flow. The application of hydrodynamic drag directly results in a displacement of nuclei downstream. This hydrodynamic drag applied to the nucleus is resisted by the acto-myosin cytoskeleton to exert an overall control over the response to shear flow. Downstream translocations of the nuclei were instantaneous when air bubbles were passed through a flow chamber that contained Human Umbilical Vein Endothelial Cells (HUVECs). Given the large extent and rapid time of response of these nucleus translocations, the involvement of intracellular transport was ruled out. As a result of this downstream displacement of the nuclei, the HUVECs were polarized against the direction of passage of the bubble.

Figure 3.4 below shows (a) the system to generate the air bubble-based shear stress on HUVECs; (b) The air bubble (from top to bottom) causes displacements of nuclei [The positions of nuclei are shown before (red) and after (brown) the bubble passes Microtubule Organization Centers (MTOCs) are shown by yellow dots; The cell polarizations are shown against the direction of the air bubble passing in (c) and the polarity seen in endothelial cells against the flow due to the nuclear displacement is shown in (d) (Tkachenko et al, 2013; *This is an Open Access article distributed under the terms of the Creative Commons Attribution License (http://creativecommons.org/licenses/by/3.0), which permits unrestricted use, distribution and reproduction in any medium provided that the original work is properly attributed*):

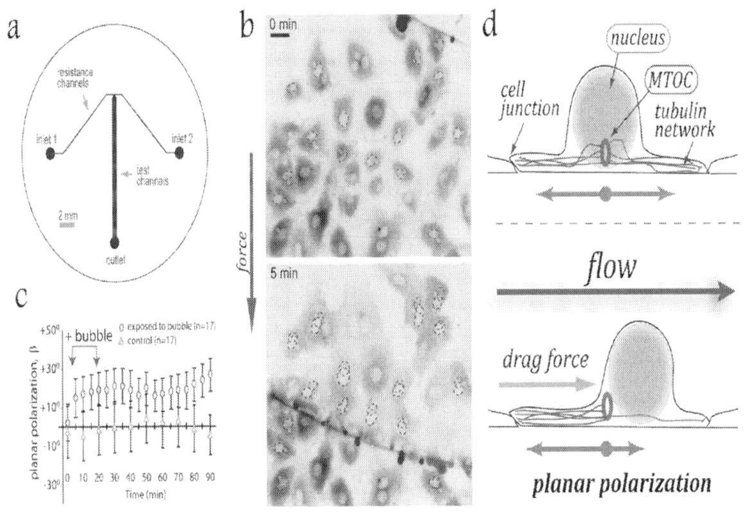

Figure 3.4

These observations in microfluidic perfusion chambers and mice open up another function of the nucleus in sensing flow and play a role in planar cell polarity.

(Tkachenko et al, 2013)

3.4 LIGHT, CHROMATIN AND THE NUCLEUS

Light is important for photosynthesis however when there is too much light, the outcome is reactive oxygen species (ROS) formation that can wreak havoc on the cell. The cell and genome are adversely affected by ultraviolet (UV) radiation from strong sunlight. This brings in a delicate balance between harvesting light for photosynthesis and responding to changing light. This brings in the role of photoreceptors (in the cytoplasm and nucleus) that are associated with chromophores to sense specific wavelengths. There are several intricate cascades that are triggered by these photoreceptors to control the expression of light-responsive genes. A factor called DE-ETIOLATED-1 (DET1) is part of the signaling in the Arabidopsis model that has been linked to nucleosomes opening up the connection between light and chromatin.

The expression of light-responsive genes involves an amalgamation of factors bound to chromatin and post-translational modifications of histones. Light signaling has been shown to be associated with changes in specific genes and chromosomes into the nuclear space to thus allow plants to bear solar radiation.

Figure 3.5 below shows the harvesting of photosynthetically active radiations (PAR) by photosynthetic pigments. Nuclear photoreceptors can set off pathways to adapt to the environment (Bourbousse et al, 2020; *this is an open-access article distributed under the terms of the Creative Commons Attribution License (CC BY). The use, distribution or reproduction in other forums is permitted, provided the original author(s) and the copyright owner(s) are credited and that the original publication in this journal is cited, in accordance with accepted academic practice. No use, distribution or reproduction is permitted which does not comply with these terms*):

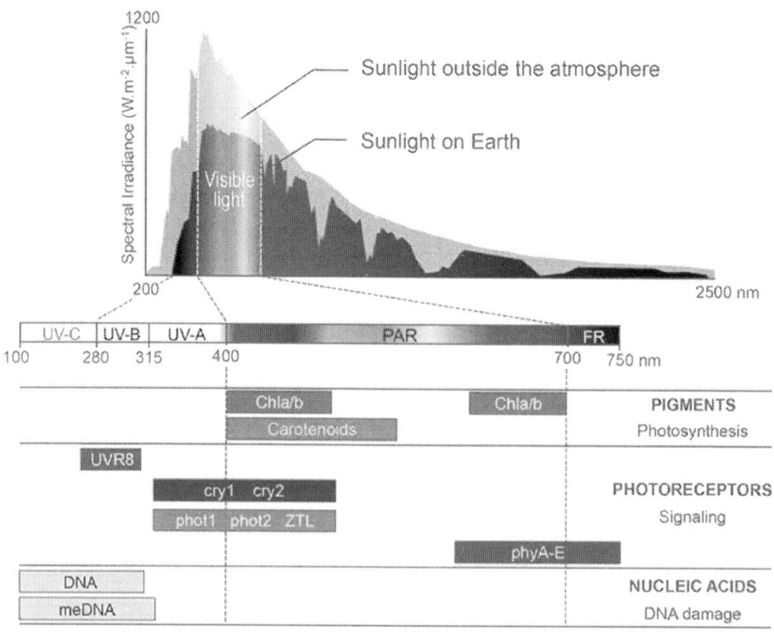

Figure 3.5

Studies on what is termed as the ***nuclear architecture*** in Arabidopsis have revealed that the process of cotyledon germination and the establishment of seedlings are based on light. The nuclei in dry seed cotyledons are tiny with the heterochromatin arranged as compact two to four subnuclear foci. This is suggested to be protective of drying conditions.

This heterochromatin is subjected to massive relaxation followed by an average expansion of the nucleus during the process of imbibition (Phase 1), either in dark or light. 3 days post-imbibition, Phase 2 is dependent on light (mainly blue) where the nucleus increases further in size with the formation of chromocenters or distribution of the heterochromatin as 8-10 sub-nuclear regions. If there is no light, CONSTITUTIVE PHOTOMORPHOGENIC 1 (COP1) and DE-ETIOLATED-1 (DET1) proteins suppress this compaction of the heterochromatin (Phase 3). The process of de-etiolation (conversion of etioplasts into chloroplasts) releases this arrest to be suggestive of the sensing of light vital for the development of the nucleus.

Further, during the process of de-etiolation, there is a repositioning of light-inducible genes such as *CAB* (chlorophyll a/b-binding proteins) and *RBCS* (*Rubisco small subunit*) from inside to the periphery of the nucleus. This is indicative of light influencing the reorganization of the nucleus for not only heterochromatin but also euchromatin.

Figure 3.6 below shows Phase 1 where there is an expansion of the heterochromatin and the nucleus during the process of seed imbibitions (Note the dispersion of the signal in methylated DNA immunolabeling). This is followed by Phase 2 where there is a continued expansion of the nucleus due to light perception and decondensation of heterochromatin if there is no light (Phase 3). Phase 4 is the development of a photomorphogenic organization during the process of de-etiolation that shows the aggregation of photoreceptors with the downstream factors and relocation of light-inducible genes to the periphery of the nucleus [d; days post-imbibition] (Bourbousse et al, 2020; *this is an open-access article distributed under the terms of the Creative Commons Attribution License (CC BY). The use, distribution or reproduction in other forums is permitted, provided the original author(s) and the copyright owner(s) are credited and that the original publication in this journal is cited, in accordance with accepted academic practice. No use, distribution or reproduction is permitted which does not comply with these terms*):

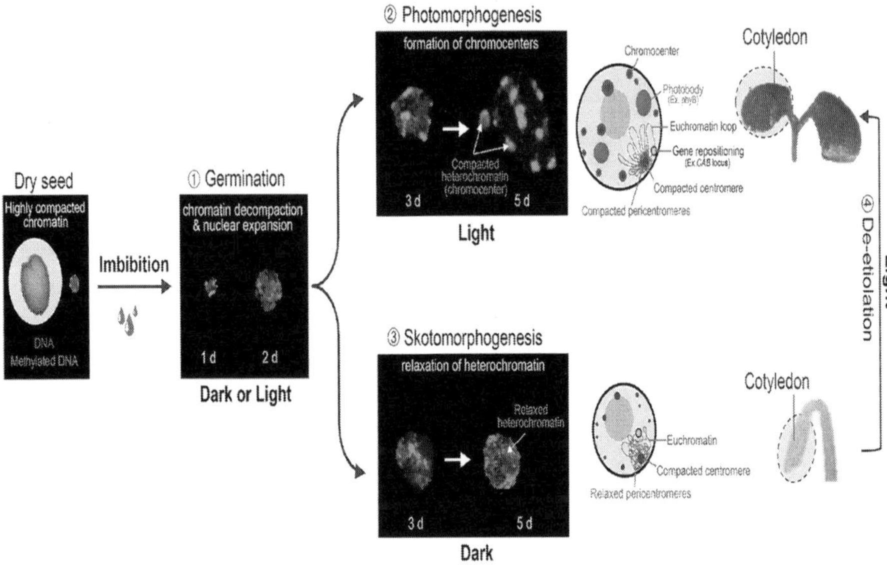

Figure 3.6

(Bourbousse et al, 2020)

3.5 THE SPECIALTIES OF THE NEUTROPHIL NUCLEUS

Described as a migratory and ameboid type of cell, a neutrophil is capable of high-speed migration and undergoing deformation. The average velocity of these cells is described as 100-times more rapid than that of invasive cancer cells and fibroblasts. This is accounted for their ability to reach a site where inflammation or infection is present. While the majority of other cells are not capable of passing through spaces that are lesser than 1.5 μm, neutrophils can pass across endothelial cell layers (around 1 μm) in order to reach tissues.

Interestingly, the migration of these cells is not from the "front" but from the back of the cell. There are pseudopodia at the front end and a contractile uropod at the rear. There is a squeezing motion where the cell is not pulled but pushed forward. It has been indicated by studies that the nucleus of these cells is not experiencing the force but maintains the force to allow efficient movement of the neutrophils.

The nuclear envelope proteins in a mature neutrophil nucleus are distinctive. For example, the expression of LBR (Lamin B receptor) is higher while that of LINC, LaminA/C, and LaminB1 are on the lower side. In mature neutrophils, the lamin present the most is LaminB2. The composition and development of the neutrophil nucleus show conservation across species. This is suggestive of the optimum functioning of neutrophils based on its flexibility and nuclear features.

The profile of the lamins is connected to the ability of the nucleus of neutrophils to become flexible and facilitate the movement of the cell. Research shows that the deficiency of LaminA/C results in the softening of the nuclei by half while overexpression causes impairment in the migration of cells in *in vitro* experiments. These are overall suggestive of the aforementioned lamin profile to facilitate flexibility of the nucleus that in turn allows the migration of these immune cells.

(Manley et al, 2018)

Figure 3.7 below shows the link between the expression of lamins and the nucleus of neutrophils. The number of lobules shows an increase during the maturation with parallel changes in the lamin profile (Manley et al, 2018; *This is an open-access article distributed under the terms of the Creative Commons Attribution License (CC BY). The use, distribution or reproduction in other forums is permitted, provided the original author(s) and the copyright owner(s) are credited and that the original publication in this journal is cited, in accordance with accepted academic practice. No use, distribution or reproduction is permitted which does not comply with these terms*):

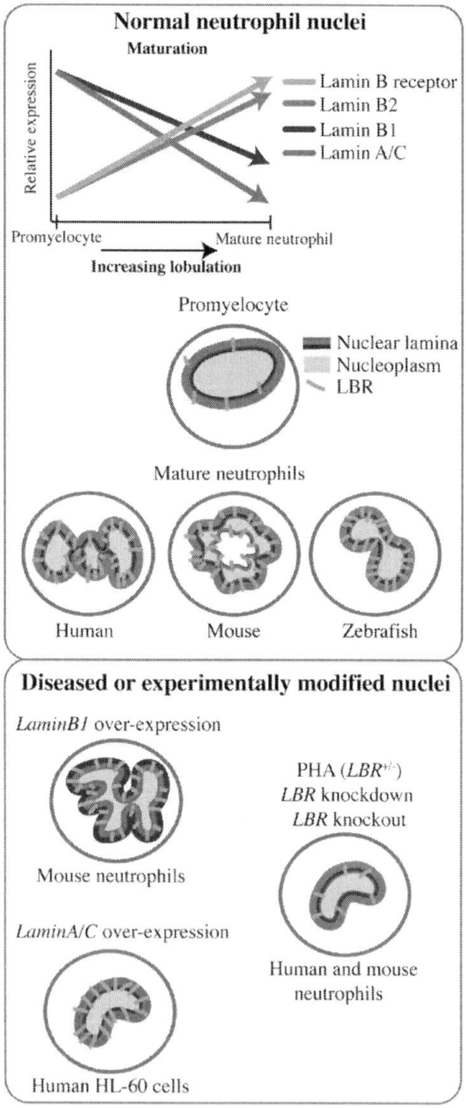

Figure 3.7

Table 3.1 below highlights the profile of the nucleus of a neutrophil and its connected function(s) (Adapted from Manley et al, 2018):

Table 3.1

Component	Level	Indicated role
LaminA/C	Downregulated	Allows for nuclear and cellular flexibility to pass through narrow spaces
LaminB1	Downregulated	The high expression has been shown to cause hyper-lobulation of the nucleus
LaminB2	Average expression	This is the major lamin in a mature neutrophil nucleus
Lamin B receptor (LBR)	High	Lowered expression reduced the number of nuclear lobes in Pelger-Huët Anomaly (a rare genetic disorder)
LINC proteins	Downregulated	The exact functioning and significance requires further research
Chromatin	Heterochromatin	The compactness of chromatin is suggested to allow for nuclear flexibility

Further, the role of the nucleus in these cells has also been extended to the formation of neutrophil extracellular traps (NETs) by neutrophils. As part of the immune response, the release of NETs made of antimicrobial peptides and chromatin webs allows for pathogens to be targeted. Yet, too much of NETs can manifest as diseases such as rheumatoid arthritis or lupus or even metastasis. As a response to microbes or chemical signals, the stimulation of neutrophils results in decondensation of chromatin, breakdown of the nucleus, and combining the chromatin to granule proteins that are then released from the cell. The process can be "vital" where the viability of the cells remains or "suicidal" where the chromatin is released violently with rupture of the cell membrane and cell death.

The process of the release of chromatin necessitates the partial disassembly of the nucleus. The exact functioning remains to be ascertained. The nuclei first (whether the pathway is suicidal or vital) are rounded from their multiple lobules. This is followed by decondensation of the chromatin associated with *in vitro* and *in vivo* nuclear swellings. In the suicidal pathway, the swelling of the chromatin reaches a tipping point to break open the nuclear membrane and later the cell membrane. The swelling and decondensation are associated with the translocation of cytoplasmic granule proteins into the nucleus like neutrophil elastase, myeloperoxidase, and SerpinB1. The involvement of LBR and lamins in the detachment of the chromatin from the nuclear envelope has been outlined by research.

In the case of "suicidal" NETosis, there is an explosive extrusion of the decondensed chromatin and cytoplasmic granules from the neutrophil. For "vital" NETosis , vesicles with granule proteins and DNA bud off from the nucleus to fuse with the cell membrane and lyse. These enucleate neutrophils retain their viability and functions of chemotaxis and phagocytosis!

Figure 3.8 shows the scheme of NET formation and the role of the nucleus (Manley et al, 2018; *This is an open-access article distributed under the terms of the Creative Commons Attribution License (CC BY). The use, distribution or reproduction in other forums is permitted, provided the original author(s) and the copyright owner(s) are credited and that the original publication in this journal is cited, in accordance with accepted academic practice. No use, distribution or reproduction is permitted which does not comply with these terms*):

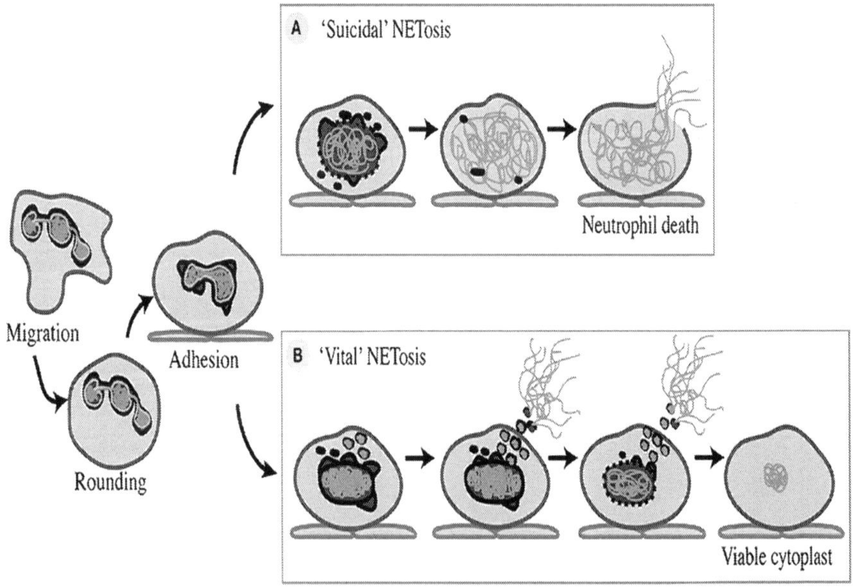

Figure 3.8

The plasticity of the nucleus of cells such as neutrophils is attributed to the envelope proteins such as lamins. These proteins play a role in influencing the flexibility of the cells to allow the process of immune protection and migration across small pores.

(Manley et al, 2018)

3.6 REFERENCES

1. Farshid Guilak, John R. Tedrow, and Rainer Burgkart. Viscoelastic Properties of the Cell Nucleus. Biochemical and Biophysical Research Communications 269, 781–786 (2000) doi:10.1006/bbrc.2000.236.
2. Calero-Cuenca FJ, Janota CS, Gomes ER. Dealing with the nucleus during cell migration. Curr Opin Cell Biol. 2018 Feb;50:35-41. doi: 10.1016/j.ceb.2018.01.014. Epub 2018 Feb 20. PMID: 29454272.
3. Manley, H. R., Keightley, M. C., & Lieschke, G. J. (2018). The Neutrophil Nucleus: An Important Influence on Neutrophil Migration and Function. *Frontiers in immunology*, *9*, 2867. https://doi.org/10.3389/fimmu.2018.02867.
4. Eugene Tkachenko, Edgar Gutierrez, Semion K. Saikin, Per Fogelstrand, Chungho Kim, Alex Groisman, Mark H. Ginsberg. The nucleus of endothelial cell as a sensor of blood flow direction. Biology Open 2013 2: 1007-1012; doi: 10.1242/bio.20134622.
5. Bourbousse, C., Barneche, F., & Laloi, C. (2020). Plant Chromatin Catches the Sun. *Frontiers in plant science*, *10*, 1728. https://doi.org/10.3389/fpls.2019.01728.

CHAPTER 4

ORIGIN OF THE CELL NUCLEUS, MITOSIS AND SEX: ROLES OF INTRACELLULAR COEVOLUTION

Thomas Cavalier-Smith

Department of Zoology, University of Oxford, South Parks Road, Oxford, OX1 3PS, UK

ABSTRACT

Background

The transition from prokaryotes to eukaryotes was the most radical change in cell organisation since life began, with the largest ever burst of gene duplication and novelty. According to the coevolutionary theory of eukaryote origins, the fundamental innovations were the concerted origins of the endomembrane system and cytoskeleton, subsequently recruited to form the cell nucleus and coevolving mitotic apparatus, with numerous genetic

Citation: Cavalier-Smith: Origin of the cell nucleus, mitosis and sex: roles of intracellular coevolution. Biology Direct 2010 5:7

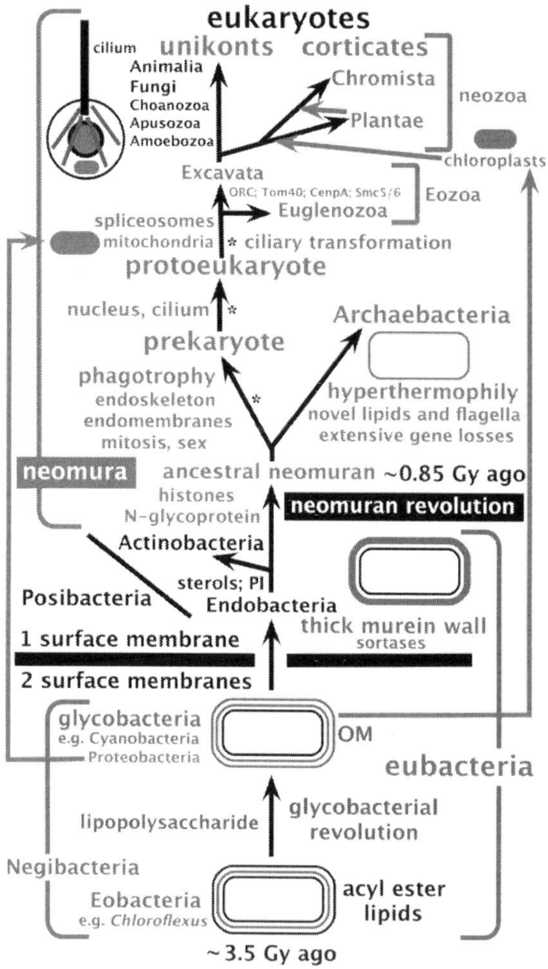

Figure 1: The tree of life and major steps in cell evolution. Archaebacteria are sisters to eukaryotes and, contrary to widespread assumptions, the youngest bacterial phylum [6,13]. This tree topology, coupled with extensive losses of posibacterial properties by the ancestral archaebacterium, explains (without lateral gene transfer) how eukaryotes possess a unique combination of properties now seen in archaebacteria, posibacteria and α-proteobacteria. Eukaryote origins in three stages indicated by asterisks probably immediately followed divergence of archaebacteria and eukaryote precursors from the ancestral neomuran. This ancestor arose from a stem actinobacterial posibacterium by a quantum evolutionary shake-up of bacterial organization - the neomuran revolution

[6,12]: surface N-linked glycoproteins replaced murein; ribosomes evolved the signal recognition particle's translational arrest domain; histones replaced DNA gyrase, radically changing DNA replication, repair, and transcription enzymes. The eukaryote depicted is a hypothetical early stage after the origin of nucleus, mitochondrion, cilium, and microtubular skeleton but before distinct anterior and posterior cilia and centriolar and ciliary transformation (anterior cilium young, posterior old: [3]) evolved (probably in the cenancestral eukaryote [9]). Kingdom Chromista was recently expanded to include not only the original groups Heterokonta, Cryptista and Haptophyta, but also Alveolata, Rhizaria and Heliozoa [9], making the name chromalveolates now unnecessary. Excavata now exclude Euglenozoa and comprise just three phyla: the ancestrally aerobic Percolozoa and Loukozoa and the ancestrally anaerobic Metamonada (e.g. *Giardia*, *Trichomonas*), which evolved from an aerobic *Malawimonas*-related loukozoan. Sterols and phosphatidylinositol (PI) probably evolved in the ancestral stem actinobacterium but the ancestral hyperthermophilic archaebacterium lost them when isoprenoid ethers replaced acyl ester lipids.

Evolution of complex characters typically involves preadaptation, radical mutational innovation, and different selective forces acting in succession [3,6,10]. Here I paint an integrated picture of how the nucleus, sex, and the eukaryotic cell cycle originated and congealed into a novel, unified, and very conservative cellular lifestyle during later stages of the conversion of a bacterium into a eukaryote. In addition to establishing the phylogenetic context (Fig. 1) there are three crucial problems for understanding the origin of the nucleus [5]: (1) assembly of endomembranes around chromatin (the DNA-histone complex); (2) evolution of the nuclear pore complex (NPC), which crucially allows a channel between nucleoplasm and cytoplasm; and (3) origin of centromeres and mitotic spindle, without which nuclear chromosomes cannot be stably inherited. As first argued 30 years ago [11], origin of the cell nucleus cannot be understood in isolation from other major innovations of the eukaryotic cell; intracellular coevolution among different cell constituents that interact physically or that profoundly affect selective forces acting on each other is the key to understanding eukaryote origins [3,4]. Elements of the present synthesis were presented then [11], e.g. that sex began even before the nuclear envelope, i.e. in a prekaryote phase of evolution (see Fig. 1) and the dominant selective advantages. However, the phylogenetic context has changed dramatically with our now much more robust understanding of cell phylogeny (1) [3,7,8,12,13]. Moreover, genomics has enabled molecular origins of many key eukaryotic constituents, including NPCs, to be traced [14-17], whilst advances in molecular cell biology tell us how nuclei actually assemble [18,19] and function. Building on these insights, I now propose the first specific physical mechanism for evolving nuclear envelope architecture and explain its major genetic consequences and why other theories are inadequate.

As the field of eukaryogenesis has been confused by a plethora of contradictory ideas, some not compatible with established evidence, before presenting the novel explanations I summarise two areas to put them in context: (1) the phylogenetic origin of the eukaryotic components, and (2) the origin of the endomembrane system and cytoskeleton. I only outline the conclusions, giving references for details, as most of the evidence and arguments is not new, being already published. Because the nature of molecular changes during major evolutionary transitions is more diversified and complex than some molecular evolutionists have realised, I also preface my original explanations of the origin of the nucleus with an outline of some basic but widely neglected evolutionary principles that apply to all such major innovations in body plan. This background is rather long because the proper evolutionary context is so important: the nucleus did not evolve on its own; explanations of its origin make no sense without understanding the prior evolution of the endomembrane system of which its envelope is a specialised part. Intracellular coevolution of about a 100 novel properties is at the core of understanding eukaryogenesis.

Phylogenetic Context for Eukaryogenesis

Eukaryote cells are all evolutionary chimaeras of an ancestrally phagotrophic host cell with nucleus, endomembranes, and endoskeleton [3] and an enslaved α-proteobacterium converted into a mitochondrion close to the time when the nucleus itself originated, i.e. prior to divergence of any extant eukaryotic lineages (Fig. 1) [20]. Contrary to some assumptions [17,21], the host for that symbiogenesis was not an archaebacterium, but an otherwise fully developed early eukaryote with NE and cilium (a protoeukaryote) or else an intermediate stage (prekaryote) that had already evolved rudiments of phagocytosis (the likely means of engulfing the α-proteobacterium) and internal membranes already differentiated into a primitive ER and peroxisomes, endoskeleton, centrosomes and mitosis (see [3,6,20,22] for further explanation). Fig. 1 emphasizes the key importance for early cell evolution of ancestral groups like Posibacteria and Eobacteria that are necessarily paraphyletic (in contrast to derived holophyletic groups like archaebacteria and actinobacteria), but which are phylogenetically perfectly respectable, the 'arguments' against them being fundamentally flawed (see [23]).

Figure. 1 differs from many widely discussed views of the tree of life in three major respects: the position of the root of the whole tree, the position of the eukaryotic root, and in the idea that both archaebacteria and eukaryotes evolved from Posibacteria. Though these topics are explained in detail in other papers, many readers may not have assimilated the evidence therein that rather strongly supports them, so I shall begin by outlining the evidence for these interpretations and add a few novel arguments and new evidence for them and explain the flaws in alternative ideas on the rooting and topology of the tree.

Clade Neomura and Its Posibacterial Origin

Archaebacteria are clearly related to the eukaryote host (together forming a clade called neomura [4,12]). But there is no sound evidence that archaebacteria are directly ancestral to eukaryotes. Instead several arguments show they are their sisters [6,12,13]. Thus the >20 features shared by both groups but absent from eubacteria (e.g. N-linked glycoproteins, more complex RNA polymerases, core histones) are not specifically archaebacterial, but neomuran characters that evolved in their common ancestor during the neomuran revolution [4,6,12,13]. Purely archaebacterial characters (notably unique isoprenoid ether lipids and flagella) evolved in the ancestral archaebacterium after it diverged from the prekaryote lineage [12,13]. Moreover, genes shared by eukaryotes and eubacteria, but not archaebacteria (e.g. MreB that became actin [3,6], and eubacterial surface molecules that became NE lamin B receptors [14], and enzymes making acyl ester phospholipids), were probably lost by the ancestral archaebacterium, which apparently underwent massive gene loss during its secondary adaptation to hyperthermophily [12,13]. In addition to those earlier arguments, the most comprehensive multigene analysis to date convincingly places archaebacteria as a holophyletic clade that is sister to eukaryotes, not ancestral to them [24]. However, these authors confusingly refer to the 'deep archaeal origin of eukaryotes' despite their strong evidence that all extant archaebacteria form a derived clade not a paraphyletic ancestral group. The phrase 'archaeal origin' wrongly implies that the common ancestor of eukaryotes and archaebacteria had the specific positive attributes of archaebacteria that distinguish them from both eukaryotes and eubacteria, of which there are very few: notably the isoprenoid ether lipids, archaeosine modified rRNAs, flagella, and duplicate versions of DNA polymerase B [25].

It is unparsimonious to assume that such characters were present in and then lost by the ancestors of eukaryotes. Though the replacement of archaebacterial lipids by acyl ester lipids derived from the enslaved proteobacterial ancestor of mitochondria is a formal possibility [26], it would be evolutionarily extremely onerous and thus unlikely, and phylogeny gives no convincing reason to assume it in the first place. Moreover, the hypothesis of replacement by archaebacterial lipids by eubacterial lipids from the α-proteobacterial symbiont totally fails to explain the origin of phosphatidylinositol, which played a key role in eukaryogenesis [27] and is present in all the actinobacterial relatives of neomura but never in archaebacteria or proteobacteria. Thus, it is far more likely that both archaebacteria and eukaryotes evolved from a common ancestor that was a prokaryote with acyl ester lipids including phosphatidylinositol, but which had not yet evolved either the specifically archaebacterial properties like isoprenoid ether lipids or any eukaryotic properties.

Sterol evolution even more strongly refutes the idea that eukaryotes evolved from archaebacteria and independently shows that neomura are most closely related to actinobacteria. Sterols in actinobacteria and eukaryotes are synthesised from squalene, as are the hopanoids of eubacteria. In all posibacteria squalene is produced from isopentenyl diphosphate (IPP), which is also the precursor for the isoprenoid tails of archaebacterial lipids; in posibacteria, archaebacteria, and eukaryotes that never have plastids (which use instead the cyanobacterial DOX isoprenoid pathway) IPP is generated by the mevalonate synthetic pathway, the enzymes of which were clearly in place and inherited vertically from the last common ancestor of Posibacteria and neomura [28,29]. As the enzymes that convert IPP into sterols are entirely absent from archaebacteria and mostly absent from α-proteobacteria, this simultaneously refutes the popular but totally erroneous ideas that archaebacteria were directly ancestral to eukaryotes [26,30,31] and that eukaryotes got sterols from the enslaved mitochondrion [26,31-33]. Actinobacteria are the only bacteria in which many genes needed for making sterols are phylogenetically widespread and of ancient origin within the group. Sequence trees for four major enzymes of sterol synthesis refute the idea that any of these genes entered actinobacteria by lateral gene transfer [34] and are totally consistent with the vertical descent of sterol biosynthesis from an actinobacterium-like posibacterium to the first eukaryote (and their loss in the ancestral archaebacterium when replacement of acyl esters by isoprenoid ethers provided an alternative and superior means of making membranes more rigid). Oddly, though recognising that their trees rule out lateral transfer from eukaryotes to actinobacteria, Desmond and Gribaldo [34] evade the obvious conclusion that Posibacteria were indeed ancestral to neomura by postulating lateral transfer (LGT) of these genes from a stem pre-eukaryotic lineage into actinobacteria, despite there being no evidence whatever for that implausible and unparsimonious scenario, which would require that Actinobacteria are younger than pre-eukaryotes. The first enzyme of sterol synthesis for squalene monooxygenation (making squalene epoxide) is so widespread in actinobacteria that it must have been present in their last common ancestor [34]; elsewhere in prokaryotes it is known only from a few gamma and delta proteobacteria and one planctomycete (all members of the clade Gracilicutes [13]); as the trees do not require any LGT it probably evolved in the last common ancestor of Posibacteria and Gracilicutes after the prior divergence of Cyanobacteria and the oxygenation of the atmosphere; it is entirely absent from archaebacteria and α-proteobacteria. As sterol synthesis requires oxygen its loss by secondarily or facultatively anaerobic lineages is unsurprising (the likelihood that the ancestral archaebacterium was largely anaerobic [12] is another reason why it lost sterols).

The second enzyme of the sterol synthesis pathway, oxidosqualene cyclase, catalyzing cyclisation of squalene epoxide to make lanosterol and/or cycloartenol is even more widespread in eubacteria, being present in both posibacterial subphyla (Actinobacteria, Endobacteria) as well as Proteobacteria (including even α-proteobacteria), Planctobacteria, and Cyanobacteria, so probably evolved even earlier before Cyanobacteria diverged from the other groups, and was presumably never present in Eobacteria and lost by Sphingobacteria, Spirochaetae, and Archaebacteria. The tree suggests that one planctobacterium (*Stigmatella*) replaced its own enzyme by one from eukaryotes, but gives no evidence for LGTs amongst eubacteria, contrary to the authors assumption [34]. Such replacement by LGT of one enzyme within a pathway is mechanistically simple, but there is no evidence for LGT of the whole pathway at any time in the history of life (by contrast symbiogenetic replacement by whole cell enslavement did allow the mevalonate part of the pathway to be replaced by that of cyanobacteria). The third enzyme in the pathway that catalyses C14 demethylation of lanosterol is known only from the order Actinomycetales (widespread) within Actinobacteria and from one delta and one gamma proteobacterium; as the tree does not support the idea of LGT, most likely it evolved at the same time as the first enzyme but was lost (or evolved beyond bioinformatic recognition) more often. The enzyme DHCR24, which makes the more complex sterols ergosterol and cholesterol, is present widely and phylogenetically deeply in Actinomycetales within Actinobacteria and is sister to its eukaryotic homologue [34] if the tree is rooted between them and the β-proteobacterium *Rhodoferax* in accord with Fig. 1, suggesting that this enzyme also originated at the same time as enzymes one and three but was lost even more often. Homologues were detected in only one other bacterium: *Methylococcus*; its sequence branches well within opisthokonts and was therefore probably acquired by LGT from an animal; however there is no evidence for LGT for that gene provided one roots the tree correctly. The simplest interpretation of the alternative lanosterol and cycloartenol pathways in eukaryotes [35] is that the first eukaryote inherited the posibacterial oxidosqualene cyclase vertically and that it was mutationally modified in plants at the time of origin of plastids and to make cycloartenol preferentially and later transferred to other eukaryotes by secondary symbiogenesis (i.e. to chromists and photosynthetic euglenoids); the sequence tree [34] is consistent with the rooting of eukaryotes within Eozoa (Fig. 1) and refutes my old idea that the plant enzyme came from the cyanobacterial ancestor of plastids [4].

Thus sterol and phosphatidylinositol evolution independently refute the idea that eukaryotes evolved from archaebacteria and both strongly indicate that the closest relatives to neomura are actinobacteria (in agreement with a dozen other characters [12]). However, the evolution of archaebacterial lipids and neomuran glycoproteins suggests that neomura may have evolved from the other posibacterial subphylum, Endobacteria. Homologues of the glycosyl transferases that make N-linked glycoproteins were detectable only in Endobacteria among eubacteria [13] and geranylgeranylglyceryl phosphate synthase (GGGPS) the enzyme that attaches isoprenoid tails to *sn*-Glycerol-1-P to make the membrane lipids of archaebacteria is known only from Endobacteria (specifically Bacillales) and the sphingobacterium *Cytophaga*, making it likely that Endobacteria rather than Actinobacteria were ancestral to neomura. This evidence for an endobacterial origin of neomura can be readily reconciled with the more extensive evidence for their actinobacterial affinities by the posibacterial tree topology of Fig. 1, where Endobacteria are shown as ancestral to both neomura and their sister Actinobacteria. We need only postulate that the cenancestral actinobacterium lost glycosyl transferase and GGGPS after it diverged from neomura and that phosphatidylinositol evolved immediately prior to that bifurcation and was lost only by archaebacteria (together with other acyl esters). This topology also allows the extra 5' Alu domain of the neomuran signal recognition 7SL RNA to have been inherited directly from Endobacteria [12], making it unnecessary to postulate that the positionally equivalent domain present in some Endobacteria (alone among eubacteria) is convergent [13] - assuming that 5' domain was lost in the ancestral actinobacterium. As previously discussed [13], the other key enzyme for the archaebacterial replacement of eubacterial lipids, *sn*-glycerol-1-phosphate dehydrogenase, which makes their unique *sn*-glycerol-1-phosphate, almost certainly evolved from a known posibacterial homologue (also present in *Thermotoga* and Proteobacteria [28,29]). The idea that archaebacterial lipids evolved independently of eubacterial biosynthetic pathways and the idea that their cells evolved independently of eubacterial cells [36,37] are both utter nonsense.

If actinobacteria are holophyletic (Fig. 1), there is also no need to assume that any of the five unique proteins of actinobacteria [38] or any of the actinobacterially unique paralogues of more widespread proteins like the iron uptake regulator Fur [39] were lost by the ancestor of neomura. However, one would have to assume that the most divergent actinobacterial branches had lost 20S proteasomes, as they are restricted to Actinomycetales [13]. Skophammer et al. [40] suggest that archaebacteria plus Endobacteria are a clade because of two claimed shared indels; however, it is evident that one gene pair they considered are not really paralogues and the other is self contradictory [41] so there is no convincing evidence against the topology shown in Figure 1. A quaternary structure argument for dihydroorotate hydrogenase (PyrD) evolution [41] supports a common ancestry for archaebacteria and Endobacteria; but that does not mean that they alone form a clade, for we all accept that the ancestral eukaryote was cladistically closer to Archaebacteria than Endobacteria, so it must have lost the PyrD 1B paralogue; an additional loss by the ancestral actinobacterium reconciles their argument with Fig. 1. An indel argument to exclude the root of the tree of life from Actinobacteria [42] actually excludes it only from the orders Actinomycetales and Bifidobacteriales, as their analysis included no DNA gyrase GyrA proteins from the three most deeply branching orders. But that limitation of the argument does not matter, as there was never any reason to think the root was within Actinobacteria in the first place. My own alignment indicates that the only available GyrA from the deepest branching actinobacterium (*Rubrobacter*) does not have the four amino acid insertion found in other actinobacteria, suggesting that it evolved after the first internal divergence, possibly substantially later (incidentally the insertion region seems incorrectly aligned in [42] and the gap should probably be moved by five amino acids). One cannot use this indel to argue against the topology or rooting of Fig. 1 because when histones evolved in the neomuran ancestor DNA gyrase was replaced by DNA topoisomerase VI [12], whose B subunit probably evolved from GyrB, but whose A subunit is so radically different that they cannot be aligned with GyrA [43,44]. A eukaryote-specific topoisomerase (IIA) probably also evolved from DNA gyrase by fusion of GyrA and GyrB to make a chimaera also so different from its eubacterial ancestors that one cannot apply the indel argument to it. Given the ancestral neomuran transformations of gyrase into novel topoisomerases, the very few archaebacterial GyrAs that can be aligned with those of eubacteria almost certainly entered archaebacteria by LGT from eubacteria [43], so the absence in them of the higher actinobacterial 4-amino acid insertion [42] must not be used to argue against actinobacteria being sisters of neomura (Fig. 1).

The above arguments from eukaryote and archaebacterial lipid evolution strongly contradict (and are more compelling than) a recent 53-gene analysis in which, in contrast to standard phylogenetic methods that show archaebacteria as holophyletic sisters of eukaryotes, a theoretically superior heterogeneous method shows archaebacteria as paraphyletic ancestors to eukaryotes [45]. Several statistically strongly supported branches within eukaryotes on that tree are topologically incorrect, so it cannot safely be concluded (as the authors did) that the grouping of eukaryotes as sisters to crenarchaeotes alone is not also a phylogenetic reconstruction error - as I think is likely. The same problem applies to their analyses of rRNA and 41 protein genes [46]. Virtually all these genes underwent marked episodic accelerated evolution in the eukaryote stem, making them so substantially different from those of archaebacteria that reconstructing the correct tree with confidence is extremely difficult. None of these trees adequately test my thesis about the sister relationship of neomura and actinobacteria as no actinobacteria were included and taxon sampling was generally too sparse to get the best trees. Given that methods and datasets conflict and that the topology within eukaryotes is suspect (and contradictorily different) in all the trees, even though it ought to be easier to reconstruct, I do not share the authors' hope that this type of analysis can establish the historical truth by itself unambiguously enough to be trusted. Comparison of their overall tree with their eukaryotes only tree shows that including the prokaryotic outgroups changes the topology within the eukaryotes and misroots the eukaryote part of the tree. As the stem at the base of neomura is even longer than the stem at the base of eukaryotes it is likely to cause even worse problems of misrooting within the basal neomuran branches. Therefore one can have no confidence in the conclusion of this analysis. The difficulty of deciding by even the best current sequence tree methods whether archaebacteria are holophyletic or paraphyletic emphasizes two things: (1) we should give more weight to other phylogenetic evidence for establishing the correct phylogeny, as I and Lake [40,42,47-51] have, and (2) it is very unlikely that archaebacteria can be several times as old as eukaryotes, as the assumption that archaebacteria are as old as eubacteria (e.g. [36]) would require given the fossil evidence that eubacteria are at least 2.5 and more likely 3-4 times older than eukaryotes [6]; if archaebacteria were really 2.5-4× older than eukaryotes, sequence trees should place eukaryotes clearly relatively shallowly within one or other of the two archaebacterial subphyla, which they never do. By far the most parsimonious interpretation of the overall evidence concerning the topology and root of the tree of life is that the common ancestor of neomura was neither an archaebacterium nor a eubacterium (to label it as either is seriously confusing), but a transitional intermediate between these two major groups, which had eubacterial/eukaryotic type lipids, but neomuran N-linked surface glycoproteins and DNA-handling enzymes [3,4,12].

As this transitional 'missing link' arose somewhat earlier than eukaryotes themselves, during what I named the 'neomuran revolution', the reader is referred to earlier discussions of this enabling revolution in cell structure, and of the compelling evidence from the fossil record and transition analysis that eubacteria are paraphyletic and ancestral to neomura and very much older [6,13]. These papers refute the unwarranted, widespread assumption that the root of the tree of life is between neomura and eubacteria (assumed for example by [52], notwithstanding their mistaken assertion that their Fig 1. tree was 'unrooted'). They detail why that is incorrect, and why the root is within photosynthetic gram-negative eubacteria (Negibacteria: Fig. 1, where ancestrally photosynthetic taxa are green or purple) [6,12,13]. The conclusion that the root of the bacterial tree is within Negibacteria has been questioned on the basis of indel distributions [40,42,47-51], but Valas and Bourne [41] have re-examined these indels critically in the light of protein three dimensional structure and show that the supposed contradictions to my thesis cannot be substantiated and that a root of the tree within Negibacteria, specifically beside or within Chlorobacteria, remains the best interpretation for rooting the tree of life.

As eubacteria are the basal, most ancient group of cells from which neomura evolved, far more genes than often supposed were inherited by the ancestral prekaryote vertically from the eubacterial ancestor of neomura and were already present before mitochondria and nuclei evolved. The absence of numerous eukaryotic gene homologues in archaebacteria, and their presence in many eubacteria, is often used to suggest that they were acquired from the enslaved α-proteobacterium or by independent lateral gene transfer (LGT) [17]. However, that conclusion is probably wrong, being based on dubious assumptions: that LGT is easier than multiple gene losses; that the host was an archaebacterium not a prekaryote derived from the neomuran ancestor. The eubacterial ancestor of neomura could not have been a negibacterium with two bounding membranes, but was a posibacterium with a single surface membrane, like neomura; probably a stem actinobacterium, i.e. an early intermediate between Endobacteria and crown Actinobacteria [4,6,12,13]. As such, it was probably extremely gene-rich. Thus archaebacteria are probably secondarily simplified, their ancestor having lost many hundreds of eubacterial genes during its novel adaptation to hyperthermophily after its divergence from the eukaryote ancestor, e.g. the loss of most genes for aerobic metabolism, including the cytochrome P450s that were precursors of ER respiration, which probably was not derived from the proteobacterial symbiont (see [22] for discussion of the likely gene numbers of the putative intermediates and of the relative contributions of the two major gene donors in eukaryogenesis, which corrects some widespread misconceptions). Like mycoplasmas, archaebacteria are highly derived and specialised, not primitive, bacteria. By contrast their large-celled aerobic sister group, the prekaryote lineage became far more genetically complex through the greatest burst of gene (and probably genome) duplication in the history of life. By the time the nucleus began to evolve, members of this lineage had ceased to be bacteria, having already evolved ER and rudimentary endoskeleton [6,12].

The Eozoan Root of the Eukaryote Tree between Euglenozoa and Neokaryotes

Just as one must work upwards from bacteria to eukaryotes with a reliable phylogeny rooted in cellular and palaeontological reality, not unfounded speculation, so one must work downwards systematically from the known diversity of eukaryotes to infer the nature of their last common ancestor. Only when both inferences are sound can we hope to explain the transition from prokaryote to eukaryote realistically. Past reasoning has been hampered by the root of the eukaryote tree also often being misplaced. Though not as inherently difficult as rooting the whole tree of life, correcting these errors has not been easy. We can now rule out the earlier idea that any premitochondrial lineage of eukaryotes survives as there is good evidence that all extant lineages have relics of a mitochondrion [53]; thus the root of the tree cannot lie within the secondarily anaerobic excavate phylum Metamonada that includes *Giardia* and *Trichomonas*, as was often supposed in the past, and must lie amongst ancestrally aerobic protozoan phyla. That the double-membrane mitosomes of *Giardia* indeed evolved from mitochondria and did not evolve separately from α-proteobacteria (as some have speculated) is shown by the presence of Tom40 [54], the outer-membrane protein that in all mitochondria of neokaryotes (i.e. all eukaryotes except the early diverging Euglenozoa [9]) mediates protein import into the mitochondrion from the cytosol.

A few years ago evidence suggested that the eukaryotic root lies between two major groups differing radically in cytoskeletal and ciliary organization: unikonts and bikonts [3,7,8]. Unikonts, comprising animals, fungi, Choanozoa, and Amoebozoa, were postulated ancestrally to have had one cilium and centriole and interphase microtubular cytoskeleton in the form of a cone of single microtubules emanating from the centriole, like a half-spindle [3]. Bikonts, comprising the plant and chromist kingdoms plus 10 protozoan phyla that form four clades (alveolates, excavates, Rhizaria, Apusozoa), ancestrally had an asymmetric cortical skeleton of bands of multiple microtubules forming the roots of two dissimilar cilia and centrioles [3]. However, it now seems that that interpretation was mistaken [9]. One argument for this bifurcation being fundamental was based on a misinterpretation of ciliary and centriolar development of the unikont slime mould *Physarum* that suggested that these were fundamentally different in the unikont Amoebozoa from the prevailing pattern in bikonts, but the discovery of an overlooked correction to the earlier interpretation now makes it likely that most, possibly all, eukaryotes share fundamentally the same pattern of cilia transformation and evolved from an early ancestor that had already evolved two centrioles and ciliary and the widespread pattern of ciliary transformation from a younger to an older cilium [55]. Recent evidence that the biciliate gliding Apusozoa are phylogenetically within unikonts (apusomonads at least probably being sisters to opisthokonts [9,56]) makes it likely that the simple cytoskeletal patterns of Amoebozoa and opisthokonts are secondarily derived following the independent loss respectively of the posterior and anterior cilium. A second argument based on a derived fusion gene against the root being within bikonts [8,57] is also now invalidated by the growing evidence that apusomonads, which have that fusion gene [57], belong within unikonts as sisters to opisthokonts [56].

Another favoured position for the position of the root was within the excavates beside the jakobid flagellates because of the primitive nature of their mitochondrial genome [58]. However, stronger arguments stem from the numerous distinctive features of Euglenozoa, which are most simply explained if the root is between Euglenozoa and excavates [9] as shown in Figure Figure1.1. Two euglenozoan characters in particular, the absence of mitochondrial import protein Tom40 and of the DNA replication preinitiation 'origin recognition complex' (ORC), both of which are likely to be ancestral characters of the most primitive eukaryotes rather than secondarily derived simplifications [9]. Previously Tom40 and ORC were assumed to have originated in the ancestral eukaryote; I now think it more likely that both arose somewhat later in the common ancestor of all eukaryotes other than eukaryotes (collectively called neokaryotes to contrast them with the very different Euglenozoa [9]). It seems possible that RNA polymerase II transcription factors IIA, F, and H also originated only in the ancestral neokaryote not the first eukaryote, which are absent from trypanosomatid genomes, unless they were lost by Euglenozoa when their cenancestor replaced the original neomuran transcriptional regulation by posttranscriptional gene regulation [59].

We can confidently eliminate most other formally possible positions of the eukaryotic root, and infer with high confidence that the last common ancestor of all eukaryotes was a phagotrophic protozoan with nucleus, at least one centriole and cilium, facultatively aerobic mitochondria, sex (meiosis and syngamy) and dormant cyst with cell wall of chitin and/or cellulose, and peroxisomes (Fig. 2); these conclusions all follow whether the root is beside the jakobids or the Euglenozoa. This last ancestor was probably non-photosynthetic, unless cyanobacteria enslaved chloroplasts simultaneously with mitochondria, as has sometimes been proposed [60] but which is unlikely if the root is beside or within Euglenozoa or excavates. Importantly for the present paper, the eukaryotic cenancestor (last common ancestor) already had a highly developed nuclear envelope with complex NPCs with all proteins shared by animals, plants, and Euglenozoa and at least eight different karyopherins to mediate nucleocytoplasmic exchange. Those NPC proteins apparently missing in the parasites *Giardia* and *Plasmodium* [14,17] have either diverged sharply beyond recognition (most likely) or been lost, and do not represent a simpler primitive state as some suggest [17]. This is certainly so for *Plasmodium*, which diverged from plants after the origin of red algae, whose plastid their ancestor enslaved (given chromalveolate monophyly: [61,62]). However, if the root is indeed between Euglenozoa and the rest of eukaryotes, at least some of those apparently absent from the trypanosomatid *Leishmania* [17] might be genuinely, primitively and universally missing from Euglenozoa. The last common ancestor of all eukaryotes (cenancestor) was almost certainly sexual and probably haploid, undergoing syngamy prior to encystment, and meiosis during cyst germination (excystment), but further study of ploidy in early diverging lineages is needed to test this [3,63]. Though most euglenoids have been considered to be asexual (*Scytomonas* is the only one with proven syngamy, but even for it meiosis has not been seen), evidence for relatively normal sexual mechanisms is growing for the euglenozoan *Trypanosoma brucei* [64,65].

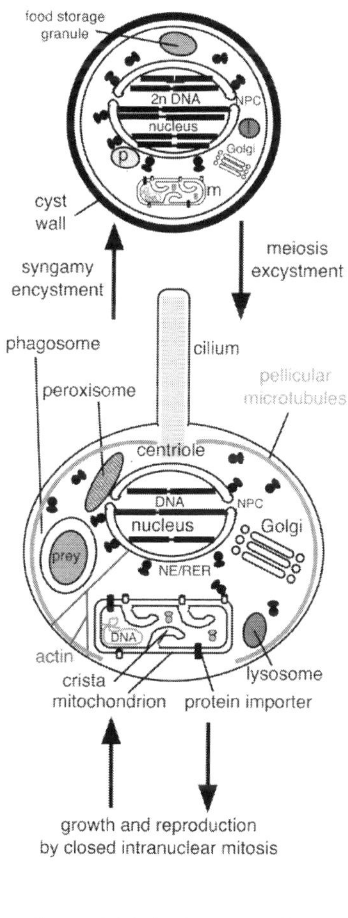

Figure 2: Inferred life cycle and high degree of organellar complexity of the last common ancestor of all extant eukaryotes. This reconstruction assumes that

the root of the eukaryotic tree is between Euglenozoa and excavates [7,8]. If so, every homologous character present on both sides of the neokaryote/euglenozoan split must have evolved prior to the cenancestor, provided that its later lateral gene transfer from one to the other can be ruled out, as it can for the complex characters shown. The major uncertainty is whether there were only one centriole and cilium as shown or more likely two of each [9]. In addition to the pellicular microtubules there would also have been centriolar roots consisting of bands of microtubules (probably two if the ancestor was uniciliate and three if biciliate) and a specialized anterior cytostome and cytopharynx for prey ingestion (all not shown for simplicity). The peroxisome (p) was probably attached to the nucleus and the Golgi was probably attached to a centrin body; centrin would also have been associated with the centriole and intranuclearly at mitotic spindle poles. The mitochondrion (m) was probably actually attached to the centriole and/or nucleus. A branched actin cytoskeleton permeating the cytoplasm was linked to nuclear envelope (NE) via KASH/Sun integral membrane protein complexes and to the plasma membrane via membrane-embedded integrin proteins. Syngamy involved fusion of plasma membrane, NE, and probably mitochondria.

This rerooting of the tree is important for thinking about the origin of eukaryotes, as it means that several characters often assumed to be general for eukaryotes and to have evolved in the first eukaryote actually evolved later, e.g. in the ancestor of neokaryotes or neozoa. Thus it now seems that eukaryotic N-linked glycoproteins were probably initially somewhat simpler than in animals and plants. If other Euglenozoa resemble trypanosomatids in lacking glucose termini in the glycosyl group that is added to proteins cotranslationally in the RER [66], then the enzyme making dolichol-phosphate-glucose the donor in neokaryotes for adding three extra residues may have evolved only in the ancestral eozoan, not the first eukaryote; probably quality control over glycoproteins was also simpler as they lack two of the four enzymes that Neozoa use to digest faulty ones (Mannosidase I and peptide-N-glycanase) [67].

Unless the root of the eukaryote tree were within Euglenozoa between the euglenoid *Scytomonas*, which (possibly primitively) has only a single centriole and cilium [68], and all other Euglenozoa that ancestrally had two (a possibility that cannot currently be excluded), then the last common ancestor of eukaryotes almost certainly had two centrioles and cilia per daughter cell. Centrioles would probably have been duplicated at the beginning of S-phase and the two parental centrioles would have separated prior to division, each associated with one new daughter centriole, as in all well studied ciliated eukaryotes. As the closest group to Euglenozoa on unrooted trees is the excavate phylum Percolozoa (Heterolobosea and their relatives) [69], if the root is between Euglenozoa and excavates, the earliest branching excavates would have been the discicristate phylum Percolozoa. This is important for understanding the origin of mitosis, as both Percolozoa and Euglenozoa have intranuclear mitosis with an intact nuclear envelope and a nucleolus that divides, very unlike the open mitosis of animals and plants where the nucleolus and nuclear envelope both disperse prior to metaphase. This probable root position between Euglenozoa and Percolozoa means also that in the ancestral eukaryote (as in both these phyla) the centrioles will not have been be directly at spindle poles but were indirectly attached to them by a cytoplasmic fibrillar cytoskeleton. Moreover, as neither phylum was ancestrally amoeboid, their common ancestor would have had a well developed semi-rigid cell cortex supported by longitudinal cortical microtubules; thus both mitosis and cell division probably evolved in a cell with semi-rigid surface, this rigidity probably stabilising the earliest eukaryotes and allowing fairly accurate DNA segregation following the loss of the eubacterial cell wall; as previously argued [70] the widespread assumption that the earliest eukaryotes were soft-surfaced amoebae is probably a myth; such formless intermediates would have exacerbated the problems of maintaining efficient DNA segregation during eukaryogenesis when the bacterial connection of chromosomal DNA to the cell surface and rigid wall was lost.

Geometric order is essential for DNA and organelle segregation. The important point for this paper is that the eukaryotic cenancestor had at least one centriole and cilium in daughter cells (possibly two) and at least two centrioles (possibly four) in predivision cells, probably attached to the nucleus during interphase to form a karyomastigont complex. Probably the cortical microtubular skeleton that persists during the whole cell cycle and is divided amongst daughters, with new elements being inserted into each, coevolved with the purely temporary mitotic spindle; the origin of the first protozoan pellicle is important for understanding eukaryogenesis as is discussed below. Furthermore, the cenancestral eukaryote had already evolved the coupling of centriole duplication to the onset of DNA replication at the beginning of S-phase [71-76], and had fully eukaryotic cell cycle controls [77] based on cell cycle kinases, phosphatases, and proteases, plus cyclin-mediated anaphase proteolytic resetting of hundred of proteins, as well as growth control over the G1 to S transition and post S-phase involvement of ε-tubulin in centriole duplication [78]. This temporal ordering is as important as geometric order for accurate cell reproduction. All these novelties were evolving simultaneously during the prokaryote-eukaryote transition, effectively at the same time as the nucleus, our main subject.

Symbiogenesis: An Accessory to Eukaryogenesis, Not the Primary Instigator

Although symbiogenesis explains the origin of mitochondria, Mereschkowsky's theory of a symbiotic origin of the nucleus [79], and recent attempts to modernize it, are all decisively refuted by the NE being three subdomains of the endoplasmic reticulum (ER) (to say it is 'connected to' ER [17] is wrong; it is ER, invariably having ribosomes on its outer surface); analogies with mitochondria or bacteria are extremely naïve [80]. The really distinctive steps in eukaryogenesis - all much more radical than the helotic origin of mitochondria - were the integrated origins of phagocytosis, endomembranes, endoskeleton, mitosis, nuclei, centrioles, cilia, cell cycle controls, meiosis, and syngamy [3,4,11]. As stressed above, these arguably evolved by the rapid autogenous structural transformation of a bacterial cell through entirely novel selective forces and drastic intracellular structural transformations caused by the onset of phagotrophy [3,4]. Although mitochondrial symbiogenesis involved transfer of many α-proteobacterial genes to the nucleus, whose proteins were often (not always) retargeted to the mitochondrion [20], these genes were probably inessential for any non-mitochondrial major innovations, except for supplying in some transferred genes group II self-splicing introns that evolved into spliceosomal introns and RNAs [3,81]. It is important to realise that the mitochondrion itself is an evolutionary chimaera with many key proteins of host origin being imported, e.g. the inner membrane carrier proteins [20]. Recent trees indicate that the inner membrane proteins Oxa1 (which inserts respiratory chain proteins from the mitochondrial matrix) and Oxa2 (which assembles cytochrome oxidase) evolved in the ancestral eukaryote by duplication from the host's YidC gene, rather than from the α-proteobacterial YidC (as crenarchaeotes lack YidC they cannot have been ancestral to eukaryotes) [82]. On the YidC tree [82] neomuran sequences are a clade (with archaebacteria and eukaryotes sisters) that branches between Endobacteria and Actinobacteria and there is a clear bipartition between the neomuran/posibacterial sequences and those of negibacteria plus chloroplasts, consistently with Fig. 1.

Chloroplast symbiogenesis was probably after the unikont/corticate split (Fig. 1), shortly followed by secondary enslavement of a red alga to yield chromalveolates [83]; both major photosynthetic symbiogeneses replaced host fatty acid and other enzymes, but did not significantly affect basic nuclear properties, except that in chromists among chromalveolates membrane fusion placed the enslaved red alga within the perinuclear cisterna [61]. In contrast to mitochondria the AlbC protein that inserts proteins from the stroma into the thylakoids probably evolved from the cyanobacterial YidC, the host YidC being unavailable by then for such co-option as it had already been modified for mitochondrial function as Oxa1 and 2; its weakly supported failure to group with cyanobacterial rather than other negibacterial sequences is probably artifactual [82]. But even chloroplasts are chimaeric having inner membrane carriers of host origin, and having like mitochondria had their original outer membrane lipopolysaccharides replaced by host phosphatidylcholine.

An important, insufficiently appreciated, feature of symbiogenesis is that it supplied several novel genetic membranes to the eukaryotic cell. In many ways addition of genetic membranes was more important than that of DNA, genes or genomes, because without them genes for oxidative phosphorylation would be useless. Lateral gene transfer had enabled foreign genes to be acquired by bacteria since life began, but for 3.5 Gy never succeeded in transferring oxygenic photosynthesis from one bacterium to another. In free-living prokaryotes cell lineages have been strictly vertical throughout history. By acquiring phagotrophy, eukaryotes could acquire whole cells and novel genetic membranes as well as genomes, not just genes from other organisms, so the inheritance of membranes has rarely been horizontal among unrelated taxa. Of course, sex also involves the horizontal transfer of membranes as well as genes. Membrane heredity is at least as old as DNA heredity - probably older [84] - and just as important for understanding cell evolution [20,62,84-88]. All membranes have been inherited from those of the first cell and the origins of novel mechanisms of protein targeting into and across membranes is central to eukaryogenesis, which involved a marked increase in the number of genetic membranes, some initiated in association with and enabling (not caused by) foreign cell enslavement [85] and some not, i.e. being purely autogenous.

Coevolutionary Origin of the Endomembrane System and Cytoskeleton

The endomembrane system and cytoskeleton are coadapted and interact in numerous ways. Branching networks of actin attach to plasma membrane, endomembranes and organelles by specific protein links. The endomembrane system fundamentally depends on coat-mediated budding of vesicles from one compartment, uncoating, and fusion of smooth vesicles with target compartments. Both budding and fusion are mediated by suites of mechanical effectors, targeting specificity factors, and controlling proteins among which GTPases play a major role. Vesicles are transported along the cytoskeleton by molecular motors absent in bacteria: myosins for actin filaments, dyneins and kinesins for microtubules. Probably three different functionally specialised myosins were present in the cenancestral eukaryote [7] (this estimate is not changed by recent rerooting the tree [9], though one must now regard the addition of the most widespread tail domains to two of them as neokaryote synapomorphies only, not shared by the earlier diverging Euglenozoa). They evolved during eukaryogenesis by successive gene duplications after their common ancestor became the first myosin by the radical transformation of a former bacterial GTPase by a shift in nucleotide specificity to ATP [89] and various domain fusions [90] to make a complex motor. Likewise the cenancestral eukaryote probably had 11 different heavy chain kinesins [91], with multiple roles in mitosis, ciliary motility and vesicle transport, including interactions with dynein (note that the recent rooting of the tree between Euglenozoa and neokaryotes [9] reduces the previously estimated number in the cenancestral eukaryote because one can now treat kinesins 4-8 and 15 as having evolved later in the ancestral neokaryote only; the rerooting also invalidates the earlier suggestion that kinesin-17 is a synapomorphy for bikonts [91]; instead it probably originated in the ancestral eukaryote and was lost by the ancestral unikont, as it was independently several times within Plantae and Chromista). Kinesins and myosins are structurally related ATPases, probably having a common origin in the prekaryote by a primary gene duplication. Dynein motors are related not to them but to midasin and arose independently from a bacterial ATPase [92]. Duplications produced nine different heavy chain dyneins in the cenancestor [93], most involved in ciliary motility [94].

When I proposed that the origin of actin for a role in phagocytosis was the primary molecular invention that triggered eukaryogenesis [11] it was not even known that actin and myosin were thus involved. Since then a universal role has become apparent for actin and myosin, not only in phagocytosis but in other forms of endocytosis retained even by eukaryotes like yeast, whose ancestors abandoned phagotrophy [95-99]. Thus, an intimate association of actomyosin with endocytosis in all its forms and with endomembrane vesicle trafficking is central for eukaryote cells. Actin is a key player in endomembrane biology and not just involved in general cell motility and cytoplasmic division. It originated from the bacterial membrane skeleton protein MreB, which like actin helps mediate both cell shape and division [100-105]. Prekaryote gene duplications produced not only actin, but actin related proteins (Arps) that nucleate actin filaments, Arp2/3 being essential for branching to make a 3D skeletal network for the first time and for endocytosis. Similarly, six tubulins (α, β for microtubules, γ-tubulin for centrosomal nucleation, and δ, ε, η for centrioles) must have arisen prior to the cenancestor by gene duplications of a relative of FtsZ, the filamentous bacterial GTPase which dates back to the last common ancestor of all life and is the general marker for the site of bacterial division (despite being secondarily lost by the ancestor of many crenarchaeotes and also within Planctobacteria).

Novel Proteins and Eukaryogenesis

Thus prior to the origins of the nucleus and mitochondrion the prekaryote underwent massive gene duplications that created characteristic eukaryote structures [106]; notably of small GTPases involved in vesicle budding or fusion [107]; of actin, tubulins, and molecular motors; of proteins of the proteasome making this digestive cylinder immensely more complex than in bacteria because of the novel ubiquitin-linked proteolytic controls over the cell cycle and removal of faulty ER lumen proteins by the ERAD system [108]; and - of special significance for the origin of the nucleus - coated vesicle proteins (e.g. COPII, COPI, clathrin). These gene duplications, and origins of novel protein domains (the most extensive since cells began) [6], had a key role in eukaryogenesis, the central logic of which is summarized in Fig. 3 in six major stages. (a). Ancestrally an FtsZ ring between daughter DNA termini (T) divided bacteria, their shape being controlled by cortical skeletal MreB (blue) and rigid murein wall (brown). (b). The origin of phagotrophy then disrupted this. In the neomuran ancestor flexible glycoproteins (yellow) replaced murein, allowing MreB in the ancestor of eukaryotes to become actin (blue) and power phagocytosis, which internalised DNA-membrane attachments (centre); soon thereafter evolution of COP-coated vesicle budding, and fusion with plasma membrane after uncoating, made permanent endomembranes (EM: precursor of ER, NE, Golgi, lysosomes; peroxisomes (P) separated earlier) and disrupted bacterial DNA segregation. (c). Hypothetical origin of simple mitosis in a prekaryote cell where FtsZ gene duplications evolved stable microtubules and γ-tubulin-containing centromeres (probably also containing centrin) still attached to the surface membrane. Centrosome binding to the chromosome origin and microtubule attachment parallel to the cell membrane throughout the cell cycle prevented phagocytosis in that region (upper) and thus avoided chromosome internalization as in b but allowed it elsewhere (e.g. lower example shown). Actin and myosin that originally evolved for phagocytosis were recruited to form a contractile ring for cytokinesis, replacing the lost function of FtsZ. Probably they were thus recruited before FtsZ duplicated to α-, β- and γ-tubulin and the prime function of microtubules was to prevent actomyosin dividing the cell before replication finished or internalising the chromosome attachment sites. This stabilising function would be less demanding for the initial origin of microtubules and centrosomes than many more complex later ones. Kinesin (a myosin relative) may have evolved by gene duplication and divergence from myosin at this stage to push apart the antiparallel microtubules attached to the daughter centrosomes, only then would slight microtubule shortening allow the actomyosin ring to cleave the cell. This phagotrophic prekaryote I, with both kinesin and myosin motors that could segregate its DNA and divide the cell, had a centrosome duplication cycle, but no nuclear envelope and perhaps not yet even permanent endomembranes. (d). Accidents in centrosome duplication and phagotrophic membrane internalisation generated a more complex prekaryote II in which stable endomembranes differentiated into peroxisomes (P) and protoendomembranes (EM, i.e. the ancestors of ER and Golgi; see [27]), some of the latter remaining associated with the internalised centrosome and DNA, whereas another centrosome remained at the cell surface and continued to stabilise it and control the site of cytokinesis by the actin ring. Ultimately the surface centrosome generated the pellicular microtubules and the centriolar and ciliary microtubules of the cenancestral eukaryote, whereas the ER-associated MNC nucleated its intranuclear spindle. (e). The first eukaryote. EM attachments around DNA made a primitive NE, and connectors (brown, possibly containing centrin) evolved to ensure that both nuclear and cytoplasmic organelles were properly segregated at division, which was constrained to be between sister cytoplasmic centrosomes and their attached pellicular microtubules. The Golgi apparatus (G) was probably by then differentiated from the ER and attached to the centrosome.

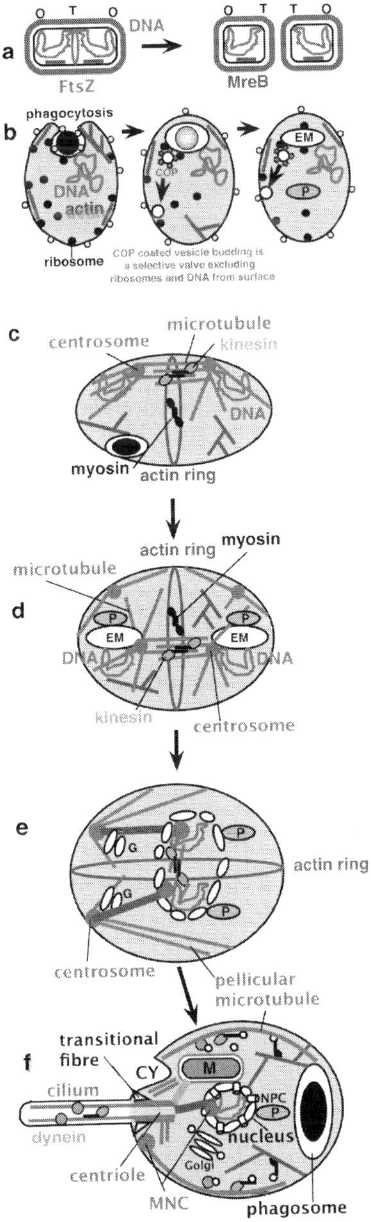

Figure 3: Evolution of eukaryotes from a posibacterium, emphasizing changes in DNA segregation caused by internalization of DNA-membrane attachments. (a). An FtsZ ring between daughter DNA termini (T) divides bacteria; cortical skeletal MreB (blue) and rigid murein wall (brown) control cell shape. (b). Dis-

years before any eukaryotes could diversify and leave modern descendants is evolutionarily ridiculous.

Population genetics of large microbial populations and modest selective forces could achieve this in much less than 80,000 years, even 8,000 years (well over 3 million generations; ten times the number that separates us from our common ancestor with other apes), so very likely stem lengths of paralogue subtrees are commonly inflated 10,000-fold or more by quantum evolutionary divergence consequential upon divergence of sister paralogues, compared with rates of change deduced by comparisons among derived groups where purifying selection dominates and keeps change slow. As argued previously [12], a comparable gross rate inflation probably applied to cytoplasmic rRNA temporarily during eukaryogenesis as a consequence of the evolution of the nuclear envelope and nucleolus with novel processing and transport (but not to mitochondrial rRNA, which evaded these novel coevolutionary selective forces by being kept entirely within the mitochondrial matrix). Thus, extreme quantum evolution can occur not only following divergence of paralogues but in unduplicated molecules given a sufficiently intense shift of function. Overlooking the fundamentally different evolutionary principles that apply to megaevolution compared with ordinary macroevolutionary divergence has led to much misinterpretation of rRNA and paralogue trees. These misinterpretations have most seriously affected interpretations of the timing of evolutionary events and the rooting of trees, but quantum evolution can also introduce such serious long-branch artefacts that no algorithm can reconstruct the correct topology. Molecules like rRNA and RNA polymerase are not evolutionary chronometers, but evolve extremely heterogeneously in speed across time; rates can be roughly estimated only by considering both the fossil record and the degree to which quantum evolution may be involved, which cannot be statistically modelled - unique events like the origin of eukaryotes or the vertebrate skeleton violate all naïve statistical assumptions of regularity and repetitiveness. It is fundamentally wrong to assume that all molecules evolve the same way throughout time and to ignore unpredictable irregularities. A related kind of misinterpretation concerns the source of the different components of eukaryotes. For example Esser et al. [115] concluded that among bioinformatically tractable proteins, the majority in yeast were more similar to those of eubacteria than to archaebacteria. From this truth they concluded a probable falsehood - that most nuclear genes came from the α-proteobacterial ancestor of mitochondria.

There are three reasons why that conclusion is unsound. First, the evidence that eubacteria are ancestral to archaebacteria implies that ancestral archaebacteria probably lost about 1000 genes compared with eubacteria [6,13,22]. If most of those genes were lost after they diverged from the prekaryote lineage, 1000 extra genes of host origin would have been present in the host to leave modern eukaryotic descendants, which would radically increase the proportion deduced to have come from the host, by itself completely reversing the authors' conclusion. In other words, the genes now in archaebacteria grossly underestimate the much wider spectrum present in the neomuran ancestor; many eubacteria-like eukaryotic genes probably came not from the mitochondrion (as others undoubtedly did) but from the host, their sister homologues being lost by the archaebacteria. Secondly, proteins well enough conserved during the prokaryote/eukaryote transition to allow such sequence comparisons are a small, non-representative sample of the total. Most of those host proteins that evolved radically novel, specifically eukaryotic, functions (such as nuclear pore complexes and vesicle coat proteins, and molecular motors) were so transformed by quantum evolution that their sequences are too divergent to be recognisable by BLAST analysis. Evidence of the precise phylogenetic origin of the majority of eukaryotic genes has simply been erased by the quantum evolution that made eukaryotes. Probably the vast majority came vertically from the host, not the α-proteobacterial symbiont. The third major bias to the proportions is symbiont replacement of functionally equivalent host genes. This well-known evolutionary phenomenon applies not only to the primary symbiogenetic origin of mitochondria and chloroplasts, but to secondary symbiogenesis, e.g. the acquisition of plant fatty acid biosynthesis genes from the enslaved cyanobacterium [83,116] and later from the enslaved red alga by the ancestral chromalveolate [117]. Symbiogenesis introduces thousands of symbiont genes into the host nucleus. If the encoded proteins were exactly equivalent to host proteins, there is a good chance that some will replace host proteins, since among two equally well expressed and well controlled proteins it does not matter which is lost by functionally advantageous elimination of wasteful duplicates [60]. If the symbiont is a bacterium, as was the protomitochondrion, the host versions of many of the relatively unaltered metabolic genes they share will be eliminated. But novel host eukaryotic genes that have become so different from their bacterial progenitors that the protomitochondrion had no close relatives are less likely to be lost. Thus, host gene replacement by symbiont genes is biased towards a small subset of genes - relatively conserved and bioinformatically tractable metabolic genes and against those very genes that played the most important role in the origin of eukaryotes and whose ancestry cannot be traced from their sequence.

The value of symbiogenesis was that it created a novel organelle with minimal structural change to the enslaved bacterium and no fundamental genetic change simply by adding one novelty: inner membrane carriers and associated protein-import machinery [20,22]; the elimination of thousands of symbiont genes and retargeting of some were inevitable consequences of this one key innovation. Symbiogenesis probably contributed very little, if anything, to the key initial steps in eukaryogenesis, the concerted origin of the cytoskeleton, endomembranes, and mitosis. Two major effects on the host were the origin of spliceosomal introns from α-proteobacterial group II self splicing introns, and the loss of its own plasma membrane/endomembrane ATP-generating, oxidative phosphorylation enzymes - assuming that the host, like the symbiont was a facultative aerobe able to live both in fully aerobic and temporarily anaerobic environment [22].

RESULTS: THE ORIGIN OF MITOSIS AND THE NUCLEUS

Megaevolutionary Principles Apply to Cell and Molecular Evolution

From the classic megaevolutionary perspective of Simpson [10,113], understanding the origin of eukaryotes (eukaryogenesis) is much less difficult than is often supposed, despite it being the most dramatic molecular, cellular and genetic upheaval in the history of life. The central requirement for a convincing explanation is to specify correctly the adaptive shift that caused the innovation in body plan and the nature of the ancestral organism that made that innovation - in this case predation by phagotrophy. Thereafter it is relatively straightforward to reconstruct a biologically sensible and mechanistically plausible sequence of likely successive changes.

Most megaevolution occurs without any symbiogenesis; only five megaevolutionary events in the history of life involved symbiogenesis. In the origin of the kingdoms Plantae and Chromista (now = chromalveolates [9]), of chlorarachnean Cercozoa and Euglenia, the internal symbiogenetic enslavement of a photosynthetic cell was the central determinative event; in every case phagocytosis engulfed the slave, proving that phagotrophy preadapts a cell for symbiogenetic enslavement of other cells. Symbiogenesis was of the essence, but phagotrophy was the essential preadaptation. In the fifth case, the origin of eukaryotes, phagotrophy was causally primary and mitochondrial enslavement purely secondary. The fundamental novelty in body plan and way of life of eukaryotes involved the unique origin of phagotrophy, internal cytoskeleton and endomembranes. Very likely the host that acquired a mitochondrion was already a facultative aerobe with oxidative phosphorylation [3]. Adding mitochondria was an important extra refinement to increase energy efficiency, but probably did not fundamentally change the cell's way of life or body plan or impose any radically new selective forces on it as did the evolution of phagocytic prey capture and internal digestion. Oxidative phosphorylation had been around for ~2 billion years before eukaryotes evolved [12]; phagotrophy only arose with them. Margulis [32,33] claimed that mitochondrial enslavement preceded and enabled phagotrophy. But neither she nor recent adherents to this unsubstantiated view (e.g. [26,36]) ever convincingly explained how such an implausible reversal of the above logic could have worked selectively or mechanistically to produce the cytoskeleton and endomembrane system.

Simpson emphasised that the magnitude of megaevolutionary steps ensure that they occur only if the ancestral lineage was in many ways preadapted for the transition, both by its structure and physiology and its ecological accessibility to the empty adaptive zone that it alone has a realistic potential to invade. Evolution occurs not in abstract sequence space, but in the real world of organisms with specific body plans and ecological contexts. Preadaptation does not imply evolutionary foresight or planning; it is just an historical accident - not a lethal accident or uncaused accident, but one culminating a unique stream of historical events that gave one lineage uniquely suitable properties for being a viable vehicle for mutations of radical effects that would have extinguished less fitted lineages of different background. Fish with lungs and lobed fins happened to be preadapted for land life and becoming tetrapods, in a way that fish with spiny fins and no lungs, or still worse echinoderms or jellyfish, which never made it to land, were not. Phylogenetic preadaptation of a unique lineage is a necessary part of the explanation of every megaevolutionary event. Exceptional preadaptation and discontinuity invariably apply to body plan innovation. Thus, the origin of eukaryotes was necessarily an intrinsically abnormal event, not one understandable simply by extrapolating the trivial tinkering that occupies most evolution. To understand it we must invoke something quite exceptional. Actually four things, none miraculous but all uniquely innovative: one is preadaptation of one lineage for phagotrophy; secondly the novel selective forces that phagotrophy brings; thirdly the disruptive effects that phagotrophy had on the cell division and DNA segregation machinery by internalising the DNA-membrane attachment sites [4,11]. Not only did this disruption necessitate the adoption of novel systems (cell division by actomyosin and DNA segregation by microtubules/kinesin), but also because these were inevitably initially inefficient it directly caused repeated polyploidy (whole genome duplication in modern jargon) in the transitional intermediate. This would make the cells large, as in the giant bacterium *Epulopiscium* [118], probably advantageous for ingesting others as food. Every gene probably duplicated many times in just a few days. The intermediate was almost certainly multigenomic and in consequence also larger than its ancestors and better able to engulf other cells and also survive some otherwise harmful mutations. Interestingly, large complex filamentous multigenomic cyanobacteria and actinomycetes can segregate their DNA randomly, and thus remain viable without MreB, unlike most bacteria [58,119]. Advanced large, metabolically diversified and multigenomic prokaryotes, not the tiny degenerate ones implausibly postulated by Margulis [32,33,120] as our ancestors, could best have made the transition to eukaryotes.

Key Novelties and Preadaptations for Eukaryogenesis

There were three key preadaptive features of the ancestor of eukaryotes. First it had one bounding membrane (like Posibacteria and Archaebacteria alone among prokaryotes), not two as in all other eight eubacterial phyla, which would have prevented the evolution of phagocytosis and endomembrane budding. Thus of eubacteria, only Posibacteria were preadapted to become the ancestor of either eukaryotes or archaebacteria. This first preadaptation was the origin of Posibacteria by the loss of the negibacterial outer membrane, probably at least 700 million years before the origin of eukaryotes, thus not a triggering event [121]. Secondly, the three dimensionally covalently bonded murein wall of most eubacteria makes both the origin of phagotrophy and sex (cell fusion and ploidy reduction by meiosis) impossible. The essential preadaptation was the replacement of the murein cell wall by individually potentially mobile surface N-linked glycoproteins. This neomuran revolution was I argued the triggering event for the origin of eukaryotes [4] and equally of their archaebacterial sisters; each evolved divergently, one by genome reduction and cell compaction and the other by gene multiplication and cellular expansion. The third preadaptation for eukaryogenesis was a combination of ideal cellular factors: large cell size to enable engulfing of other bacteria, large genome size with many potential gene ancestors, a diverse set of lipids to favour endomembrane differentiation, a large secretome including many digestive enzymes as precursors for internal digestion (about a third of eukaryotic proteins are secretory); and a facultatively aerobic/anaerobic metabolism to allow them most simply to become hosts to a facultative aerobic/anaerobic, probably photosynthetic, α-proteobacterium [3,4].

Of the two posibacterial subphyla, Endobacteria and Actinobacteria, no Endobacteria were thus preadapted, but many actinobacteria were. Actinobacteria have the largest secretome of any bacteria - over 800 secretory proteins, mostly digestive enzymes, large genomes and cells, complex life cycles with protein phosphorylation controls, differentiated resting spores, phosphatidylinositol, proteasomes of the neomuran type for intracellular protein digestion, and H1-like histone, and enzymes related to those that make eukaryotic sphingolipids, and many more cytochrome P450 (precursors of the ER oxidation system) than other bacteria. Regulation of actin polymerisation by phosphatidylinositol 4,5-bisphosphate activated by a small GTPase [107] can only have evolved in an actinobacterium-derived ancestor, as no other prokaryotes whatsoever have phosphatidylinositol [13]; the inositol phospholipids of myxobacteria are chemically different (dialkyl ethers not acyl esters; see [13]); citing them as evidence for a myxobacterial rather than an actinobacterial ancestry [122] is a spurious argument. Duplications of the same GTPase gene family provided Ran GTPase for NE assembly and nucleocytoplasmic transport and GTPases for endomembrane differentiation during the multiplication of novel genetic membranes [85,107]. Among actinobacteria, actinomycetes are of special significance for eukaryogenesis (and the origin of archaebacteria) as they not only have eukaryote-like small GTPases, crucial for endomembrane differentiation and nuclear origins, and a neomuran type proteasome, but also biosynthetic pathways for polyketides like fungi, protozoa, chromalveolates and plants; mycobacteria can also make cholesterol. Contrary to Margulis [32], who suggested that mitochondrial enslavement by a mycoplasma-like host introduced sterol synthesis into prekaryotes thereby enabling phagotrophy, α-proteobacteria do not make sterols, and mycoplasmas are too small to have been the host; they evolved as intracellular parasites inside eukaryote cells after phagocytosis had already evolved, enabling a posibacterial ancestor to lose its murein wall in an osmotically protective environment and for mycoplasma cells and genomes to be miniaturized. Mycoplasmas lost their posibacterial walls independently of the neomuran ancestor. The genomic reduction of the ancestral archaebacterium, associated small size, lack of extensive secretome or suitable lipid precursors, more anaerobic lifestyle, a likely hyperthermophilic specialisation, absence of phosphatidylinositol or other phospholipids and sterols, all make archaebacteria entirely implausible as ancestors of the eukaryote host, being devoid of most preadaptations that would have been present in the neomuran ancestor. Though in lipid biology mycobacteria seem preadapted for the evolution of eukaryotes, they are so specialised that the actual ancestor of neomura was more likely a less specialised early diverging relative with similar metabolic capabilities.

Jékely [107] suggested that eukaryogenesis was initiated not by phagocytosis, but by the evolution of secretion by exocytosis, with phagocytosis arising later, simply because this appeared more compatible with his small GTPase sequence tree. While I once briefly held that view [70,123], it has immensely less explanatory power than phagotrophy as the initiator [3,4,6]. Moreover, the GTPase tree that fuelled his proposal (see Fig. 4a) was misleading through concealing branch lengths and misinterpreted. It is evident from other work that the Arf branch is the longest among the eukaryote paralogues and that the small GTPase tree is essentially unresolved at its base [124]; therefore rooting of the tree with the far longer eubacterial outgroup could have wrongly attracted it to the base. It is as unwise as for other paralogue trees to assume that the rooting is accurate. Even if the position of the root is correct, it was wrong to infer that the Arf containing branch arose prior to the Rab-containing branch. In such a basal divergence one cannot say that one sister branch is earlier than the other. The worst argument, however, was that the Arf-1 branch is involved in secretion and the Rab one in phagocytosis, and to combine these two bad arguments to conclude that exocytosis evolved before phagocytosis [107]. In fact, there is ample evidence that Rab paralogues are extensively involved in secretion in early, middle and late stages [125,126], and conversely that Arf1 is involved on phagocytosis [127]. Thus Arf and the Rab paralogues are both required for secretion and exocytosis and equally for endocytosis and phagocytosis, decisively invalidating any attempt to infer the relative timing of the origin of these complementary processes from GTPase paralogue trees. Since his work additional prokaryotic small GTPases paralogue classes have been discovered, which complicate the story and disrupt the branching pattern of the tree [128]. In my view these authors also are overconfident of their (somewhat different) tree and misinterpret it as evidence for separate origins of Arf/Sar and Rab/Ras/Rho from different bacteria; more likely both these major classes of eukaryotic GTPases have a common ancestry in the RarD paralogue found only in archaebacteria and actinobacteria [87]. Rather than making dubious deductions from possibly distorted sequence trees, it is preferable to use general evolutionary and cell biological principles and as much relevant knowledge as possible to infer the overall most likely sequence of events. Fig. 4b therefore proposes a plausible and simple scenario for endomembrane differentiation prior to the origin of the nucleus; almost certainly endocytosis and exocytosis coevolved, recruiting from a common pool of enzymes when assembling their toolkit.

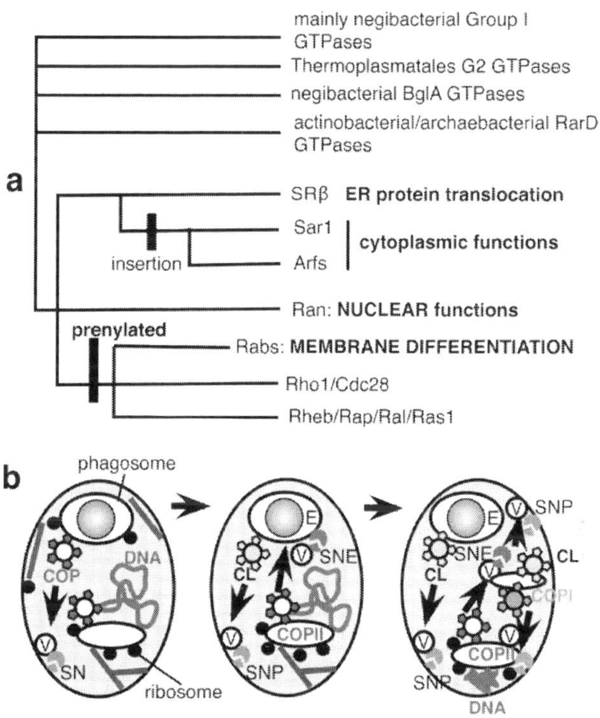

Figure 4: Evolutionary differentiation of endomembranes. (a). Schematic tree for controlling small GTPases [124,128]. Sar-1 and Arf-1 have an extra, derived insertion, so the root cannot be in that branch. Because of disparate rates of evolution among paralogues and the shortness of the molecules it is unclear from trees whether the seven eukaryotic clades (lower) are all mutually related as shown and which of the four bacterial clades (upper) are their closest relatives. (b). After endomembranes, peroxisomes, and plasma membrane became distinct genetic membranes (Fig. 3b) most secretory ribosomes were on old DNA-bearing cisternae; the first COP/adaptin coats generated vesicles (V) from the

protoendomembrane/phagosome; early SNAREs (SN, left) fused them with the plasma membrane. Endomembrane differentiation improved digestion by targeting digestive enzymes specifically to phagosomes, mediated by successive concerted duplications and divergence of coat proteins, cognate SNAREs able to bind to them, and associated small GTPases. Primary specialisation between digestion and synthesis involved clathrin vesicles (CL) associated with plasma membrane SNAREs (SNP) and COPII vesicles associated with endomembrane SNAREs (SNE). Interpolation of Golgi, by mutual fusion of uncoated COPII vesicles, stabilised by COPI-mediated recycling (right), allowed specialisation between lysosomes and surface growth. For a fuller discussion of endomembrane origins see [91].

Endomembrane differentiation required not only duplications of vesicle coat proteins and GTPases controlling budding but also of SNAREs controlling targeting of each type of uncoated vesicle [85] (and GTPases controlling fusion) [107]. Even the first step, separation of rough ER (RER) from plasma membrane probably needed SNAREs initially. 20 different SNAREs arose before the eukaryotic cenancestor [129]. The genetic identity of RER was secured by insertion of SRP-recognising docking protein (pentagons Fig. 3a, b) in which small GTPase SRβ participates [107]. It was more efficient for digestive enzyme-laded CopII vesicles budding from proto-ER to fuse with phagosomes not the cell surface. Thus was born the primary divergence between ER outwards secretory traffic and plasma membrane inwards endocytic traffic seen in GTPase [107] and SNARE [130] trees. So we should not ask 'did phagocytosis or exocytosis evolve first?' Both evolved together, with phagotrophy being the entirely novel selective advantage, as De Duve [131] and Stanier [2] first argued. I refer the reader to [27] for a more detailed discussion of the origin of the endomembrane system, which had to precede the origin of the nucleus.

The Overlapping Origins of Mitosis, Centrioles and the Nucleus

As Fig. 3 indicated, a simple form of mitosis probably originated before the nucleus (Fig. 3c), but the modern version is more complex and its elaboration must have overlapped and coevolved with the origin of the nucleus. A nucleus could not evolve without a relatively efficient means of DNA segregation, making it misleading to consider the origin of the nucleus on its own, as many do. I shall argue that NPCs and nucleocytoplasmic transport and the chromatin-attachment of NPCs and the inner envelope membrane - the key features of the cell nucleus - evolved by the structural and functional integration of novel duplicates of pre-existing key proteins whose basic functions originally evolved separately for endomembrane and chromatin-associated mitotic functions. Before explaining this integration I must outline the origins of endomembranes and of mitosis, the essential prerequisites for the origin of the nuclear envelope, NPCs and nucleocytoplasmic compartmentation.

As explained in the first detailed discussion of the logic of the transition from bacterial to eukaryotic DNA segregation [70], this involves three major changes. First a change in the timing of DNA replication relative to mechanical separation; both processes are spread over almost the whole cell cycle in bacteria and replication origins are separated (as then postulated, but now known [132,133]) well before replication is complete, whereas in eukaryotes replication is completed before sister chromatid separation begins. Second was the origin of the eukaryotic chromosome condensation cycle in which DNA is relatively dispersed for replication and transcription but typically more tightly condensed during mitosis. Third, was a topographic change from chromosomes being attached to and geometrically ordered on the bacterial cell surface membrane and wall to chromosomes being attached to the nuclear envelope in interphase during replication but attached instead to the mitotic spindle microtubules during cell division. Spindle microtubules consist of α- and β-tubulin and are nucleated by γ-tubulin at the core of the centrosome; when these and the other universal tubulins evolved by repeated gene duplications of a bacterial GTPase they underwent such radical changes in a short time that they now have only a slight sequence similarities to their bacterial relatives FtsZ [134] and TubZ [135]. It has usually been assumed that tubulin evolved from FtsZ [136], whose single filaments form the core of the Z-ring which marks the site of division in most bacteria and chloroplasts and the more primitive mitochondria. However, the recently discovered GTPase TubZ, a double filament protein that is important for the segregation of DNA in certain plasmids of the posibacterium *Bacillus*, is a better analogue of tubulin, because it seems to have similar treadmilling and capping properties and might actually be involved in pushing apart sister DNA molecules [135,137], unlike FtsZ which is more a morphogenetic marker to guide the membrane scission machinery into place. The fact that TubZ coexists in cells with FtsZ means that its DNA segregation properties can evolve without at the same time abandoning FtsZ's marker function. I suggest that tubulins evolved from either an early relative of TubZ itself (quite likely as it evolved like tubulin in the posibacterial lineage) or an independently evolved paralogue of FtsZ that functioned similarly in plasmid DNA segregation. TubZ evolves much faster than either tubulin or FtsZ, presumably because it interacts with fewer other proteins, and is extremely divergent in sequence from both. During the origins of α-, β-tubulin γ-tubulin numerous new features evolved to allow α- and β-tubulin to form dimers and their protofilaments to form microtubules and γ-tubulin to nucleate them, so much so that sequence trees cannot be expected to establish which bacterial paralogue tubulin evolved from [134]. However, an origin from a treadmilling posibacterial TubZ involved in plasmid DNA separation would offer the easiest transition for the origin of microtubules and the mitotic spindle, with the least trauma for the evolving phagotroph.

By contrast there was a much less radical change in the organization of a fourth key component of the DNA segregation system, whose role and universal importance to DNA segregation was unknown even 20 years ago: SMC proteins and kleisins [138]. SMC (acronym of structural maintenance of chromosomes) proteins have homologues in all organisms and are essential for accurate DNA segregation. SMCs are giant proteins with a globular domain at each end and a long helically coiled rod in the middle. They exist as dimers in which the rods are antiparallel and twisted around each other as coiled coils, and hinged in the middle enabling them to form a V shape in which the end of each arm contains one C-terminal and one N-terminal globular domain. This close association of the contrasting domains enables the globular head end of each arm to bind ATP and function as an ATPase. Each head interacts with kleisin proteins; when these are bound to both, the whole complex forms a ring known as a cohesin or condensin (depending on its function and which SMCs and kleisins it contains). Cohesin rings enclose sister DNA molecules and prevent their separation till the cell is ready to divide, because they are loaded onto the parent DNA molecule before replication is completed (in eukaryotes before it even starts). When the cell is ready to divide proteases digest the kleisin (at anaphase in eukaryotes), converting the ring into an open V, thereby allowing the mechanical separation of DNA sister molecules (chromatids) [139,140].

Eukaryotic cohesins are loaded into DNA at the G1-S phase transition. Bacterial DNA segregation broadly follows the principles adumbrated long ago [141], but we now know the molecular machinery [132,133]. Bacterial SMC condensins are loaded onto replication origins by the DNA partitioning protein ParB and thereby mediate chromosome compaction of daughter chromosomes [132,133,142], which is essential for accurate DNA segregation as the origins are moved to opposite poles by an ATPase (Soj), i.e. actively as postulated [141]. Thus DNA gyrase in eubacteria and histones in eukaryotes are not sufficient for chromosome compaction as was originally assumed [70]. The ring-shaped condensins that hold DNA loops together to allow proper chromosome folding [142] clearly evolved in the last common ancestor of all life, as SMC is known from most bacterial phyla [143] and I have found them by BLAST in the earliest diverging Chlorobacteria (their greater sequence divergence within this phylum compared with most others is consistent with it being the oldest [6,13]). Moreover SMC had also undergone two gene duplications in the last common ancestor to generate two proteins for DNA repair: Rec9 and SbcC. RecN is a eubacterial protein involved in repairing double-strand breaks in DNA; neomura have a homologue (also for break repair) called Rad50 that probably evolved from it. SbcC helps in other forms of DNA repair and seems to have been lost by the ancestral neomuran. A small subset of γ-proteobacteria of the family Enterobacteriacae (including *Escherichia, Salmonella* and *Shigella*) replaced SMC by a larger but related protein, MukB that interacts with different kleisins, and groups on trees with SbcC and must have evolved either from it or from SMC by radical changes. Possibly these unique drastic changes in these enterobacterial SMC-like proteins occurred when their ancestor (alone among prokaryotes?) evolved the capacity to initiate several successive rounds of chromosome replications, yielding branching chromosomes that allow the cell cycle to become even shorter than the replication fork travel time [141].

SMC proteins are so long (over 100 amino acids) and well conserved that they make better sequence trees [143] than most proteins commonly used for prokaryote phylogeny (e.g. [45]); the SMC tree shows archaebacteria, neomura, and 'neomura plus Posibacteria' all as monophyletic and places neomura between actinobacteria and Endobacteria as in Fig. 1 (though as sisters of Endobacteria not Actinobacteria). Prokaryote SMCs are all homodimers, whereas in eukaryotes they are always heterodimers. During the origin of eukaryotes two gene duplications generated four paralogues [143]: Smc1 and 3, which form cohesin by binding the kleisin Ssc1; and Smc2 and 4, which form condensin I by binding the kleisin CAP-H (or CAP-H2 in condensin II which evolved later in an early animal only). As Smc1 and 2 group together on trees, as do Smc3 and 4, it is likely that stem eukaryotes initially evolved an ancestral general purpose heterodimer containing one of each of these pairs, but that further duplications differentiated it into cohesin (loaded onto chromatin in interphase) and condensin (loaded during prophase of mitosis). This differentiation must have gone hand in hand with the new temporal controls via cyclins and phosphorylation. The first heterodimeric

SMC was probably a condensin as in bacteria; the newer function of cohesin depended on the coevolution of novel anaphase proteolysis of Scc1 (see below). Though Euglenozoa have condensin I, and cohesin, Ssc1 and the separase enzyme that cuts it are essential for chromosome segregation [144], they lack another Smc heterodimer (Smc5/6), which may act as a cohesin for rDNA segregation [144] as well as having functions in DNA repair or recombination, suggesting that it evolved somewhat later (in the ancestral neokaryote, being present in *Naegleria* and *Giardia* genomes).

Chromosome condensation, separation and all other aspects of mitosis and the cell cycle are regulated by four related families of eukaryotic serine/threonine (S/T) protein kinases: Cdk, Nek, aurora and polo-like. All are present in the most divergent eukaryotes (Euglenozoa) so all originated by gene duplication in the ancestral eukaryote. Their closest relatives and putative ancestors of these and numerous other eukaryotic families of S/T kinases is the PKN2 family of eubacterial kinases which are widely and mainly present in Posibacteria (both subphyla) and Cyanobacteria but never in archaebacteria [145]. Almost certainly they were vertically inherited from Posibacteria to the ancestral neomuran and immensely multiplied by eukaryotes but lost by the ancestral archaebacterium. The idea that they originated from nowhere in the first eukaryote and were repeatedly laterally transferred into eubacteria [145] is bizarre, implausible, and not supported by trees (presumably proposed because the authors accepted the erroneous dogma that eubacteria are not ancestral to neomura). In two groups of actinobacteria the S/T kinases regulate cell growth and division including DNA segregation [146]. I suggest that this was also true of the stem actinobacterium from which neomura probably evolved, so there was even some continuity of cell cycle regulation between these posibacteria and eukaryotes. These posibacterial kinases have a binding site for peptidoglycan [147], which would have been lost when the ancestral neomuran replaced murein by glycoproteins allowing their recruitment for new cell cycle functions when the segregation machinery in the ancestral eukaryote was internalised as a consequence of phagocytosis (see next section below).

Though archaebacterial S/T kinases are more widespread [147] than previously thought [145], the apparent absence of cell cycle regulating PKN subfamily S/T kinases is yet another reason why archaebacteria, unlike Posibacteria, cannot be ancestral to eukaryotes. It is important to stress that as well as chromosomes mitosis segregates other organelles. Golgi duplication and centriole duplication are also regulated by some of these cell cycle kinases; so all segregation mechanisms must have coevolved.

Polo-like kinases are essential for centriole and Golgi duplication (the latter by phosphorylating Golgi-nucleating centrin blobs [148]) and separase is important for centriole as well as chromosome separation in Euglenozoa [149], so these functions all evolved in the early eukaryote at the same time as the nucleus. Aurora and polo-like families are probably sisters; on the polo-like paralogue-rooted aurora kinase tree Euglenozoa are the most divergent eukaryotes [150], as they are for each of the four SMC subtrees [144] and on Fig. 1, giving further support to the conclusion that Euglenozoa are the earliest diverging eukaryotes [9]. In trypanosomes aurora kinase has more diverse and basic cell cycle functions than polo-like [151,152] suggesting that it might have been the original mitosis related kinase; the involvement of polo-like kinase in centriole duplication, mitochondrial DNA segregation and cytokinesis [153] suggests an early division of labour between the two kinases with respect to chromosome segregation (aurora) and cytoplasmic organelle segregation (polo-like). Proper geometric localization of the segregation machinery, exemplified by that of polo-like kinase along the presumptive division plane [154] would have been important in the early evolution of eukaryote cell division and would have been facilitated by the complex and semi-rigid character of the cell cortex supported by an extensive array of cortical microtubules; the rooting of the tree between Euglenozoa and Percolozoa (both of which ancestrally had rigid microtubule supported pellicle) plus a distinct cytostome for ingestion [9] confirms that the importance of cell surface rigidity for the evolution of organelle segregation that I have long emphasized; contrary to early ideas dating back to Haeckel the cenancestral eukaryote was not a formless amoeba but a phagotrophic flagellate with rigid cell cortex and localised ingestion apparatus for phagocytosis. That is the only way to reconcile the conflicting demands of surface rigidity for geometric control of segregation and surface fluidity for ingestion during the prokaryote to eukaryote transition. The transitional organism had simultaneously to perfect ingestion and digestion and maintain viability by reasonably accurate organelle segregation. Mutual attachment of almost every cell organelle is seen in such modern cells as trypanosomes [155-157] and the cercozoan *Sainouron* [158].

The contractile calcium-binding protein centrin and proteins that bind it play key roles in organellar positional control that are often more disparate and pervasive than its core centrosomal function. Trypanosomes have five centrins [156] with diverse functions including coordinating cell and nuclear division [159] and ciliates far more, mainly to construct their uniquely complex and pervasive contractile cortical infraciliary lattice [160];

microtubules first thus evolved at the cell surface [4], but because the tree was misrooted I did not then think (as I now do) that their cytoplasmic/cytokinetic/pellicular microtubular division mechanism probably descended directly from the hypothetical prekaryotic surface system, and is not just an analogy.

The two key constituents of centrosomes are gamma tubulin, the nucleator for microtubule assembly, and centrin for positional control and architecture. It is centrin, the positional controller that underwent gene duplication to several paralogues to enable the complex structure of the cenancestral eukaryote. However, in the prekaryote before the origin of nuclei and cilia a simple centrosome with just one centrin paralogue would have sufficed. When I first speculated about the origin of eukaryotic chromosomes [70] it was generally assumed that bacterial DNA segregation was passive, by wall growth between DNA attachment points, so I proposed that centromeres evolved from the termini of bacterial replicons, because of the requirement that DNA replication be completed before division and the idea that replication termination could physically directly signal this to a adjacent segregation machinery and wall division site. Later I decided that bacterial DNA segregation must be by active pulling apart of replicon origins by ATPase motors [141], now known to be true, and argued instead that the first microtubule nucleation site therefore probably evolved in association with the pre-existing binding sites on the cell membrane of bacterial replication origins, as shown in Fig. 3c. We now know that Par genes that are central to bacterial DNA segregation cluster close to bacterial chromosome origins and that ParS marks the site of loading Smc condensins [132,133]; the bacterial Par system is widely considered a centrosome analogue, so the idea that centromeres originated from the origin region of the bacterial chromosome is common place, albeit seldom made explicit. However at the prekaryote stage when chromosomes were still attached to the cell surface membrane and there was only a single MNC, the distinction between a primitive centrosome and primitive centromere was less clear; both functions could have been assumed by a single macromolecular complex at one point on the cell surface [141]. I considered the postulated single MNC to have been a proto-centrosome (not a centrosome) because microtubule nucleation and the division of MNCs had to evolve before there could be any selective force for the evolution of kinetochores to attach chromosomes to the growing end of microtubules. MNCs are much more fundamental and general that kinetochores because they are essential for the segregation of many other organelles as well as chromosomes and because all interphase cell architecture is organised around the various cytoplasmic MNCs and centrin foci and their attachments to the nucleus.

I assumed above that when the nucleus evolved there was only a single MNC at each spindle pole. However that might not be true, because in euglenoids there are several distinct sub-spindles in the nucleus [110] so they must have multiple MNCs at each pole. If the eukaryotic root were actually within euglenoids (e.g. between the unicentriolar *Scytomonas* and more typical bicentriolar genera) then this multiple MNC condition would have been ancestral for all eukaryotes. Even if the root is between Euglenozoa and neokaryotes it could still have been the ancestral condition and the uni-MNC condition of neokaryotes and Kinetoplastida/Diplonemea (the main other putative euglenozoan clade than euglenoids) could have arisen independently by polar MNC merger. Previously I argued that the polarity seen in animal epithelial cells, and to a lesser extent also in migratory mesenchymal cells [167] stems fundamentally from the cell polarity that evolved in the first phagotrophic zooflagellate [58]. With the root being beside or within the ancestrally zooflagellate Euglenozoa [9] we can now say that the basic pattern and mechanisms of such cell polarity go right back to the last common ancestor of all eukaryotes and that it is a universal principle of eukaryotic cell biology. Haeckel's idea that ancestral eukaryotes and cells generally were amorphous amoebae without centrioles or cilia [168] has mislead thinking for over 150 years. Skeletal organisation is fundamental to understanding eukaryogenesis; in the cytoplasm the cytoskeleton consists of cross-linked actin filaments and microtubules. The nucleoskeleton is primarily chromatin itself [169,170], heterochromatin in particular whose origin I next discuss. Actin is being found to be increasingly important also in nuclear functions of animals at least [171-173]. But as its role in Eozoa, if any is unknown, and even in animals actin's contribution to basic nuclear architecture is much smaller than that of chromatin, I omit discussing it, beyond pointing out that if it has similar roles in trypanosomatids, these must be of universal significance and must eventually be taken into account in fuller discussions of nuclear origins.

Origin of Heterochromatin and Centromeres

Heterochromatin evolution is of key importance for the origin of the nucleus because the nuclear envelope is generally bound to at least a thin layer of condensed chromatin that would usually be called heterochromatin and bacteria have no equivalent structure [174]. Traditionally, constitutive heterochromatin was defined as chromatin that was condensed and microscopically visible throughout the cell cycle in contrast to euchromatin that dispersed in interphase and condensed only in mitosis as visible chromosomes. It has bizarre and largely unexplained patterns of variation in abundance and intranuclear distribution among organisms and developmental stages, but is often associated with centromeres, telomeres and the nuclear envelope. Until recently its function was a mystery; it was assumed not to be transcribed. All these ideas were oversimplified. Some organisms like budding yeasts lack visible chromatin condensation in both mitosis and interphase and others like euglenoids have visibly condensed chromosomes throughout the cell cycle. Heterochromatin generally has a low density of ordinary protein-coding genes; but much is transcribed at some stage of the life history to make transient non-coding RNA of varied functions. Long ago I postulated that heterochromatin is really DNA that has a largely structural role, such that its volume and degree of folding or unfolding can be modulated according to its needs - both in the interphase nucleus (where volume needs to be modulated at different developmental stages) and during mitosis where the compaction of DNA into centromeres and the geometric ordering of their kinetochore microtubule binding sites must be critical [169,175,176]. I also suggested that multicellular organisms may often transcribe it at stages when maximal nuclear volume was needed [169,170].

Recent ideas about heterochromatin have been dominated by the discovery that it is typically formed by the covalent modification of histones and is associated with an elaborate gene silencing machinery involving small 20-30-nucleotide non-coding RNA (microRNA or miRNA [177]) and associated enzymes that cut RNA that miRNA recognises by base pairing [178-181]. The significance of this vast body of work has been unclear for nuclear origins because although known in all higher eukaryotes (specifically Neozoa - all descendants of the last common ancestor of animals and plants) it was uncertain which phenomena and molecules were in the first nucleus. It is best known in the fission yeast fungus *Schizosaccharomyces pombe*, where RNA polymerase II transcribes heterochromatin (especially centric) only briefly during S phase and the RNA silencing machinery chops up the transcripts (recognised via double stranded loops) to make miRNAs which direct the histone deacetylating and methylating machinery for transmitting the heterochromatic state to the same locations in newly replicated DNA [182-185]; a form of epigenetic positional control. Plants use different specialized RNA polymerases (IV and V) for transcription and methylate DNA as well as histones to make heterochromatin, but the principle of

RNA silencing and the machinery for generating and using small RNA is similar across all neozoa. Just enough is known about gene silencing and histones in trypanosomes to say that the basic gene silencing machinery was already present in the eukaryote cenancestor, but probably with substantially lower molecular complexity than in animals. As in neozoa, all four core histones are acetylated and methylated (sometimes at homologous sites) in trypanosomes [186], and a limited number of homologous acetylating, deacetylating, and methylating enzymes (mostly targeting lysines) exist; there are also a few chromatin remodelling enzymes and histone-binding proteins with bromodomains, PDH fingers and Tudor domains, so this basic machinery all evolved in the ancestral eukaryote. There is also clear evidence for telomere silencing, which presumably evolved in the ancestral eukaryote when telomeres arose making chromosomes linear. But it is unclear whether their core histones are phosphorylated or ubiquitinated; though core histone phosphorylation by aurora kinase can occur in vitro in trypanosomes, it is likely that the mechanisms and their uses are simpler in Euglenozoa.

In neozoa, heterochromatin assembly at the centromere is typically a prerequisite for the loading of CenpA onto kinetochores and therefore for binding spindle microtubules (i.e. in all but budding yeasts (*Saccharomyces*), which secondarily evolved point centromeres where a different histone variant Cse-1 is loaded directly onto centromeric DNA in association with proteins able directly to recognize the centromeric DNA sequence [187]; model systems are often the exception to the rule, as for *E. coli* and SMCs). Neozoan heterochromatin is marked by heterochromatin protein 1 (HP1) which recognises lysine 9 (K9) in histone H3, directing methyltransferase Clr4 to methylate it. However trypanosome H3 lacks lysine K9, having four fewer amino acids in this region, so like *Saccharomyces* they must also use different machinery for assembling heterochromatin. As archaebacterial histones [188] lack the N- and C-terminal tails that in eukaryotes are important for the chromosome condensation cycle and which are modulated by acetylation and methylation, the slightly shorter N-terminal tails of trypanosomatids plausibly represent the ancestral eukaryotic condition, but until we know more about H3 and H4 in other Euglenozoa we cannot be sure that they are not secondary deletions. Although archaebacteria do not acetylate histones, they do acetylate their other major chromatin protein, Alba [188] which is a prerequisite for Mcm function in replication [189]; Alba probably evolved from a more universal RNaseP protein [190]; deacetylation is done by a homologue of the eukaryotic Sir2 histone deacetylase, so the acetylation/deacetylation machinery was in place in the neomuran ancestor

for suppressing transposons [205] and silencing their own genes [206]. However their canonical case of gene silencing is of genes located near telomeres that probably exploited a pre-existing telomeric gene silencing machinery (see next section). In trypanosomes one Argonaute protein is used for controlling chromosome segregation and also mRNA in polysomes for posttranslational control [205], while another targets exogenous RNA [207]. Probably the first eukaryote had a single Argonaute [208] primarily concerned with heterochromatin assembly. The heavy emphasis on translational control in trypanosomes is not likely to be a primitive character, but is probably a secondary consequence of the evolution of polycistronic messengers and universal trans-splicing of mini exons to pre-mRNA in the ancestor of Euglenozoa [59,209-211], coupled with the virtual abandonment of transcriptional control, which must have occurred after Euglenozoa diverged from neokaryotes. The fact that gene control in neokaryotes is largely transcriptional as in prokaryotes means that this must have been true also of the eukaryote cenancestor. Likewise the use of histone variants to label the boundaries of polycistronic transcripts in Euglenozoa [212,213] is almost certainly a secondary innovation for them, but it could have evolved from a more basic early eukaryotic use of histone variants to label functional chromosome domain boundaries.

I suggest that gene repression by small RNAs (miRNAs) and suppression of exogenous viral or transposons transcripts by destruction via analogous small silencing RNAs (siRNAs) could both easily have evolved and be applied to a great variety of targets according to the specific needs of different lineages after the basic small RNA-based machinery evolved for centromeric assembly and was modified for telomeric chromatin. Both latter functions are general for all eukaryotes, but specific gene repressions such as mating type silencing in yeast or variant surface antigen repression in the parasites *Giardia* and trypanosomatids are lineage-specific exploitations of a more basic pre-existing machinery. Various eukaryotes have secondarily lost the capacity for suppressing exogenous RNA by siRNA (also called RNA interference), e.g. *S. cerevisiae* ([214], but related budding yeasts have retained it [215]) and *Trypanosoma cruzi* (but *T. brucei* retains it). That such organisms still have some heterochromatin suggests that the origin of heterochromatin was more complex than I have implied; study of its types and functions in Euglenozoa broadly are important for clarifying its origin. The secondary origin of centromeres by direct binding of centrosomal proteins to specific DNA sequences (bypassing typical heterochromatin) that evolved in budding yeasts may have predisposed some of them (e.g. *S.*

cerevisiae) to lose the standard siRNA mechanism, as purifying selection for efficient centromere function would no longer retain it.

A more broadly distributed form of miRNA that probably evolved almost as early as centromeric small RNAs is miRNAs derived from snoRNAs, which arose at least as early as excavates, being found in *Giardia* as well as across neozoa [216]. One potential hypothetical function for snoRNA-derived miRNA that might be universal would be the positional control of nucleolar heterochromatin which is present in most eukaryotes and likely to be important in nucleolar architecture. snoRNAs evolved in the ancestral neomuran and come in two classes, boxC/D snoRNAs that recognize sites for methylation (or in one case - U3 - proper folding of prerRNA for cleavage [217]) and boxH/ACA snoRNAs that recognize prerRNA sites for isomerization to pseudouridine; each interacts with a different 4-protein catalytic complex. They underwent little change during the origin of the nucleus; trypanosomatid and euglenoid snoRNAs are notably simpler than those of neokaryotes (including the excavate *Giardia*) and more closely resemble those of archaebacteria [218-220], giving additional strong support to the primary eukaryote divergence being between Euglenozoa and neokaryotes (Fig. 1). Unsurprisingly on this view of the root, U3, which is far more variable in length than other eukaryotic snoRNAs, is shortest in trypanosomatids [217].

Origins of Telomeres and Telomeric Heterochromatin

The primary reason d'être for telomeres was to solve the end replication problem of linear chromosomes [221] and the telomerase machinery could in principle have been recruited either from host enzymes or from exogenous selfish genetic elements [70,222]. Linear chromosomes have evolved several times in viruses, bacteria, and mitochondrial DNA as well as during the origin of eukaryotes and the solution to the end replication problem has been varied. Probably in most cases the original fragmentation of the once circular chromosomes was accidental and all mechanisms were ad-hoc rescues. The origin of linear eukaryotic chromosomes may have occurred in the prekaryote prior to the origin of the nuclear envelope if meiosis (see below) began then, since odd numbers of meiotic crossovers or mitotic sister chromatid exchanges in circular chromosomes produce circular dimers, resulting in accidental breaks during DNA segregation [11,70]. Even before the origin of the nucleus telomeres would have acquired additional proteins to help block accidental covalent joining of ends of different chromosomes; a key protein involved in this end protection is Rap1 [223] that must have been present in the ancestral eukaryote as it present in trypanosomatid telomeres, where it is also essential for silencing genes located in telomeric heterochromatin [224,225]. Telomeric and centromeric gene silencing may both be by-products of the assembly of heterochromatin. Telomeric heterochromatin depends not only on region-specific proteins like Rap1, but also on others shared with centromere biogenesis; histone acetylases are involved [203] and the mi-RNAs derived from snoRNAs [216] potentially target the telomere-located variant surface antigen proteins. If this is also true of Euglenozoa it could mean that snoRNA derived miRNAs were recruited for telomere heterochromatin formation even in the first eukaryote. However, when the nuclear envelope evolved (see next two sections) telomeres would have acquired additional functions for attaching to it. Both in interphase and meiotic prophase telomeres are normally attached to the nuclear envelope. Probably the compacted nature of both centromeric and telomeric heterochromatin were prekaryotic preadaptations that would have facilitated the attachment of endomembranes to the surface of chromatin, the central evolutionary innovation in the origin of the nuclear envelope. The nucleolus also typically has associated heterochromatin, which also might have evolved before the nucleus and might as suggested above depend on small RNAs for its positional control.

Phagocytosis, Endomembranes, Mitosis and Novel Cell Cycle Controls

It is often overlooked that internalisation of DNA-attachment sites by primitive phagocytosis would automatically produce an endomembrane vesicle with the chromosome attached to its surface by bacterial membrane proteins that bind DNA [4]. Almost certainly some such proteins were retained for similar DNA-binding duties in the inner membrane domain of the NE during eukaryogenesis. Likely candidates are Man1 (its MSC domain is related to the DNA-binding helix-turn helix fold of many bacterial transcription regulators) and RfB [14,17]. There had to be much continuity in DNA attachment and segregation processes during eukaryogenesis, despite its radical nature, or cells would die [4]. This DNA-bearing endomembrane was the NE precursor [4]. Such internalization of DNA must immediately have caused segregation problems [4,70], for any membrane division machinery carried with it might divide the endomembrane after DNA replication, yielding two daughter DNAs attached to separate cisternae, these would not be attached to the cell surface, which initially might still divide by an FtsZ ring. Without direct physical coordination of division of surface and endomembranes, daughters lacking DNA or having multiple chromosomes would inevitably be produced at a fair frequency until protoNE/DNA segregation became less random [3,4]. Avoiding this was the primary selective force for the radical origin of mitosis [3,4,70]; as suggested above the difficulty of the transition may have been somewhat mitigated by the transitional intermediate being multigenomic, so unequal divisions would produce DNA-free cells with lower frequency than if there were only one genome per cell. Provided that the transitional cell had a uniquely strong compensating advantage that completely set it apart from all bacterial competitors (eating other cells) it could still enjoy net reproductive success and rapid population growth despite the handicap of generating a proportion of DNA-less daughters.

Once it achieved this degree of success the main competition would be among its offspring, leading to increased segregational efficiency of the best lineage, with all the less efficient ones dying out. This same principle would apply to all innovative aspects of eukaryogenesis, ensuring that there was probably only one eventually successful lineage surviving the transition, with no half-evolved lineages persisting long enough to become ecologically important. Mis-segregation could be avoided only by coordinating division of both membranes through novel indirect physical connections to the cell surface. As the surface skeletal bacterial protein MreB had already undergone duplication and evolved into actin and Arps for phagocytosis, other gene duplications produced Arps to nucleate a contractile actomyosin ring for dividing the surface membrane [6]; even in eubacteria the divisome includes an actin relative, FtsA [226]. Membrane division also needs membrane deforming and membrane scission proteins. It now appears that novel proteins of this sort evolved during the neomuran revolution when the peptidoglycan wall was replaced by surface glycoproteins. In eubacteria it is presumed that the actual membrane scission might involve the growth of the wall and formation of the new murein septum, though all the machinery involved has not been identified [227]. Obviously that could not continue when murein was lost. Instead ancestral neomura evolved a novel membrane deforming and scission machinery using proteins ancestral to both the ESCRT III complex of eukaryotes [228] and the CdvA, B proteins of (mainly) crenarchaeote archaebacteria [229,230]. In eukaryotes this is involved in membrane division during cytokinesis and endosome division and in archaebacteria in cytokinesis and membrane blebbing into the environment [231]. The mesophilic former crenarchaeotes now called Thaumarchaeota [232] retain both the old FtsZ and the new ESCRT-like division machinery (which probably have complementary functions), but hyperthermophilic crenarchaeotes lost FtsZ (and so could not be ancestral to eukaryotes unless eukaryotes got their tubulin ancestor from FtsZ mitochondrial ancestor - very unlikely) and retained only ESCRT-like division machinery (Thermoproteales evolved a third unknown mechanism for their odd snapping division), as did eukaryotes (assuming tubulin evolved from TubZ not FtsZ). Conversely euryarchaeotes except *Thermoplasma* lost ESCRT-like division proteins. Another membrane division novelty, the large GTPase dynamin, which promotes division by forming a helical collar around the neck of constricted membranes, evolved later in the ancestor of eukaryotes only. Many dynamin paralogues evolved for different functions: cytokinesis, vesicle budding from the Golgi, endocytosis, peroxisome (probably all before the origin of the nucleus during endomembrane diversification [27]) and also for nuclear fusion (see below); and later mitochondrial division, and even later a plant cytokinetic one was further duplicated to help chloroplast division [233].

Another likely consequence of murein loss was a basic change in the termination of DNA replication. The ancestral mechanism in eubacteria is DNA site-specific and tightly linked to murein septation, and involves a recombinase (Xer in proteobacteria) that unlinks catenanes by site-specific recombination at the terminus [234,235], together with a DNA translocase (FtsK in proteobacteria) that strips proteins from DNA and helps unlinking [236]. This machinery was lost in the ancestral neomura, which all use non-site specific DNA termination mechanisms and different ways of controlling DNA replication in relation to the cell cycle, including novel topoisomerase II enzymes that can separate intertwined sister DNAs. As argued before [12], many novel features of neomuran DNA replication machinery and its control stem from the origin of histones rather than the loss of murein in the cenancestral neomuran. This is arguably so for the new preinitiation machinery involving Mcm proteins and Cdc6 discussed below.

Triplication of a TubZ-like gene in the ancestor of eukaryotes alone to yield γ-tubulin (centrosomes) and α- and β-tubulins making microtubules [3,6] provided a new machinery for DNA segregation, but as Fig. 3 indicated an initial key role may have been surface stabilisation to prevent DNA internalisation by phagocytosis. Initially there was no need for kinetochores and microtubule attachment to DNA since centrosomes and chromosomes were both bound to the membrane (Fig. 3c). DNA and endomembrane would automatically be cosegregated if microtubule polymerisation pushed daughter centrosomes apart [3,6], as is postulated for TubZ. Later, efficiency increased by evolution of kinesin motors to actively slide apart antiparallel centrosome-to-centrosome microtubules (Fig. 3c), and probably later still of dyneins to move vesicles and chromosomes towards the minus end of microtubules (Fig. 3e, f). Thus all major elements of mitotic spindles, possibly remaining through interphase as stable half-spindles, probably evolved before the nucleus [3,4,6], though I suspect that dynein, whose sister paralogue is midasin (a giant protein involved in rRNA export from the nucleus) may only have evolved at the same time as the nuclear envelope; most dyneins are concerned with ciliary function. I reserve discussion of the origin of centrioles (whose basic function is generating cilia) and cilia themselves to a later paper because of its complexity and because they are not fundamental to mitosis, being just 'there for the ride' [237-240], even though they originated prior to the eukaryote cenancestor. That FtsZ starts to assemble at the time of initiation of DNA replication [241] is intriguingly similar to the synchronization of centriole duplication at the beginning of S-phase; might the timing of γ-tubulin's initial assembly have been inherited from bacteria? By contrast to the more optional centrioles, as soon as protoER/NE became a distinct genetic membrane [85], its segregation without loss was as important for viability as that of DNA [163].

Efficient segregation and avoidance of DNA breakage (whether by entanglement with molecular motors effecting chromosome or vesicle movement or by contraction by the new actin contractile ring) required greater chromosome compaction [4,11,70]. Thus histone H3 and/or H4 (both arose in the neomuran ancestor [12,13]) underwent duplication and major modification, yielding histones H2a and H2b, forming octomeric nucleosomes, and an H1 precursor already present in the actinobacterial ancestor (lost by archaebacteria) [12], was recruited to link them as solenoidal 30 nm chromatin fibres. Methylating, phosphorylating, and acetylating enzymes were recruited and temporally coordinated to effect a chromosome condensation cycle (aurora kinase phosphorylates opisthokont H3 and Cdk1 does for H1 [242]), with looser interphase transcribed chromosomes alternating with inactive mitotic chromosomes of 30 nm fibre loops tightly folded around a proteinaceous core, including especially DNA topoisomerase II, essential to relieve positive supercoiling during transcription and to separate tangled sister DNAs. A further histone duplication generated CenpA, the core constituent of centromeres. It was previously assumed that this allowed evolution of specific microtubule binding directly to chromatin for the first time and thus fully developed mitosis and that it arose in the cenancestor of eukaryotes [12]. However, the rerooting of the eukaryote tree between Euglenozoa and other eukaryotes makes it possible that the apparent absence of CenpA in kinetoplastid Euglenozoa is the primitive state for all Euglenozoa and eukaryotes generally and that CenpA evolved only in the ancestor of neokaryotes (all eukaryotes other than Euglenozoa: [9]); CenpA homologues are present in excavates, including the earliest diverging Percolozoa, e.g. *Naegleria* [212]. Given that minichromosomes or yeast and trypanosomes can be segregated accurately without all the machinery associated with typical large chromosomes the ability to move organelles bipolarly on spindles is probably more basic and general than the specific modes of attachment. As suggested above the most fundamental innovation for centromeres may have been the origin of properly folded centromeric heterochromatin assembled at positions controlled by the base pairing of small RNAs. The requirement of Argonaute in trypanosomes for segregation of both mini and ordinary chromosomes [118], which do not have the same kind of kinetochores (and neither has CenpA), supports this.

That separase is needed but cohesin is not for trypanosome minichromosomes whereas both are needed for the large chromosomes [149] confirms the multiplicity or redundancy of the segregation machinery, as does its ability to segregate other organelles such as centrioles, Golgi, mitochondria and the nuclear envelope ER; in many protozoa the singleness and nuclear attachment of peroxisomes means that they also must be attached to the segregation machinery [158]. That segregation of these organelles may depend on similar mechanism to chromosome disjunction is shown by mammal cohesin being involved in centriole adhesion and separase cleavage of Scc1 (perhaps recruited to centrosomes by protein Aki1) causing their separation [243]. As separase mediates centriole separation in both trypanosomes and mammals, cohesin was probably recruited for attaching parent and daughter centrioles when centrioles first evolved prior to the eukaryote ancestor (between stages e and f on Fig. 3). The evidence that eukaryotes were ancestrally biciliate and that even the cenancestor had evolved ciliary transformation from a younger anterior to an older posterior cilium [9] means that proteolytic separation of mutually attached old and next generation centrioles must date from then. It is remarkably simplifying that chromosomes and centrioles use the same mechanism for adhesion and separation despite their fundamentally different structures and modes of duplication.

Very little is known about the molecular mechanisms of mitosis in Euglenozoa, except that mitosis differs cytologically in several respects from that of neokaryotes as well as among the euglenozoan classes, always lacking a chromatin condensation cycle (permanently condensed in euglenoids, permanently diffuse in kinetoplastids); in trypanosomatids there are often both minichromosomes (often more than the number of spindle microtubules so their segregation is unconventional); euglenoids also can have large numbers of small chromosomes; localised sites for binding DNA topoisomerase II may correspond with centromeres in some but not all trypanosomatid chromosomes [212,213]. Comparative studies of molecular mechanisms in Euglenozoa should greatly clarify the origin and early evolution of mitosis.

Two other key features of the eukaryote cell cycle [244] probably arose at about the same time as centromeres: DNA replication licensing and pervasive anaphase proteolysis and cycle resetting. Internalisation of DNA by phagocytosis would have disrupted not only the spatial coordination of DNA segregation but also the temporal coordination of replication and cell division as practised by bacteria [70]. This would have immediately been exacerbated by the origin of centromeres, because efficient mitotic DNA segregation totally removed the ancestral stabilizing selective force requiring only a single replicon per chromosome so replication termination could directly signal division to occur directly between the only two daughter replicon termini [70]. Thus inevitable recurrent mutations multiplying replicon origins or termini previously stringently removed by purifying selection for >2 Gy would suddenly be harmless and spread like wildfire [70] (and be beneficial, as multiple replicons would ensure more rapid completion of replication long before division might try segregating incomplete daughters and abolish wasteful overlapping rounds of replication [70]). Likewise mutations making several separate chromosomes would spread automatically, as mitosis, unlike bacterial systems, can cope with many [70]. Furthermore, rapid replication of numerous replicons removed any constraint for circularity arising from the need for replication forks to converge on a single terminus and signal completion, making mutations for linearity less harmful [70]. Thus, the characteristic multi-replicons of eukaryote chromosomes (and in part their linearity) are consequences of the origin of mitosis plus mutation pressure, not positive selection [70]. Having multiple replicons caused problems for ensuring that all replication was complete before division. The two solutions were evolution of Mcm-based DNA licensing to ensure all replicons initiated together, and of cyclins and recruitment of protein phosphorylation, dephosphorylation and proteolysis to coordinate this with cell growth, so that on average a round of replication was initiated at every doubling of cell size, with mitosis being inhibited till its completion. Mcm proteins themselves evolved earlier in the neomuran ancestor as they are also present in archaebacteria [245], but not apparently eubacteria, and work as a six-protein DNA helicase that moves replication forks. In archaebacteria and in the early diverging kinetoplastid Euglenozoa [246] replication origins are recognised by the protein Cdc6, which with the help of a partially related protein Cdt1 loads the Mcm hexamer onto replication origins. In neokaryotes the system is more complex: an additional prereplication complex consisting of up to six different origin recognition complex (ORC) proteins is also required. A major domain of five ORC proteins (Orc1-5) is related to Cdc6. I proposed that at least two of them (Orc1 and Orc2) evolved by gene duplication from Cdc6 in the ancestral neokaryote after it diverged from Euglenozoa [9], but that Orc4 (apparently missing from *Naegleria*) evolved only after the divergence of Percolozoa from other eukaryotes. Orc6 evolves faster and its origin is less clear. It thus appears that not all features of the ORC system that are universally present in higher eukaryotes (Neozoa) were present in the first eukaryotes and that the neozoan cell cycle control system probably evolved in stages during the successive divergences of Eozoa. Direct biochemical studies of excavate and euglenozoan cell cycle controls are essential to test this and elucidate the various steps.

Proteasomes, which probably arose in thermophilic actinobacterial ancestors of neomura to degrade denatured proteins [13], were complexified in the ancestral eukaryote by numerous gene duplications and access to them for destruction became controlled by the novel ubiquitin system and they were recruited for anaphase proteolysis. Numerous proteins, including especially the cohesin loops [247-249] that hold sister DNA molecules together from replication onwards (even in bacteria; they are also Smc proteins related to condensins) [63], were thus destroyed at the anaphase transition: the key switch of the eukaryote cell cycle [244]. It may be that this system is simpler in Euglenozoa and did not evolve all at once; one difference is that unlike in higher eukaryotes trypanosomatid Cdc6 is not degraded at S phase and persists in association with chromatin throughout the cycle [250].

Later, additional checkpoints blocking mitosis until all was ready were added. Attempts to deduce the order of evolution of cell cycle kinases from paralogue trees [251] are hopeless, as trees are obviously dominated by systematic biases caused by the radically different ways each family adapted immediately after gene duplication, against which statistical tests provide no protection whatever. Kinase trees do not remotely reflect time, are almost certainly misrooted (a pervasive problem with paralogue trees [13]), and likely to be topologically wrong. Fig. 3 of [251] has no valid outgroup and is simply 'rooted' among the three longest branches; cdc28 may appear late merely because since the cenancestor it evolved 2-3 times slower than them. All we can safely conclude from those trees is that all major mitotic and meiotic controls existed before the cenancestor (and the subsequent relative evolutionary rates of each paralogue). However gene duplication order is less important than the principle that the pervasive selective force of avoiding mis-segregation and DNA breakage made the eukaryote cell cycle, which was a rescue operation after the genetically immensely disruptive effects of DNA/membrane internalization by phagocytosis, which arose because of its immense trophic benefits despite this harm. Thus much of eukaryogenesis occurred to find a new equilibrium in which the benefits of phagocytosis could persist without its severe genetic costs. It is not intrinsically better to divide cells the eukaryote rather than the prokaryote way, but if DNA is attached to endomembranes instead of the cell surface there is no other option but a radically new solution, which was clearly constrained by the possible precursors that happened to be available and the new cell structure.

Thus, prior to compartmentation that finally made the nucleus, many features of eukaryotic chromosomes, including the chromatin condensation cycle and novel replication controls and novel segregation machinery, had probably evolved as an indirect consequence of the changeover from surface membrane to mitotic DNA segregation. As soon as the novel replication and other cell cycle controls were in place (necessitated to complete replication well before mitosis which is sudden at anaphase and not gradual and spread out over the whole cell cycle as in bacteria) accidental duplications of replicon origins inevitably spread them across the whole chromosome, but this would also have been positively selected as simultaneous replication at many points could compensate for the much slower movement of replication origins though nucleosomes and allow shorter cell cycles than otherwise); overall replication time would no longer limit genome size. There is no reason to think that an increase in genome size per se, which in eukaryotes is independent of organismal complexity [17], or quantitative population genetic factors such as effective population size were determinants of such a radical change in genome organization, as is sometimes claimed [79]. The above concentrated on the cell's trophic phase, but the cenancestral eukaryote also had resting cysts [3]. The evolution of syngamy and meiosis were probably associated respectively with encystation and excystation; a later section argues that the fundamental reason for the origin of sexual cycles with haploid and diploid nuclei was the conflicting selective forces during growth and dormancy (cell multiplication and survival respectively). However, I first consider the next steps in the origin of the nucleus of trophic cells, even though in reality sex-related chromosomal evolution was probably concurrent with the evolution of interphase nuclear structure.

Nucleocytoplasmic Separation: A Two Phase Evolution

NPCs must have evolved in two stages with different selective advantages [5]. Initially the basic octagonal cylinder embedded in the NE evolved, primarily to prevent complete fusion of ER cisternae (Fig. 3d) around chromatin, which would have been lethal by preventing transcripts reaching the cytoplasm and stopping growth [4,5]. Later this wide cylinder, allowing nucleocytoplasmic exchange by passive diffusion (Fig. 5a), was constricted by inserting the inner FG repeat nucleoporins (Nups) to exclude ribosomes from the nucleus and DNA and RNA polymerases and RNA processing complexes from the cytoplasm, and active transport machinery evolved to export and import them (Figs 3e, 5b). FG repeat proteins have long domains consisting of numerous repeats in which the basic amino acid phenylalanine (F) and the acidic amino acid glutamic acid (G) predominate, leading to a structure that is far less ordered than in most proteins. While probably a little more ordered than random coils, the FG-repeat Nups appear to form a hydrophilic meshwork with dynamic properties that impede the spontaneous diffusion of large macromolecular complexes through the NPCs, but can be modulated by the temporary binding of karyopherins [252] that can thus force large macromolecular complexes bound to them through the FG repeat putative hydrogel. FG repeat Nups not only dominate the central plug of the NPCs, but are also prominent in the nuclear basket that caps the NPCs on their nucleoplasmic face and in the long cytoplasmic filaments that help to control ingress to the nucleus nuclear on the cytoplasmic side. A gradient in binding affinity between karyopherins and the FG repeat Nups from weak outside to strongest on the nuclear basket is instrumental in the efficient polarised import of karyopherins and their cargos [253]. A basic principle of the evolution of the NPCs is that karyopherins and FG repeat Nups must have coevolved, as must karyopherins and their disparate cargos.

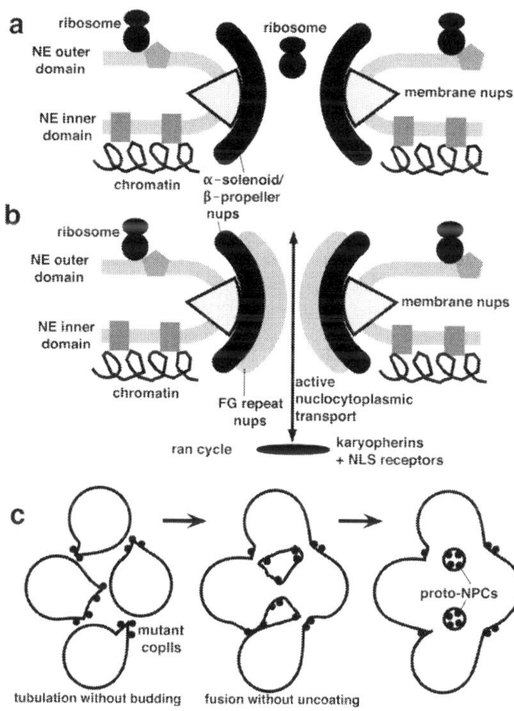

Figure 5: Two-phase origin of the nuclear envelope and trans-envelope transport. (a). Nucleoporins (Nups) forming the octagonal cylindrical scaffold evolved by duplications of coat proteins of COPII secretory vesicles with α-solenoid and/or β-propeller domains, being attached by integral membrane Nups descended from actinobacterial membrane proteins; (b). NPC lumens were narrowed by plugs of FG-repeat-rich Nups, which form a dynamic gel-like meshwork that prevents passive diffusion of macromolecular complexes and mediates active specifically-targeted nucleocytoplasmic exchange by carrier complexes, typically consisting of large karyopherin proteins and their cargo either bound directly or by adaptors. (c). Phase I surface view, showing complete Ran GTPase-mediated fusion of RER cisternae prevented by COPII coat proteins (black blobs) remaining in place to become octagonal NPC scaffolds.

Extensive gene duplication and domain shuffling generated the two extensive and structurally different but functionally complementary protein families, the trans-envelope shuttling karyopherins and the FG-repeat Nups. The selective force for the coupled restriction of free diffusion of larger molecules and of active transport across Npcs was the benefit of compartmentation and specialization; higher concentrations of protein synthesis enzymes in the cytoplasm only and, especially nucleic acid synthesis enzymes in the nucleus could be maintained at much lower cost [4,5]; especially important in reducing cost was the effective exclusion of nuclear proteins from the much larger cytoplasmic compartment, thereby reducing the biosynthetic load on the cell.

I previously suggested that NPC transport machinery might in part have arisen from that for secretory vesicles [5,70]. This seems partially true. The basic octagonal scaffold Nups have α-solenoid and/or β-propeller domains clearly related to those of coated vesicles (COPII from ER; COPI from Golgi; clathrin-associated adaptins from plasma-membrane and endosomes) [15-17]. I now suggest that the NPC scaffold evolved from COPII coats, probably before COPI evolved; at that time COPII vesicles would be uncoated and fuse directly with the plasma membrane or protoendomembranes, processes mediated by small GTPases, that themselves underwent differentiative duplications in concert with those of vesicle coats that multiplied the number of genetically distinct endomembranes. Two β-propeller proteins (Sec13/She1) actually form a subcomplex in both COPII coats and NPCs [15-17]. β-propellers have several vanes composed of WD repeats that are widely present in proteins involved in binding other proteins (some in bacteria and many in eukaryotes). The combination of β-propellers and α-solenoid domains that is found in several core Nups is unique to eukaryotes; as such proteins are centrally involved not only in NPCs, and coated vesicles, but also in ciliary transport particles that supply growing ciliary axonemes with newly made proteins (and had also already evolved prior to the eukaryote cenancestor), their origin and diversification by gene duplication and domain shuffling were centrally important for eukaryogenesis. Since this paper was written the close structural relationship between the structural scaffold of the nuclear pore complex and the COPII outer coat lattice has been directly demonstrated by solving the crystallographic structure of the Nup85 20 α-helix/Seh1 β-propeller complex, placing their evolutionary relationship beyond reasonable doubt [254].

The universal binding of the nuclear envelope to chromatin during interphase is a really fundamental feature of the nucleus of profound importance for the evolution of eukaryote genome size and for the explanation of why eukaryote genome size correlates with cell volume [18], but has been largely overlooked by geneticists and theorisers about nuclear origins unfamiliar with cell biology. It was crucial for nuclear envelope evolution, because its inner membrane and components of the NPC are bound to chromatin and because the binding of RanGTP to chromatin, and the exclusion of the ancestrally soluble RanGAP that mediates its conversion to RanGDP provide a directional polarity to nucleocytoplasmic transport and nuclear assembly. RanGTP is concentrated in the nucleus because of the chromatin-attachment of its cofactor RCC1. Nuclear RanGTP binds to karyopherins, promoting the disassembly of cargo from karyopherins involved in import (e.g. importins α and β and the attachment of cargo to those involved in export of which the exportin Crm1 (exportin 1) that exports ribosomal subunits and many other cargos is most important) [255]. RanGTP remains bound to karyopherins till they exit the nucleus but is converted to RanGDP in the cytoplasm with the help of the activator protein RanGAP that promotes GTP hydrolysis to GDP; free RanGTP is concentrated in the cytosol because of the cytoplasmic location of RanGAP. In contrast to RanGTP, RanGDP promotes the attachment of cytoplasmic proteins to karyopherins for import into the nucleus, but it does not accompany them. Thus nucleocytoplasmic transport continually depletes the nucleus of Ran; it is replenished by transport to the nucleus of RanGDP bound not to karyopherin but to a completely unrelated carrier, an NTF2 dimer, which binds to FG-Nups in a different but partially similar way [256,257]. Inside the nucleus, RCC1 charges Ran with GTP, dissociating it from NTF2, which is transported back into the cytosol. The complementary distribution of the two forms of Ran [258] must have been a pervasive spatial backcloth against which the nucleocytoplasmic exchange machinery evolved [259].

Its importance, however, goes way beyond the origin of the nucleus, because RanGTP is intimately involved in spindle assembly and mitosis through promoting kinetochore and centrosome functions. If as I have repeatedly argued the origin of mitosis was so crucial to the survival of the earliest prekaryote cells that its main features must have evolved before the nucleus [3,4], it is reasonable to suppose that the initial function of the Ran system was for mitosis [260], thus preadapting the prekaryote for the subsequent and probably immediately following origin of the nucleus. RanGTP promotes the assembly of microtubules at the kinetochores, and its binding to chromatin may have originally evolved for this reason. Its putatively pre-existing chromatin location explains how the assembly of the envelope (itself under RanGTPase control) was spatially organized around chromatin from the outset. If the basic features of the RanGTP/GDP cycle existed prior to the origin of the nucleus, it both shaped and facilitated the latter. As proposed by Jékely [260], RanGTPase probably originated for its mitotic functions by gene duplication and divergence from Rabs. Prior to the origin of the nuclear envelope I suggest RanGTPase was already associated with chromatin by its RCC1 exchange factor that is essential for the conversion of RanGTP to RanGDP (the cytosolic form). Thus the RanGTP/GDP cycle and the chromatin attachment of RanGTP via RCC1 probably evolved for the control of mitosis including its spatial aspects in the prekaryote. Interestingly RCC1 is another β-propeller protein with 7 WD-40 domain blades like so many other such proteins (e.g. trimeric G proteins, the vesicular traffic protein Sec13) that proliferated during eukaryogenesis. Thus even before the origin of NPCs the cell was spatially polarised into Ran-GTP-rich chromatin and RanGTP-rich cytosol [259]. RanGTPase and a suite of Ran binding proteins played key roles in both phases of NPC evolution.

Phase One: From Coated Vesicles to Nuclear Pore Complex

The key step thereafter for the origin of the nuclear envelope, I propose, involved duplication and modification of some COPII component(s) that allowed cisternal fusion without prior total separation of vesicles from the donor membrane and therefore necessarily without uncoating (Fig. 5c left). Thus the essential key innovation for making the nucleus was a modification of COPII coats. As the cisternae fused together, the COPII-derived coatomers would remain on the membrane (Fig. 5c centre) and automatically become clumped as cylindrical aggregates, shown in surface view in Fig. 5c. Thus, a single mechanistic innovation in the known precursor would at a stroke evolve both a fenestrated nuclear envelope and NPC scaffolds allowing passive nucleocytoplasmic exchange. Cytoplasmic motors actively moving COPII, and later other, secretory vesicles would be prevented from causing DNA damage and interference with transcription, by getting entangled with DNA and nascent RNA, especially pre-rRNA, which being so long (comprising 3 molecules before cleavage) would be especially susceptible to shearing. Homotypic fusion of ER-derived COPII vesicles occurs in modern cells to generate the preGolgi intermediate compartment [261,262]; allowing it to go ahead without prior uncoating would generate a plausible precursor of the nuclear envelope (including its pore complex) in a single step, provided that cisternae already attached to chromatin (or COPII vesicles containing chromatin-attachment proteins) were also involved in that fusion.

Thus, avoiding DNA and RNA breakage was the primary advantage of the NE and NPC, which initially was only narrow enough to exclude secretory vesicles but not ribosomes and polymerases (Fig. 5a). Avoiding DNA breakage has been criticised as a selective force, by the claim that there is no problem '...during mitosis, even in species where they are permanently uncondensed, despite the fact that the nuclear envelope disintegrates at the beginning of mitosis...' [122]. This objection is misleading and invalid. The supposed counter-example actually indirectly supports my thesis, because the only eukaryotes in which histones have been lost and DNA is truly uncondensed during mitosis are peridinean dinoflagellates in which the mitotic spindle is entirely outside the nucleus and the envelope never breaks down [263-265]. Their perpetually relatively condensed but histone-free chromosomes [266] probably evolved secondarily only after, almost uniquely, spindle and chromosome separation by evolving intra-NE kinetochores bypassed the shearing problem by a different route [267]. However, avoiding transcript breakage, a novel proposition, was perhaps an even stronger force, not only for the origin of the NE but also of the gel-like interphase nuclear matrix [163], in which DNA topoisomerase plays a key role to relieve positive supercoiling ahead of transcription complexes.

Thus, we have both a plausible physical mechanism based on known cell biology and a plausible selective advantage for the origin of the nuclear envelope. As the envelope-associated part of the NPC involved over a dozen proteins of the scaffold Nups have α-solenoid and/or β-propeller domains prior to the eukaryote cenancestor, there must have been a rapid multiplication of these following the minimal change in just one or two needed to set the process in motion. In so far as some of these core scaffold proteins are relocated to kinetochores and/or centrosomes during mitosis, it is possible that some of these duplications actually preceded the origin of the core complex and were selected initially for their functions in mitosis rather than in pore complexes. However, fruitful speculation along these lines in impeded by our knowledge of the detailed cell biology of the mitotic behaviour and possible mitotic functions of NPC components being primitive and largely restricted to one eukaryote lineage only (opisthokonts, which include animals and fungi), almost nothing being known for bikonts, which have an even greater range of mitotic behavioural diversity than opisthokonts. It is especially important to study these processes in Euglenozoa as features currently assumed to be universal for eukaryotes might only characterise neokaryotes. A related problem is that even within opisthokonts NPC behaviour differs substantially among organisms with closed mitosis (e.g. yeasts), open mitosis (animals), or semi-open mitosis (e.g. *Aspergillus*); without comparable knowledge for a good range of bikonts inferring the ancestral state even for opisthokonts, still less for eukaryotes, is hazardous. A key recent discovery is that a major core complex of about 10 scaffold Nups (the Nup107-160 complex in animals; different names in fungi) behaves as a unit during mitosis. This complex probably includes proteins with membrane curving functions likely to have been central to the origin of nuclear pores, but present knowledge also does not allow one to deduce which components would minimally have been necessary for the first phase of NPC evolution and which might have been added later in the second phase, dominated by the addition of a completely different class of Nups, the FG Nups and their interactions with karyopherins.

Compartmentation and its Consequences: Ran-GTPase, FG-Nups and Karyopherins

Karyopherins, which mediate protein import/export and the FG-rich Nups with which they interact to allow nucleocytoplasmic exchanges are both large eukaryotic gene families with extremely few bacterial homologues that became highly diversified during eukaryogenesis prior to the eukaryotic cenancestor [14,17]. The fact that karyopherins use basic amino acids as nuclear localisation signals (NLS) arguably stems from their major early cargo having been very basic histones and other DNA binding proteins [3] as well as basic proteins that bind RNA, especially ribosomal proteins [259]; early cargo probably included inner-membrane DNA-binding proteins [19], that would otherwise be impeded by the novel integral membrane Nups, unless from the start NPCs opened to let them cross the NPC NE domain, and histones. Ribosomal assembly became localised to the nucleolus probably because transcription of such long molecules is so slow that it is optimal to start assembling proteins onto them and process them by RNA cleavage, and base modification; the neomuran ancestor already had numerous RNPs and enzymes on a fibrillarin matrix associated with transcription sites [3]. Evolution normally takes the line of least resistance and changes as little as possible; stabilising selection prevents capricious 'redesign' so there was no chance of relocating ribosomal assembly to the cytoplasmic point of use. tRNAs are exported by attachment to karyopherins by proteins with NLSs, but completely separate machinery evolved for exporting ribosomal subunits and another for mRNA [17], indicating that their localisation was probably perfected in parallel not serially. Messenger RNA capping and polyadenylation probably arose primarily to prevent RNAase degradation during the much longer time it would take for mRNA to reach ribosomes than in bacteria, mitochondria and plastids, where they can attach during transcription [70]. But both probably became markers of maturity and readiness for export by the novel mRNA export machinery that arose by duplication of a bacterial GTPase to make Nug1p/Nug2p [17]. The ribosome-subunit export machinery recruited a duplicated neomuran AAA+ ATPase as Rix7P [17], possibly in the ancestral neokaryote, not the first eukaryote.

As previously argued [3,5] a phase of passive restriction of molecular exchange between the nucleoplasm must be postulated prior to active and selective import and export. Recently the likely intermediate stages have been modelled by Jékely [259], demonstrating by diffusion-reaction simulation that partial enclosure of chromatin by membranes with quite large apertures can lead to marked differences in molecular concentrations, which makes such intermediates selectively plausible; furthermore he showed that, even without any membrane boundary, direct or indirect binding of molecules to chromatin can change local concentrations of diffusible molecules. A limitation of such spatial modelling is that we do not know the size of the cell in which the nuclear envelope originated. Even though it almost certainly greatly increased in size compared with its bacterial ancestor [4] during the origin of the endoskeleton and phagocytosis, it could have been much smaller than the 20 μm he assumed, which would have made it harder to develop significant concentration gradients in the pre-envelope phase than in his model and thus increase the relative selective advantage of using a membrane and NPCs for effective compartmentation over simple binding to chromatin. The modelling has heuristic value but is notably oversimplified for the final stage with NPCs as it ignores the fact that RanGTP is exported and RanGDP imported on carriers (different ones) and do not diffuse freely.

Karyopherins have three distinct domains: an N-terminal Ran-binding loop, a central nucleoporin (Nup)-binding domain, and a C-terminal cargo-binding motif that interacts with NLS (in the case of importins) or NES (in the case of exportins). Their central region that interacts with Nups consists of HEAT repeats (comprising pairs of antiparallel α-helices) that can bind the FG repeats of FG nucleoporins. The great diversity of karyopherins and FG Nups must have arisen during eukaryogenesis from a common ancestor by repeated gene duplications, probably with some domain shuffling. As karyopherins have other functions independent of the nuclear pore complex it is likely that one of these was the ancestral one. Deciding what that was is not easy, partly because many of these functions are still poorly understood. The karyopherin Crm-1 is a member of the importin-β superfamily of transport receptors [268] that in addition to nuclear export is also needed for the intranuclear maturation and possibly export of the small ribosomal subunit. Crm-1 recognizes a leucine-rich NES [269]. This specificity might stem from the fact that RanGAP might have been its initial cargo (whether first to exclude it from the nucleus or even earlier for mitotic functions).

It was recently discovered that Crm-1 has roles in mitosis (at least in opisthokonts) as it mediates the recruitment of the RanGAP1/RanBP2 complex to kinetochores and maintains kinetochore-fibre integrity [270]. Crm-1 is also involved in centrosome duplication in animals and interacts strongly with the centrosome (spindle pole body) of fungi [271].

Domain shuffling or quantum divergence was involved in early karyopherin diversification since Crm-1 has its own special N-terminal domain that differs from the globular N-terminal domain of most importins that is a mixture of HEAT and ARM (armadillo, a related type of repeat) repeats. Likewise the tRNA exporter (exportin-t) has a unique C-terminal domain that presumably recognises tRNA cargo. Once Crm-1 became able to export its first cargo, whatever that was, by binding to leucine-rich NESs, additional cargos could be added to its repertoire by adding such residues to pre-existing proteins or by evolving binding of adaptor proteins already bearing them to cargo (notably various RNAs) that lack them, allowing them to piggy back on NES bearers.

Many importins act as cytosolic chaperones and bind strongly to basic proteins such as histone H1 and some ribosomal proteins, which would stop them binding to or interfering with cytoplasmic RNA, e.g. mRNA [272]. This would have been useful even before the origin of the nucleus. As Jékely [259] pointed out, that might have been their primary function before they were recruited as karyopherins. Such acquisition of additional functions by pre-existing molecules reduces the difficulty of evolving all components of the complex transport machinery simultaneously. Though the basic idea that some elements of the transport machinery had evolved before Nups and the envelope is attractive, as is the idea that chaperones of various sorts were probably important before compartmentation, I am unconvinced by his specific suggestion that the primary function was for binding to ribosomal proteins [259]. This is partly because different importins are used for different ribosomal proteins, which suggests secondary recruitment of variants after a primary karyopherin function was established, partly because some of these bind other basic proteins (e.g. the importin β-importin7 heterodimer binds rpL4, rpL6 and histone H1 [272]) and I do not see how one can say that ribosomal protein chaperoning was primary, partly because before the nuclear envelope evolved ribosomal proteins could have bound rather quickly to nascent rRNA, which would have been available in large amounts in the same compartment in a fairly small cell, and partly because I prefer the idea that the first nucleoporin was directly associated with loading proteins, e.g. RanGTP or histones directly onto chromatin, as Crm-1 does today for mitosis.

One class of chaperones that likely predated the nuclear envelope is histone chaperones. The two multimers from which the core nucleosome octomers are made (H3/4 and H2a/H2b) each have different chaperones (e.g. CaAF-1, Asf-1 for H3/4 or for H2a/H2b), some of which associate with them prior to import. Though they are imported as macromolecular complexes it is the core histones themselves that bear the NLS (several that can interact with several different karyopherins; there are always many different karyopherins; 15 in yeast). Core histones are imported by a monomeric importin β, in contrast to many proteins that use a more complex heterodimer in which importin α acts as an adaptor between NLS and importin β (including histone H1: [273]). I suggest that this simpler system, which is also used for most ribosomal proteins may be the ancestral one and that the classical one involving also importin α and more complex NLS is derived. Once a primitive import system evolved importin gene duplications would generate many different importins and numerous molecules that unlike histones did not have NLS as part of their intrinsic structure could have added them.

One such clear case of addition of NLS is the export adapter Nmd3; its C-terminal domain has both NLS for entering the nucleus and nuclear export signals (NES) for exiting it so it can shuttle between nucleus and cytoplasm. Inside the nucleus it binds to newly made 60S ribosome subunits and carries them out with it [274]. As the archaebacterial homologue of Nmd3 lacks this C-terminal domain it was clearly added during eukaryogenesis to allow such shuttling and ribosome export. The most important and perhaps original export receptor is Crm-1, the karyopherin that mediates export of the Nmd3-60S-ribosome complex among many others, e.g. the signal recognition particle, spliceosomal snRNAs, 5S rRNA, but not most mRNAs. Obviously karyopherins themselves have to be exported. Most exit the nucleus as they enter it under their own steam. But importin α piggy backs on importin β on the way in and on karyopherin 109 on the way out, which might be interpreted as evidence that it evolved after those two similarly ubiquitous types.

The ancestral karyopherin may have arisen by gene duplication from the adaptins that attach clathrin coats to transport vesicles as they are structurally similar with HEAT repeats [17]; if the suggestion that the closest bacterial relatives of these proteins are certain phycobilisome proteins of cyanobacteria [17] stood up to more thorough study, this could be the first example of lateral gene transfer from cyanobacterial food of the prekaryote predator being important in early eukaryote evolution. That NPCs and Sec13 are not the only nuclear pore associated proteins with affinities with coated vesicle proteins strongly supports the view that the origin of a primitive endomembrane system with various types of coated vesicles preceded and was more fundamental to eukaryogenesis than the origin of the nucleus itself [4,5,11,70,123,163].

The difficulty of deciding on the likely ancestral functions of such pervasive molecules as RanGTP and karyopherins is highlighted by the fact that in animals at least, importins α and β are both involved in spindle and nuclear envelope assembly; to be consistent both with my thesis that some mitotic functions preceded NPC-related nucleocytoplasmic exchange functions and that importins α was not part of the ancestral karyopherin importer one would have to suppose that the mitotic roles of importin α are secondary (i.e. evolved after NPCs), which current evidence cannot exclude. An analogous problem concerns the fact that RanGDP is imported into the nucleus not by karyopherins but by its own custom importer, NTF2. The suggestion that this may be because Ran import evolved before karyopherins [17] is implausible for two reasons. Such a small molecule would be the least likely to be excluded by a primitive FG Nup plug and could hardly have provided the first impetus for the evolution of active import. Secondly, if it did evolve first, why did all the other things that are imported by karyopherins not adopt that carrier. It seems to me that the diversity of karyopherins argues that they came first and were co-opted for the transport of any molecule that could acquire NLSs or could be bound by available or readily evolvable NLS-bearing adaptors to karyopherins. Perhaps the RanGDP is so small that it could not acquire a basic NLS without disrupting its function. If it did acquire an NLS this would also be present on RanGTP, which is exported (by binding to karyopherins), so its addition would contradict the need to export the GTP-bound form and import the GDP-bound one of the very same protein. It may have been easier to adopt a different carrier that would circumvent that contradiction. Its novel acidic C-terminal tail that is recognized by NTF2 was probably co-opted as the import signal because it is distinctive and absent from other Rab/Ras family proteins and thus would not lead to their accidental import into the nucleus. The suggestion that NTF2 which is essential for nuclear envelope function was acquired from the proteobacterial premitochondrial symbiont because homologues exist widely in eubacteria but not archaebacteria is fallacious, because of the likelihood that this was one of many eubacterial proteins that were lost by the ancestral archaebacterium after it diverged from its neomuran sister from which eukaryotes evolved. Such proteins are widely present in both actinobacterial and endobacterial Posibacteria so would have been present in the ancestral neomuran. Thus the phylogenetic history of this and other nuclear-envelope related proteins present widely in eubacteria but not archaebacteria do not constitute evidence that the nucleus evolved after mitochondria, a fallacious conclusion stemming from considering archaebacteria to be ancestral to eukaryotes rather than their sisters, a phylogenetic mistake which prevents many from accepting that Posibacteria and with their typically eubacterial genes were ancestral to both archaebacteria and eukaryotes.

Although there is some redundancy among karyopherins [275] some are essential for a small subset of molecules. Thus import of histone H1 is much more complex than for core histones, and some ribosomal proteins require different machinery from the majority. This complexity suggests that many molecules that were not initially imported were added later to the import repertoire in a piecemeal fashion by acquiring a diversity of NLS recognized preferentially by different karyopherins. Some ribosomal proteins show clear evidence of the later addition of NLS in expansion segments [276], an example of the quantum evolutionary impact on ribosomes of the origin of the nuclear envelope earlier postulated [3,70]. Though one might think that replication and repair enzymes and RNA polymerases were among the early molecules imported [3], it remains unclear how many of these, especially RNA polymerase are imported [277]; the case of transcription initiation factors is interesting in emphasizing that in a large macromolecular complex only some (minimally one) constituents needs an NLS - thus TAF10 a component of transcription factor complex TFII (that recognizes TATA boxes) and other transcription regulatory complexes lacks an NLS but is imported by being bound to three proteins with histone fold domains that contain NLS [278]; these histone-fold factors might themselves have evolved from core histones and thus ancestrally would have had NLSs.

Transport of multiprotein complexes is a major way in which evolution can add extra proteins to the transport repertoire without adding NLS to each; transport of ribonucleoprotein complexes in which at least one protein had an NES was key to the export of RNAs which themselves could not mutationally acquire NESs. Because NLSs and NESs are rather generalized, many proteins would by chance have had weak transport signals that could be improved by mutation, without having to add or insert an extra NLS domain.

However the generalized nature of the signals also means that numerous proteins that do not need to be transported could by chance confuse the system so there was probably also selection against such resemblances becoming too strong. There is evidence from in vivo studies that such confusion is a real factor [275], emphasizing that evolution has, unsurprisingly, not been able to evolve perfect discrimination. Though I have speculated as to what may have been imported or exported early or initially, one must emphasize that selection for compartmentation would probably not act separately on single molecular types but be a bulk selection for numerous simultaneous mutations in different genes that collectively yielded economies in protein synthesis or accelerated growth.

Origin of the NE would have complicated sex, requiring nuclear fusion before meiosis, but appropriate timing of Ran GTPase fusion control probably fixed that. Because selfish DNA, especially transposons and endogenous retroviruses, spreads much more slowly in clonal than in sexual populations [279,280], a sexual protoeukaryote would have suddenly been inundated by such genetic parasites [4]. One response was to evolve posttranscriptional gene silencing to destroy double-stranded RNA.

Nuclear Envelope Structural Proteins and Cytoplasmic Organelle Attachment Proteins

Numerous new NE structural protein and organelle attachment proteins evolved during eukaryogenesis, but little is known about them in most organisms and still less about their evolution. Of special importance for NE structure and attachments of other organelles to the nucleus are two families of NE membrane proteins: those with Sun or KASH domains [281,282]. These two protein families are jointly responsible for holding together the two (topologically continuous) NE membranes and for attaching them to the cytoskeleton and the nucleoskeleton. As Figure 6 shows, Sun-domain protein dimers are embedded in the inner membrane by membrane spanning domain(s) near their N termini with their C-terminal Sun domain protruding into the lumen of the perinuclear cisterna. Their Sun domains bind firmly to the C-terminal KASH domains of a variety of proteins similarly embedded in the outer membrane of the envelope, forming a linker complex holding the two membrane domains close together [281]. The N-termini of the Sun-domain proteins binds firmly to the DNA of the peripheral heterochromatin, whereas the N-termini of the KASH proteins are much more variable, binding to the actin cytoskeleton as well as via the molecular motors dynein or kinesin to microtubules [283]. Centrosomes are also attached to the nucleus by either KASH proteins (animals [284]) or by a Sun protein (Sun-1 in *Dictyostelium* [285]). The Sun/KASH linker complex forms a mechanically robust bridge between the inner and outer membranes of the NE that can firmly bind the nucleus to other organelles or transmit forces from outside motors (as in nuclear migration in animals [286]). In meiosis cytoplasmic microtubule motors also move telomeres together, creating the bouquet stage of prophase to help homologous chromosomes pair; Sun-KASH linkers across the NE mediate this movement [287,288]. In fission yeast Sun-KASH and other NE proteins link intranuclear centromeres through the envelope to cytoplasmic microtubules [289]. I propose that the origin of the Sun/KASH protein linkers was a key step in the original attachment of protoER membranes to heterochromatin during the formation of the nucleus.

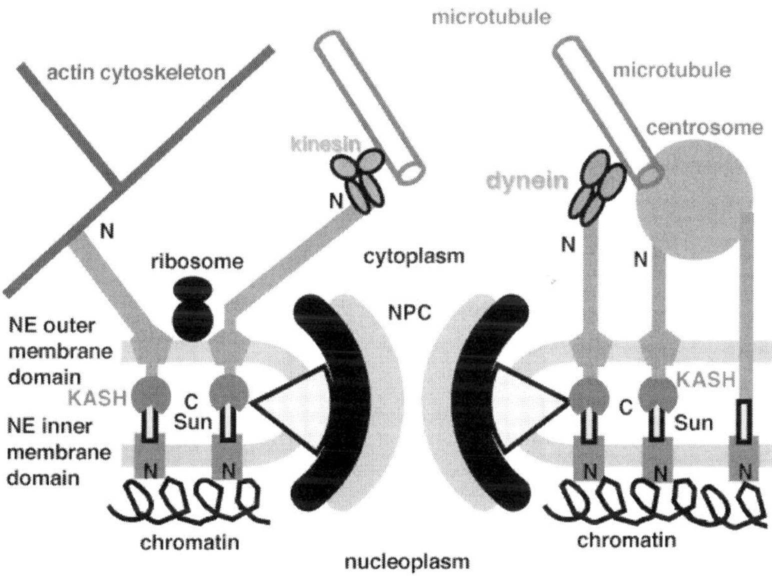

Figure 6: Role of Sun-domain and KASH-domain proteins in nuclear envelope architecture. Sun-domain proteins embedded in the inner membrane attach it directly to the DNA surface of the peripheral heterochromatin (the nucleoskeleton). Their Sun-domains (yellow) bind to the KASH domains (purple) of proteins embedded in the outer membrane, which attach it to the cytoplasmic actin filaments, microtubules, and centrosome of the cytoskeleton. Grey pentagons represent the membrane spanning domain(s) of the KASH-domain

proteins and grey rods their flexible cytoskeleton-binding N-terminal domains, which differ greatly among the various types. Microtubules may be attached to KASH-domain proteins either by kinesin or dynein motors. The firm Sun-KASH linker complex (known as LINC) within the perinuclear cisterna holds the inner and outer membrane domains of the NE together with the correct spacing and transmits mechanical forces from cytoskeleton to nucleoskeleton or vice versa without damaging it. Some eukaryotes, e.g. animals, lobose amoebae and peridinean dinoflagellates, (probably polyphyletically) evolved a proteinaceous lamina beneath the inner membrane to further strengthen the nuclear periphery, but this was probably absent in the first eukaryotes; additionally to the universal interactions shown, in animals only cytoplasmic intermediate filaments (IF) attach to KASH proteins via plectin adaptors and lamin IF proteins associate with the intranuclear domain of Sun proteins. For simplicity the fact that Sun-domain proteins are homodimers with a coiled coil domain between their two membrane-spanning and chromatin binding domains (lumped here as grey rectangles) and two Sun domains is not depicted.

Earlier I postulated that lamins were ancestrally involved at this attachment [4,5,18,163] but this is now unlikely because, like other intermediate filament family proteins, they are restricted to animals and known for sure only in animals with guts and nervous systems (Cnidaria and above). Sun proteins are much better candidates for a central role in NE origins because they are evolutionarily more widespread and because of their key role in ensuring nuclear integrity and attachment to other cell structures. The Sun domain is related to the discoidin domain; this broader domain family is present in all eukaryotes and even in some bacteria such as actinobacteria [17] so would have been present in the ancestral neomuran ancestor. The Sun-KASH interaction has only been studied experimentally and the locations of the proteins verified in unikonts, but Sun proteins have been detected through sequence homology across the neozoa [17,290]. When I studied Sun-1 and Sun-2 in *Dictyostelium* by BLASTP I could detect homologues for Sun-1 also only in neozoa, but using Sun-2 found plausible homologues with conserved Sun domains in three phyla of Eozoa including Euglenozoa, but none in bacteria. KASH proteins are more diverse in domain structure length and sequence, making such simple comparisons more difficult. Nonetheless, it is likely that Sun-domain inner membrane proteins evolved at a very early stage in the origin of the nuclear envelope to attach endomembranes directly to chromatin; that is a simpler origin for chromatin membrane attachment than one mediated by proteins like lamins that are not themselves inner membrane proteins. KASH proteins would probably have evolved almost immediately thereafter to stabilise the envelope mechanically.

Although animal Sun-domain proteins bind to lamins as well as to DNA or chromatin this is probably not their primary mode of targeting to the nuclear periphery as localization there is not prevented by the absence of lamins; some have known DNA-binding motifs. In *Dictyostelium* Sun-1 binds directly to DNA and also to a protein linker to cytoplasmic centrosomes, apparently independently of the characterized KASH protein. At least two different Sun-domain inner membrane proteins had evolved by the ancestral neozoan [290]. I suggest that a nuclear lamina composed of lamins and the intermediate filament family as a whole may have evolved in the ancestral animal (sponge) to increase the mechanical strength of the giant oocyte nucleus when the ancestor of animals evolved oogamy (giant eggs and tiny sperm) and that the family diversified into various cytoplasmic filaments when epithelia evolved in the first sponge to give added mechanical strength to them. Different paralogues were probably selected early for somatic cells and oocytes. The mechanical robustness of the lamina may have made it essential for it to be reversibly disassembled at mitosis, which may have been simplest to achieve through fragmentation of the whole envelope; thus open mitosis [291] probably evolved simultaneously with the lamina in the ancestral animal. Choanoflagellates, the closest outgroup to animals, have semi-open mitosis [292], making it likely that the direct unicellular ancestor of animals (probably a stem choanoflagellate [293]) also did.

A proteinaceous nuclear lamina evolved independently in amoebae of the amoebozoan subphylum Lobosa (e.g. *Amoeba*, *Hartmannella*), where in the giant amoebae it is far thicker and more complex than in animals [294]. Their ancestor lost cilia, centrioles and all cytoplasmic microtubules so as to focus on locomotion by actomyosin; a proteinaceous lamina was probably especially important to protect the nucleus from shearing forces which must be far greater in such amoebae that undergo repeated sol-gel transformations compared with the ancestral zooflagellate eukaryotic cells where the cytoplasm is almost permanent semi-rigid actin gel, except during cytokinesis by cell cleavage. Having evolved such a robust lamina lobose amoebae could dispense with the peripheral chromatin that underlies the NE in the zooflagellates most closely related to them (choanoflagellates and Apusozoa). Apusozoa (apusomonads and planomonads) are either sister to opisthokonts (animals, Choanozoa, Fungi) or the paraphyletic ancestors of both opisthokonts and Amoebozoa and have particularly strongly developed peripheral and internal heterochromatin in their nuclei, so must have a higher DNA content per unit nuclear volume than choanoflagellates. By contrast lobose amoebae have almost no heterochromatin visible in interphase, so their DNA per unit nuclear volume must be rather low (which goes some way to explaining why getting enough DNA for even simple PCR and sequencing from sparse cultures is often much harder for Lobosa [295] than other eukaryotes). Another group that often has giant nuclei and a proteinaceous lamina are the dinoflagellates. This coupled with the evolution of an extranuclear spindle with kinetochores embedded in the nuclear envelope probably allowed peridinean dinoflagellates to dispense entirely with heterochromatin as a nuclear skeleton and even largely also with histones, which are absent from the majority of their DNA [296] which is neutralized by divalent cations instead [297] and sparse bacterial-type DNA-binding proteins [298,299]; only one standard histone is known, perhaps involved in double-strand break repair [300]; the loss of histones entailed radically altered transcription factors that bind to TTTT instead of TATA boxes, but dinoflagellates retain standard telomeres associated with the nuclear envelope [301] and relatively normal spliceosomal intron splicing [302].

The origin of lamins and open mitosis in animals, a convergent kind of lamina and open mitosis in Lobosa, and of the dinoflagellate closed mitosis with extranuclear spindle, are all mentioned as examples of secondary changes from the ancestral state that must not be allowed to confuse us in reconstructing the first eukaryote. The peculiar mitosis of dinoflagellates in which kinetochores are embedded in the NE was once proposed as a model for an early version of mitosis before microtubules evolved, but that idea became untenable when it was discovered that they do have microtubules; I argued long ago that it was irrelevant to the origin of eukaryotes because dinoflagellates are so complex that they cannot be primitive and must be very advanced higher eukaryotes [70], which is demonstrated beyond question by their branching within the alveolates and these within the kingdom Chromista that was formed by the secondary symbiogenetic enslavement of a red alga (Fig. 1) [9,61,86,87,303]. Despite this irrelevance, dinoflagellate mitosis is oddly still sometimes mentioned as being germane to the origin of meiosis because telomeres are similarly moved on the NE [290]; though the movement of dinoflagellate kinetochores may well use the same cytoplasmic machinery (microtubules and dynein) as do meiotic pre-bouquet telomeres, and it is likely that the attachment of both depends on Sun-domain inner membrane, it is most unlikely that the ancillary attachment proteins for telomeres (e.g. Bqt1-4 [304,305]) are related to those of dinoflagellate NE plaques. How peridinean dinoflagellate mitosis may have evolved from the more normal patterns found in other alveolates is discussed in [267]. Independently of dinoflagellates, metamonad Parabasalia evolved NE-envelope embedded centromeres and extranuclear spindle, but they are equally irrelevant to the origin of mitosis or meiosis despite sometimes being mentioned in that context. Nonetheless, molecular studies of NE-embedded peridinean and parabasalian centromeres would be very revealing for cell biology and evolution.

The ratio of heterochromatin to euchromatin mass affects the quantitative relationships between genome size and nuclear and cell size [18]. However the reasons for its variation even in protists are poorly understood. I originally assumed that if a nuclear lamina was universally present euchromatin and heterochromatin ought to be able to serve equally well as a nuclear skeleton and nuclear volume determinant [169]. On that view it ought to be more economical of DNA and protein to use the less dense euchromatin for this function rather than the denser heterochromatin as less material would be needed for the same volume. I assumed that the presence of heterochromatin in so many animal and plant somatic cells mainly reflected the fact that their nuclear volumes varied so greatly that heterochromatin was only unwound fully to make dispersed chromatin in the largest cells such as giant neurons, whose large nuclei lack condensed chromatin [169,170,176]. Thus multicellular organisms would have a complete range of chromatin unfolding and nuclear volume from such giants to the most compact nuclei of sperm and red blood cell nuclei where all is maximally condensed and transcriptionally inactive. I argued that protists with multiple fission life cycles with alternation between giant cells with huge nuclei and tiny ones with compact ones should have heterochromatin, but that those with simple binary fission cycles should not, and cited much comparative supportive evidence [306]. However, as recently explained there are now several clear counter-examples to that thesis, which must therefore have been oversimplified [18]. Several examples of ordinary binary fission cell cycles without major changes in nuclear volume beyond the normal two-fold per cycle but with extensive heterochromatin, especially on the nuclear periphery, but also interstitially are now apparent. This include cryptomonads and haptophytes (interestingly now grouped together in the chromist subkingdom Hacrobia, for which extensive heterochromatin was arguably the ancestral state); Apusozoa [307,308]; choanoflagellates (probably with a lower heterochromatin/euchromatin ratio); and euglenoids. In all these cases the whole group possesses this character. It is noteworthy that in other whole groups, e.g. lobose amoebae, Excavata, and in euglenozoan Kinetoplastea and Diplonemea condensed chromatin is largely absent or relatively sparse in interphase. Thus there is strong phylogenetic constraint; for many protist groups changes from a heterochromatin-rich to heterochromatin-sparse nuclei or the converse are only made relatively rarely in protist evolution.

One way to rationalize this is to suggest that the ancestral eukaryote was a middle-sized cell in which large amounts of heterochromatin were the primary nuclear skeleton on which the nuclear envelope was assembled and cryptomonads, haptophytes, and euglenoids, for example, have retained that heterochromatin-rich state but that the protist groups with more dispersed chromatin hit upon more economical ways of making or using their nuclear skeleton or ones more suited to their individual life styles. I pointed out above that the evolution of giant nuclei in lobose amoebae and dinoflagellates was associated with cytological novelties that arguably allowed them to dispense with heterochromatin as a peripheral nuclear skeleton (the nuclear lamina of Lobosa and the permanently condensed histone-depleted chromosomes of Peridinea) and depart radically from the ancestral eukaryote condition. I now suggest that evolving especially small cells and nuclei coupled with a relatively rigid cytoplasm may also have allowed the loss of most peripheral heterochromatin (and associated metabolic and spatial economy) without mechanically harmful consequences, and a big increase in the ratio of coding to non-coding DNA. This may be why phylogenetically diverse organisms converged on a largely heterochromatin free-state (free-living budding yeasts, intracellular Microsporidia, and above all the nucleomorphs of cryptomonads and chlorarachneans (relict enslaved nuclei)) or on a sparse heterochromatin state (the intracellular parasitic coccidian Sporozoa, like the malaria parasite *Plasmodium*) and more gene-rich genomes than most eukaryotes. Somewhat intermediate relatively small-celled organisms like choanoflagellates, Filasterea or bodonids with some peripheral heterochromatin but less than Apusozoa or cryptomonads may have made some minor economy but with less radical change. In protists there are numerous secondary adaptations affecting the nuclear envelope, for example the huge macronuclei of ciliates or the large nuclei of actinophryid heliozoa that nucleate axopodia or the huge nuclei of many Retaria (Rhizaria). This proposal fits much comparative data but would require several independent reductions in heterochromatin richness in excavates and within plants and chromists, which is not unreasonable.

I have long thought it an attractive idea that both centromeres and telomeres were ancestrally attached to the nuclear envelope [4,309]. Their heterochromatin could have been a major fraction of the heterochromatin that forms the periphery of nuclei onto which the inner membrane is attached by Sun proteins. The involvement of Sun proteins as the direct binder of chromatin makes it clearer than ever that chromatin itself is really the fundamental nucleoskeleton, as I have repeatedly argued [18,169,170,176,211,309]. A major consequence of the nuclear envelope was the initiation of a new nucleoskeletal function for DNA. Elsewhere I explained how this, coupled with the adaptively significant huge variation in cell volumes of eukaryotes that the origin of the endoskeleton and endomembranes made possible, is the fundamental evolutionary reason why eukaryotic genomes are typically much larger than bacterial genomes, why their size varies so immensely, and why it correlates so well with cell volume but not with organismal complexity [18]. The (nucleo)skeletal DNA explains many aspects of the evolution of eukaryotic genome size where the still more popular selfish DNA/mutation pressure theories fail totally [18,310]. The reason why most of the non-coding DNA in eukaryotic genomes exists is fundamentally structural, which is why it is called skeletal DNA whether it is centromeric, telomeric or interstitial [18]. Parts of it are transcribed and involved in the positional control of centrosomes and heterochromatin, but at base these are structural and not coding functions; the transcription and base-pairing are needed for positional control of structure within the skeletal DNA component of the genome. As adumbrated over 30 years ago, the evolution of eukaryotic genomes are to be understood in terms of the complementary functions of genic and skeletal DNA [169]. As stressed then, the more skeletal DNA a cell has, the larger the habitat available for viruses and transposons to flourish in without harming the cell as much as they would if they if inserted themselves into genic DNA. Therefore one expects larger genomes to have more parasitic genetic elements and for these to be especially concentrated in heterochromatin because this is where they can best evade purifying selection against them.

However, many have confused these well-known correlations of selfish element abundance and location with causation. In my view they are consequences of the deeper cell biological evolutionary forces acting on nuclei, not their fundamental causes [18]. It does not matter functionally whether genomes increase in size by duplication of genes, non-coding DNA, selfish genetic elements or whole genomes - all have been important in the past - because the mutational mechanism of origin of the extra DNA is not the main or fundamental determinant of the actual spectrum of genome sizes in different organisms. To understand these we must understand the adaptive significance of the differing cell volumes in organisms with different life styles and the cell biological principles of intracellular scaling that make an approximately constant ratio of nuclear and cell volume functionally superior [18]. These cell biological considerations did not exist before the origin of the nuclear envelope and are the major reason why the evolution of genomes is so radically different in eukaryotes compared with bacteria [18,211,311]. Purely population genetic perspectives are fundamentally misleading as they fail to appreciate the key role of the major innovations in cell structure during eukaryogenesis in stimulating first a radical change in chromosome structure and then in the selective forces acting on genome size, and sometimes [121] even misrepresent what is known about the radical differences in cell biology between bacteria and eukaryotes. Though mutations in DNA initiated and promoted the changes, from a deeper perspective of cellular constraints affecting the selective forces governing the failure or spread of the causative mutations, cell biology led and DNA and genetics followed, contrary to such widespread misconceptions. Without a cell biological perspective eukaryogenesis is impossible to understand. Nonetheless, the interplay of selection acting on cell properties and that acting on the spread of parasitic genetic elements have both shaped eukaryotic genomes.

The most important evolutionary consequences of selfish genetic elements were probably in the origin of introns [81,312], and the origin of near universal trans-splicing in Euglenozoa [211], though they might also have been involved in the origin of telomeres. But they are not the fundamental reason why eukaryotic genomes are so large and variable.

Selfish DNA, Introns and Spliceosomes

Nucleocytoplasmic compartmentation facilitated the origin of spliceosomal introns from group II introns supplied by gene transfer from the enslaved mitochondrion to the nucleus. Only rapidly self-spliced introns in mRNA or rRNA or extremely short easily protein-spliced tRNA introns seem compatible with prokaryote/mitochondrial/plastid organisation; the ready access of ribosomes to mRNA during transcription and the extreme slowing of spliceosomal splicing probably prevented it from ever evolving in prokaryotes [4,5,81]. It could not have evolved until after the second phase of NPC evolution when inner Nups arose to exclude ribosomes entirely and ribosome-subunit export arose (Fig. 5b). Spliceosomal introns could not have evolved in the prekaryote phase, and recent suggestions that the NE evolved to prevent translation of unspliced messengers (latest [21,122]) are illogically back to front. This also cannot possibly have been the initial selective force for the formation of either the nuclear envelope or the initial scaffold part of the NPCs (Fig. 5a) since neither would have excluded ribosomes. This popular theory wrongly endows selection with foresight. By contrast, protection of DNA and nascent rRNA from cytoplasmic-motor shearing damage provides a strong selective force for phase I (Fig.;5a); compartmentation benefits do so for phase II (Fig. 5b). Only then could spliceosomal introns have evolved, as described previously [81]. They are consequences not causes of the NE.

I concur, however, with the suggestion that the selfish origin of spliceosomal introns caused the origin of nonsense-mediated mRNA decay [21]. This mechanism, conserved in unikonts and corticates (all key proteins and both major mechanisms are in *Arabidopsis* [313]), must have evolved prior to the neozoan cenancestor but after the origin of spliceosomal introns. It depends on ribosomes (probably attached to the outer NE) and a test translation recognising stop codons upstream of exon-exon junction sites marked by bound UF3 complexes and stimulating mRNA destruction [314]. A selective force for the origin of this junction-site specificity of UF3 is unimaginable prior to the origin of a high density of spliceosomal introns in genes and inevitable mis-splicing yielding harmful stop codons; though once evolved it could be used, as it now is in plants and fungi, also to destroy intronless similarly flawed mRNA of many intronless genes, which inevitably have sites mimicking exon junctions sequences. Thus like splicing itself, and many genomic oddities like RNA editing [211], nonsense-mediated decay is a consequence of selfish DNA evolution; correcting the bad effects of mutation pressure, not positively adaptive. Intriguingly, the cleavage enzyme itself probably evolved from one used by selfish bacterial plasmids to kill host cells losing the plasmid by mis-segregation [315]. When it originated is unclear, as it has not been demonstrated to occur in any Eozoa. If it evolved after the origin of the nucleus as is possible it could not have influenced its origin. However, the RNA helicase UPF-1 that plays a central role clearly originated in the ancestral eukaryote as it is present in excavates and Euglenozoa, as is UPF-2 with which it binds in the nonsense-mediated mRNA decay but as this helicase has multiple functions this does not prove that nonsense-mediated mRNA decay itself arose that early. Nonsense mediated decay is triggered by failure to remove the exon junction complex (EJC) from intron-containing RNA; two of the four EJC proteins are present in trypanosomatids [316], presumably involved in checking trans-splicing (as introns are extremely rare in trypanosomes [317]), so these at least are also ancient.

Origins of Nuclear Protein Modification by Sumoylation

Sumoylation is a universal eukaryote-specific process mediated by Small Ubiquitin-like MOdifier proteins (SUMO for short) that mainly modifies nuclear proteins [318,319]. It probably evolved at the same time as ubiquitination of proteasome-digested proteins very early in eukaryogenesis, and ought to have been included in Table 1 of reference [27], where I omitted to discuss its origin and importance for eukaryotes. Sumoylation is vital for heterochromatinization, maintaining the stability of eukaryotic chromosomes, for the nucleocytoplasmic exchange via the nuclear pore complex, as well as for tubulin assembly [320-323] and is thus crucial for many non-mitochondrial ('host') eukaryotic properties. Animal, fungal and plant SUMO machinery is strongly associated with the nuclear pore complex [324] and essential for viability [325] and thus was present, and probably essential for pore complex function, in the last common ancestor of neokaryotes. SUMO is also present and strongly conserved in structure in Euglenozoa [326] so evolved in the ancestral eukaryote, though little is known of its functions and ancillary proteins in Eozoa.

SUMO proteins belong to the ancient and diverse ubiquitin superfamily; ubiquitin and SUMO have similar 3-D structures and probably diverged from a common ancestral protein. Moreover, enzymes E1 and E2 that mediate ubiquitination are related to the two that add SUMO to proteins. Both ubiquitin ligase proteins probably originated during eukaryogenesis when the ubiquitin superfamily expanded massively by repeated gene duplication [327,328]. Though neither ubiquitin nor SUMO occur in prokaryotes, Iyer et al. [329] discovered ubiquitin-related genes in a scattered array of eubacteria (never archaebacteria) clustered in an operon with E1- and E2-related protein genes and also a gene for a JAB-domain protein related to the JAB-domain deubiquitinating and desumoylating enzymes. They reasonably suggest that these four proteins represent an ancestral protein modifying system that was widespread in eubacteria before eukaryotes evolved. As such operons are not known from either actinobacteria or α-proteobacteria, they could have entered early eukaryotes by lateral gene transfer from incompletely digested prey DNA. Their scattered distribution within eubacteria is suggestive of lateral gene transfer among them, so the possibility also exists of lateral transfer to the specific actinobacterial lineage that was ancestral to eukaryotes prior to eukaryogenesis.

A curious twist to the origin of this eukaryotic protein modification machinery is the unexpected discovery of an analogous system in actinobacteria, in which the small protein Pup is covalently attached to selected proteins destined for degradation by the actinobacterial proteasomes [330]. Though Pup was called ubiquitin-like by its discoverers [330], Iyer et al. [331] convincingly show it to have a different 3-D structure. It is thus unrelated, having evolved instead from a family of enzymes containing the β-grasp domain; they also showed that the putative Pup ligase (PafA) is unrelated to the ubiquitinating enzymes E1 and E2, and probably works by a different mechanism, being more closely related to glutamine synthetase. Thus the actinobacterial proteasome-related protein modification system was probably replaced by a completely unrelated one during eukaryogenesis. Why? A possible explanation could have been the transitional novel use of proteasomes for digesting externally attached prey, which I recently postulated preceded the evolution of phagocytosis [27]. As I stressed [27], it would be advantageous to digest all prey proteins indiscriminately; thus the actinobacterial restriction of digestion to a few proteins by pupylation would have been disadvantageous and was therefore lost before it could be recruited and modified for novel eukaryotic uses that arguably arose after the early endomembrane system [27]. Therefore, protoeukaryotes evolved a different pre-existing, fully functional ubiquitin-like system based on E1, E2 enzymes and JAG proteins that happened to be available for recruitment to the now more complex 26S proteasome. A 20S core of the proteasome itself, by contrast, was probably directly inherited from the actinobacterial ancestor [12], as was the eukaryotic chaperone PAC2 required for its assembly; Iyer et al. [331] show that a gene encoding a PAC2-related protein that forms a toroidal trimer suitable for nucleating the assembly of the cylindrical proteasome (also present in archaebacteria) is closely linked to PAC2-homologues in actinobacteria, but is absent from more distant species such as the planctomycete *Rhodopirellula* and the δ-proteobacterium *Plesiocystis*, which belong to groups that typically lack a pupylation system (Pup and Pup ligase) [331], suggesting that both obtained it by lateral transfer from actinobacteria.

Origin of Meiosis and Sex (Syngamy and Nuclear Fusion)

As discussed in detail elsewhere [3,4,63,332,333], the initial function of meiosis was probably ploidy reduction to correct failures of DNA segregation caused by mitotic or cell cycle control errors that would probably have been greater during eukaryogenesis. As there explained, the key step would have been a duplication of cohesin genes to create separate meiotic cohesins that would be loaded onto chromosomes at the beginning of pre-meiotic S-phase. The changed meiotic cohesins would be digested only from the chromosome arms at meiotic prophase, remaining at the centromere, thus preventing sister chromatid separation until meiosis II. Retention of meiotic centrosomal cohesin in prophase I also arguably prevented the relief of the standard block to further DNA replication initiations imposed at the beginning of each S phase, thereby letting meiosis II continue without a preceding DNA replication, halving the nuclear DNA content to produce haploid cells. Pairing was mediated by DNA hybridization during prophase I, which has a duration proportional to genome size as expected if it is rate-limited by DNA hybridization. As Cleveland first noted [334], the origin of meiosis also requires a novelty in the control of centriolar and centrosomal duplication. In normal mitotic cell cycles centrioles and centrosomes duplicate at the beginning of DNA synthesis S phase (also true of fungal spindle pole bodies, which are simply specialized centrosomes) [71,72,75,76,335-338]. But in meiosis they must also be allowed to duplicate during the abnormal interphase between meiosis I and II, which could be achieved by preventing in anaphase I the normal anaphase proteolytic digestion of the protein SAS-6, whose amount controls centriole duplication [71], the third key step in the origin of meiosis.

Conversely to meiosis, the origin of syngamy requires plasma membrane fusion and also the merger of two parental centrosomes into one or the destruction/loss of one parental centrosome or the destruction of both and the re-emergence of a single one de novo [332]. In principle two centrosomes could merge into one if they lack centrioles as in higher seed plants, but for centrioles that would be mechanistically impossible. In some animals the centriolar reduction associated with syngamy is associated with the loss/destruction of the egg centriole, only that from the sperm being transmitted. In *Chlamydomonas* all four parental centrioles from the isogamous bicentriolar gametes are disassembled in the zygote and new ones are formed in the germinating zygospore shortly before meiotic prophase I [166], as are all their associated microtubular and centrin-containing fibrillar roots. Controlling their proper numbers and assembly in the premeiotic cell would be simplest if the centrioles of one mating type were totally destroyed in the zygote but those of the other remained as two microscopically invisible procentrioles until germination. In the euglenoid *Scytomonas*, whose gametes both have a single cilium and centriole, the zygote is initially biciliate but one cilium is quickly lost and the diploid cell then swims with the other [68].

This suggests that unilateral destruction of one zygote centriole may have been the ancestral condition for eukaryotes.

Meiosis could have started to evolve as soon as cyclin-based eukaryotic cell cycle controls and cohesin evolved. Arguably it did not initially depend on synaptonemal complexes, but these were soon added to increase the mechanical stability of paired chromosomes. As synaptonemal complexes have been casually observed ultrastructurally in one bodonid euglenozoan, this stage must have been reached in the eukaryotic cenancestor. However there are no studies of meiosis in any of the three most deeply branching eukaryotic phyla (Euglenozoa, Percolozoa, Loukozoa), but recombination and Mendelian processes is established for several trypanosomatids. Therefore we do not know if any of these early branching eukaryotes have a bouquet stage of meiosis as in neozoa or whether Sun-KASH nuclear envelope protein roles in meiotic prophase evolved in the first eukaryote or only later, e.g. in the immediate ancestor of neozoa which all have fundamentally similar meiosis.

As mentioned above, meiosis is likely to have arisen prior to the origin of the nuclear envelope, probably as soon as the earliest form of mitosis evolved with primitive centrosomes and centromeres, as the selective advantage of correcting ploidy errors would have then been at its peak and cohesins, the essential molecular precursors of the meiotic machinery, would already have evolved. Explaining both its mechanistic origin and how it was selected are both much easier than if meiosis originated after the nuclear envelope and had to be preceded by the evolution of nuclear fusion, which in the absence of meiosis has no rationale or advantage. Ploidy reduction is a real, powerful, and experimentally demonstrable evolutionary force [339]. Ploidy reduction requires only chromosome pairing and meiotic cohesins and arguably evolved even before the nuclear envelope. The cytoplasmic motor driven congress of telomeres to form the bouquet stage must have followed the origin of the nuclear envelope and Sun-KASH domain interactions across it. Nuclear fusion, an essential part of sexual cycles must also have followed the evolution of the envelope. Thus the evolution of meiosis probably straddled nucleogenesis and was an integral part of eukaryogenesis, not a late evolutionary afterthought. What about syngamy?

I have argued that syngamy evolved marginally after meiosis and its primary functional significance was related to the fact that even the first eukaryotes would have had dormant resting cysts, which I argued evolved from posibacterial exospores [3,63,332,333]. Its prime role was two-fold: to make zygotes larger and increase their survival rate by being able to store more solid food reserves; to provide genetic redundancy in the dormant cysts so that accidental damage, notably double-strand breaks by radiation or other environmental insults or by physiological errors could be repaired by homologous recombination among genomes. Maximising the number of offspring when conditions improved and the cyst could germinate explains why meiosis probably ancestrally occurred then and vegetetative cells were haploid [333]. The feast and famine existence of the first predators on earth thus probably played a key role in the origin of sex, and life cycle ploidy control, its prime raison d'être. Predatory feeding came before sex. I refer readers to other discussions of the likely prime importance of circumventing deleterious mutations in the origin of sex [340,341]. It is now widely considered that removing harmful mutations may be a more important factor in the maintenance of sex than making favourable allele combinations, which sex can equally undo [342-344] and which can also be made by mitotic recombination [345]. The origin of sex cannot be considered apart from the origin of the new eukaryotic cell cycle controls, mitosis, and a wall-free cell surface with internal actin gel cytoskeleton that mechanically allow gametes the ability to fuse and the stability to survive without walls prior to fusion. The origin of syngamy did not just involve the control of plasma membrane and nuclear envelope fusion but also the merger of the cytoskeletons of the gametes. The only Eozoa where meiosis has been widely studied are parasitic Metamonada, where its hormonal synchronization with insect host moulting physiology made it technically easier. In oxymonads, unlike morphologically more conservative metamonads like the free-living *Trimastix* and parasitic retortamonads, the microtubular skeleton is no longer mainly pellicular - instead most microtubules form a giant contractile axostyle organelle deep in the cytoplasm. Remarkably, when oxymonad gametes fuse, their two axostyles also fuse into one [346], illustrating the need in all eozoan sex to halve the number of cytoplasmic organelles as well as fuse nuclei during syngamy.

Even though sex can undo favourable gene combinations as well as make them and always has some cost, the likelihood that syngamy was already evolving in the vegetatively wall-free prekaryote that periodically made dormant walled cysts at times of starvation, would have allowed different lineages of transitional forms to recombine their eukaryotic innovations and could have speeded up the transition. This is because eukaryogenesis differs radically from discussions of the maintenance of sex, which deal with a quasi-equilibrium situation where repeated (and often reversible) mutations of the same general sort may be the main factor and radical innovation is rare [342-345]. By contrast, during eukaryogenesis mutations that had never occurred before in the history of life, rather than recurrent everyday allelic mutations, were of key importance at numerous times in the process. I have given many instances where the origin of new protein paralogues by unique gene duplication and divergence, or radically new protein domains never before found in bacteria, and of totally new domain combinations to make novel multidomain proteins, were decisively important for the origin of the majority of novel eukaryotic proteins, structures and processes. Thus, once basic meiosis had evolved for ploidy reduction, accidental cell fusions between related lineages with different novel useful genes would not be so harmful; by combining independently evolved innovations it could indeed have speeded up the evolutionary transition from bacteria to eukaryotes. I do not think it reasonable to regard this as the driving force for the origin of meiosis, but it could have been a third important reason for improving the cell fusion machinery and restricting its expression to times of dormancy onset, where life was threatened already by famine and the cell had less to lose and more to gain by fusing with others. Fusion during normal growth with abundant prey when cell fusion would have been disadvantageous for a zooflagellate and therefore increasingly stringently repressed. Such condition-dependent sex is much easier to evolve than if every reproductive act is sexual, as the costs are more easily outweighed by the benefits [347]; this obviously applies to protozoa where most reproduction is clonal with cell fusion being rare at best and to bacteria where cell fusion probably almost never occurs in nature - controlled cell fusion began with the first eukaryotes.

Rerooting the eukaryote tree between Euglenozoa and Percolozoa [9] makes it much clearer than ever before that the forms of vegetative cell fusion that make multinucleate plasmodia, exhibited for example by myxogastrid slime moulds (e.g. *Physarum*) in the Amoebozoa (unikonts) and by the naked cercomonad zooflagellates in the Cercozoa (Rhizaria in the corticates) are evolutionarily advanced not primitive characters. Such plasmodial stages in the life history are entirely unknown in Eozoa, and there is now little doubt that the last common ancestor of all eukaryotes had a semi-rigid micro-tubule supported pellicle that would preclude plasmodia formation. Though amoebae did evolve secondarily within Percolozoa to form Heterolobosea [348] (and even in one instance, *Dientamoeba*, within parabasalid metamonads), the main thrusts of eukaryotic soft surfaced evolution were in advanced groups of unikonts, which probably ancestrally evolved myosin II-based pseudopodia [9] (notably in Amoebozoa), and the corticate kingdom Chromista (not only in the ancestrally amoeboflagellate Rhizaria, but also in several independently amoeboid subgroups), which by contrast were probably ancestrally strongly pelliculate zooflagellates without amoeboid stages prior to the internal enslavement of the red alga. But for understanding early sex it is the discicristate zooflagellates with strongly developed pellicles that need to be considered. Sadly, sex is unstudied in any of their free-living representatives; for almost none of them (exceptions one bodonid and the euglenoid *Scytomonas*) do we even know whether they have sex at all or whether it has been lost long ago by many or even most of them, as is perfectly possible. However, if the present rooting of the eukaryotic tree is correct, we can be sure that the cenancestral eukaryote was a pelliculate zooflagellate with dormant cysts, ciliated gametes, meiosis, and a synaptonemal complex. If, as I have argued, sex and meiosis began in a prekaryote with a single cell-surface-associated centrosome containing centrin and γ-tubulin and attached on the one hand to the chromosome replicon origin and on the other to stable microtubules that were permanently present through the cell cycle, then it is relatively easy to see how sexual life cycles could have originated despite their apparent complexity.

If the chromosome replication origin at that stage was also permanently attached to the cell surface, the lateral fusion of two such cells would have generated a zygote with two side by side chromosomes in the same compartment that could form hybrid DNA segments with each other quite easily when single stranded DNA was transiently produced, e.g. during replication or repair. By generating unequal crossover and chromosome breakage they would have been deleterious. Though this would have thereby selected against such fusions, the cell phenotype produced by syngamy, with two side-by-side chromosomes able to recombine, is essentially the same as that produced every cell cycle by replication prior to division. Thus the suppression, control or repair of such adjacent chromosomes exchanges would be important even for vegetative growth. In bacteria such dimers can be resolved by the Xer decatenation machinery at the cell surface at the presumptive division plane [235,349]; similar processes must have been in process during eukaryogenesis; given the magnitude of the changes associated with chromosome internalization on endomembrane vesicles it seems almost inevitable that at some stage chromosomes would be broken, even in vegetative cells merely undergoing binary fission and not engaging in syngamy. Odd numbers of sister chromatid exchanges for circular chromosomes have the same dimerization and consequential breakage potential as meiotic cross-overs, so the linearization of eukaryotic chromosomes and addition of telomeres need not have awaited the evolution of a primitive meiosis and sexual cycle as earlier suggested [11] but was likely as soon as the ordered segregation along the bacterial cell surface broke down during the replacement of the bacterial cell wall by a cell surface coat, endoskeleton and protozoan pellicle. Thus telomeres probably evolved soon after the origin of the first centrosomes. It has been proposed that centromeres evolved from telomeres [350], but the arguments are unconvincing. It was claimed that the fact that only linear chromosomes have evolved proper centromeres means that telomeres must have evolved first [350]. But the argument can easily be reversed; it is just as (il)logical to argue that as proper telomeres only evolved in chromosomes with proper centromeres, centromeres must have come first. As eukaryogenesis was unique, one cannot reasonably use its uniqueness per se, as did these authors, to argue the polarity of any of its component unique events. Instead one must use a logical reconstruction of the likely nature of transitional intermediates, the nature of their probable precursors, the phylogenetic context, and the selective advantage of each postulated stage, as attempted here and in previous discussions [4,70]. Those authors were unaware of those arguments that centromeres arose before telomeres [4,70], wrongly stating that the only past discussion of the origin of centromeres was [351]. The arguments in this paragraph suggest that telomeres probably slightly post-dated proto-centrosomes but could have preceded eukaryotic centromeres and kinetochores. The heterochromatic properties of centromeres and telomeres could have evolved simultaneously, but I still think it likely that centromeres came first as there would probably have been a stronger selective advantage for chromosomes to associate with microtubules than for telomeric functions. As soon as chromosome origins lost their association with centrosomes there would have been selection for secondary association with microtubules via proto-kinetochores. It seems simplest to suppose that these evolved from the bacterial origin region as both are involved in bacterial chromosome movement and that this happened soon after the origin of the nuclear envelope when the ancestral centrosome duplicated to make a separate intranuclear MNC. Telomeres might have evolved either from the chromosome terminus or random breaks elsewhere repaired by adding telomeric repeats; derivation from the replicon origin [4] now seems less likely.

Nuclear Fusion

Nuclear fusion involves both membrane fusion and the integration of the nuclear skeleton into one. If, as I have argued, the first eukaryotes had no proteinaceous nuclear lamina and the inner membrane was attached directly to chromatin by Sun-domain proteins among others, this simplifies the origin of nuclear fusion. One only has to make each membrane fuse in turn, as in *Chlamydomonas* [352] and yeast [353]; for the fusion of the outer membrane proteins involved in ER membrane fusions were probably recruited. Fusion of mitochondria from both gametes also commonly occurs in protists as diverse as yeasts, slime moulds and *Chlamydomonas*, and must also involve fusion of two membranes. I am unaware of any examples in Eozoa, so we do not know whether it evolved early in eukaryote and mitochondrial evolution. It might have occurred relatively late, for example after the ancestral bacterial FtsZ division mechanism was replaced by host dynamin division machinery [354], which could also have helped fusion. Given the multifarious roles of dynamin in membrane division (which also involves membrane fusion) [355] and that it is located on the nuclear envelope in ciliates [356] it would not be surprising if it were also recruited to help with nuclear fusion. Chloroplast fusion also occurs during syngamy in many green algae [352] and clearly evolved much later than nuclear or mitochondrial fusion, but is an important part of some sexual cycles.

Ploidy Cycle Evolution

Attentive pedants will have noticed that my saying that meiosis evolved for 'ploidy reduction' prior to the origin of the nuclear envelope is terminologically questionable. This is because the distinction between polyploidy and multinuclearity strictly only applies to eukaryotes with a NE. In bacteria with single chromosomes and lacking plasmids there is no distinction between genomes and chromosomes and the concept of ploidy does not strictly apply. In bacteria filamentous cells containing several nucleoids within one cell (quite common) are more analogous to eukaryotic multinucleate plasmodia produced by delaying cytokinesis compared with mitosis, and are not really polyploids and no special reduction division is needed to reduce their genome copy number.

The neomuran precursor of eukaryotes would conceptually have stopped being a bacterium as soon as centrosomes, microtubules, cyclin-based eukaryotic cell cycle controls, and cohesin evolved, though would not have been a protoeukaryote until the nuclear envelope evolved. But as soon as cell division of this prekaryote became obligately linked to centrosome duplication and anaphase proteolysis of cohesins, it becomes reasonable to regard cells in which this linkage has failed and which contained multiple chromosomes but only one centrosome as analogous to plant autopolyploids. With only one centrosome they would have no way of knowing that they had too many chromosomes or halving their numbers without somehow blocking DNA replication while allowing an extra centrosomal duplication, one of the hallmarks of meiosis. Thus as soon as special meiotic cohesins and partial protection of their anaphase I digestion (on chromosome arms) had evolved one can reasonably speak of prekaryotic meiosis and of ploidy reduction.

It is unclear what was the ploidy of the first eukaryotes as ploidy levels are unknown for free-living discicristates (the fact that some trypanosomatids are diploids does not allow us to infer the condition even for their closest relatives the bodonids, still less for ancestral Euglenozoa or Percolozoa), though it should be possible to deduce whether *Naegleria* is haploid or diploid when its genome sequence is properly studied. I have long supposed that the ancestor would have been haploid and only the zygote diploid, as in *Dictyostelium*, *Crypthecodinium* and *Chlamydomonas*, as that is probably the most widespread condition in free-living protozoa, except for those with very large cells like the somatically polyploid ciliates, the plasmodial myxogastrids or reticulose Foraminifera [3,63,306]. I have argued that selection for large or small cell size played a major role in evolving the diversity of protist life cycles, and that selection for small trophic cells and larger resting cells may have been of key importance in the evolution of haplo-diploid life cycles, including cell cycle variants such as multiple fission (always derived) rather than binary fission (clearly the ancestral state for eukaryotes) [306]. Experimental studies on yeasts are clarifying the selective forces acting on ploidy levels of this normally diploid saprotrophic fungus [339,357-359], but caution is needed in applying these to predatory protists with their very different ecology. Direct studies of discicristate ploidy and life cycles are needed to clarify early eukaryote ploidy evolution.

DISCUSSION

According to the present coevolutionary theory, the origin of the nucleus depended on the prior evolution of a primitive endomembrane system and a primitive mitosis, both brought about by and associated with the origin of phagocytosis. The revised multistage theory for the simultaneous origin of mitosis and the pellicular microtubule array of the first discicristate eukaryotes offered here (Fig. 3) explains the transition from posibacterial to eukaryotic skeletal, segregational, and cell division structures distinctly more smoothly and in more detail than previous ideas. The new interpretations of the origins of heterochromatin and its positional control by small RNA and the new theory for the origin of the nuclear envelope and nuclear pore complexes by coated vesicle fusion (Fig. 5) integrate a vast body of molecular and cell biological data into a coherent picture of how nucleocytoplasmic compartmentation originated, with plausible mechanisms and selective advantages for each stage. The discussion of the origin of meiosis and syngamy explains the origin of sex and of eukaryotic life cycles with alternating ploidy levels as the almost inevitable outcome of those new mechanisms of mitosis, cell cycle control, and cell compartmentation, given the need to reduce cell cycle errors in ploidy, maximize survival in dormant states induced by famine or environmental adversity that precludes growth, and the contrasting requirements for maximizing reproductive rates when food is plentiful. This is all done within a critically interpreted phylogenetic framework for both the ancestral bacterial and the derived eukaryotic parts of the tree of life, which is consistent with all major lines of evidence, molecular, cellular, and palaeontological.

In comparison with the above proposals, previous discussions about the origin of the nucleus are either unnecessarily complicated or fail to explain the most essential things. This is particularly the case with theories that invoke symbiogenesis or prokaryotic cell fusions as 'explanations' of the basic features of eukaryotic cells. To illustrate such unnecessary complexity and explanatory failure, consider the serial symbiogenetic/autogenous theory of my good friends López-García and Moreira [122]. I single this out for detailed criticism not because it is worse than other cell fusion/symbiotic theories but because it attempts to be more detailed, which is praiseworthy; most are so empty scientifically that a critic can gain no useful purchase.

Small GTPase Origins: Vertical Inheritance, Lateral Transfer or Extra Symbioses/Cell Fusions?

The ancestral eukaryotic small GTPase gene gave rise to about 10 functionally distinct paralogues [124] during the later phases of eukaryogenesis, all important for controlling the cytoskeleton, secretion, phagocytosis and nucleocytoplasmic exchange (e.g. the Ran GTPase). For some years Moreira and López-García have argued that the closest relatives of these small

GTPases, are absent not only from archaebacteria and posibacteria but also from α-proteobacteria, but are present in myxobacteria (δ-proteobacteria) [360,361]. They thought that eukaryotic small GTPases came from neither the host nor the premitochondrial symbiont. If such GTPases were truly absent from both host and mitochondrial lineages, the simplest interpretation would be that the prekaryote instead acquired a small GTPase gene by lateral gene transfer (LGT) from a negibacterium; presumably one that it ate in the earliest days of phagotrophy after actin and endomembrane vesicle traffic started but before control by small GTPases evolved [3]. However small GTPases of the BglA family are widely present in negibacteria, not just myxobacteria [3]. They have also now even been found in α-proteobacteria and several other paralogue families have been discovered in bacteria [128]. Most important is the RarD family restricted to actinobacteria and archaebacteria. This is the prime candidate for a direct vertical ancestry for all eukaryotic small GTPases making LGTs or more complex scenarios pointless.

Yet Moreira and López-García [360,361] imagined that the nucleus evolved as part of a three way symbiosis in which a myxobacterium was the host for the engulfment and enslavement of an archaebacterium and an α-proteobacterium. Supposedly, the initial stimulus for this was metabolic syntrophy between the different bacteria, followed by serial endosymbiosis, also metabolically driven. In the latest version of this syntrophy theory [122] the archaebacterium became the nucleus, but its membrane was lost, as were the outer membrane and genome of the myxobacterium! Not only is this unnecessarily and excessively complex, but its logic is fundamentally flawed, and it explains nothing important about eukaryote origins. None of the 27 novelties of Appendix 1 is explained, most not even being mentioned. The theory states that a myxobacterium phagocytosed first a methanogenic archaebacterium, to which it supplied CO_2 and hydrogen for making methane, and then a methane-oxidising α-proteobacterium. Then the myxobacterium grew surface membrane invaginations to surround the archaebacterium, supposedly to help the archaebacterium secrete proteins into the environment across the three membranes and two cell walls that then surrounded it (its own cytoplasmic membrane and cell wall and the cytoplasmic membrane, peptidoglycan wall and the outer membrane of its host myxobacterium). How this could possibly help secretion across these five barriers was not explained, nor was it stated what specific proteins were secreted or any selective advantage of that secretion mentioned. Despite these omissions that render it explanatorily empty, that membrane membranes, as in Planctobacteria, could be ancestral to phagotrophic eukaryotes is cell biological nonsense, especially for Planctomycetes as there is no known mechanism for losing the outer membrane in peptidoglycan-free negibacteria. Even their *Chlamydia* relatives, among the most reduced intracellular parasites, never lost the outer membrane despite losing murein and ATP biosynthesis; nor did mitosomes, the most phylogenetically reduced negibacterial descendants of all, which additionally lost their entire genome [369].

The discovery of α- and β-tubulin genes in some Planctomycetes is also irrelevant to eukaryote origins, as they are evidently lateral gene transfers from eukaryotes [370-372], evolving so much faster than any others that they cannot possibly be ancestral to eukaryotic genes and probably lost microtubule functions and evolved others instead with weaker stabilising selection. Just as one swallow does not make a summer, two tubulins do not make a eukaryote. Several others are needed, as are hundreds of other cytoskeletal proteins.

The latest version of the spirochaete theory of the origins of cilia, mitosis and the nucleus [120] is as devoid of phylogenetic support, explicit transformational reasoning, and cell biological plausibility as the earlier versions dating from Sagan [366], so past criticisms [366] still apply. It is imagined that a sulphur-reducing archaebacterium, like *Thermoplasma*, the smallest archaebacterium, successfully engulfed a much larger spirochaete (with two bounding membranes) and converted it into a mitotic spindle and cilium and the genomes of the two bacteria merged to become the nucleus. How any of this could have happened is totally unexplained. Early predictions of the theory that spirochaetes would have tubulins and other precursors of spindles and cilia have been firmly refuted, but instead of being properly rejected the 'theory' is now emptied of all content that might allow refutation, making it pure science fiction. The complexity of cilia with about 1000 proteins, most novel to eukaryotes and many related to non-ciliary eukaryotic proteins, always made it naïve to think of their origin primarily in terms of where tubulin came from.

By comparison, the autogenous theory [4,238-240] according to which cilia are very complex modifications of many disparate elements of a simpler cytoskeleton that evolved in the earliest eukaryotes remains valid and unrefuted, and goes from strength to strength. Though I do not agree with all details in the latest formulations [122,373], which can be further modified and improved, they are firmly founded on real cell biology and careful reasoning. An important point is that the ciliary transport particle, essential for ciliary biogenesis, contains α-solenoid and/or β-propeller domains, and so like the NPC may also have evolved from transport vesicle coats [374]. This makes it plausible that the origin of coated vesicles was not only the cause of making endomembranes permanent but also an essential prerequisite for the origin of both nucleus and cilia, whose origin is therefore subsequent to that of endomembranes. The recent rooting of the eukaryotic tree indicates that the cenancestral eukaryote had both a nucleus and an attached centriole and cilium: what classical protozoologists called a karyomastigont. The theory of the simultaneous origin of cell nuclei and centrioles and cilia is therefore correct [4]; it is accepted even by Margulis [120], who unfortunately also retains its associated postulate that some extant protozoa are primitively without mitochondria, now firmly disproved [106].

We now see that the origins of mitochondria, nuclei, and cilia were virtually simultaneous. As each was a complex series of processes it is most likely that they overlapped in time, so disputing which was first may be meaningless. One clue to relative timing is that ubiquitin is essential for spliceosomal assembly [375]; this suggests that ubiquitin was already present before the mitochondrion donated group II introns to the nucleus, allowing the origin of spliceosomes. As the role of ubiquitin in proteasomal digestion of faulty ER proteins and cell cycle proteins is likely more fundamental, these mechanisms probably evolved prior to the origin of mitochondria; thus the host was probably already eukaryotic in most respects prior to its enslavement and the origin of spliceosomes. Ubiquitination probably evolved very early in eukaryogenesis as soon as the endomembranes became stably separate from the cell surface [27]. That paper proposed that proteasomes played a key role in the origin of eukaryotes immediately before the origin of phagocytosis, by digesting prey proteins translocated directly across the ancestral plasma membrane, and that ubiquitination evolved after this hypothetical intermediate evolved phagocytosis and lysosomes, making such digestion of prey proteins redundant.

Was Ribosomal Chimaerism the Major Selective Factor in Pore-Complex Origin?

Since this paper was written a new suggestion about the selective force initiating nuclear evolution appeared [376]. This builds on my earlier idea that the symbiotic origin of mitochondria and transfer of mitochondrial ribosomal protein genes to the nucleus would have influenced cytoplasmic ribosome evolution, because of selection to avoid chimaeric cytoplasmic ribosomes with some incorporated mitochondrial ribosomal proteins [3,377]. I argued that cytoplasmic and mitochondrial rRNAs and proteins diverged to reduce the extent of heterologous misassembly. Jékely [376] proposes that this selective force would have been strong enough to favour also the origin of the nuclear envelope and pore complex. He notes that 24 mitochondrial proteins were transferred to host chromosomes before the eukaryote cenancestor, assumes that this happened before the origin of the nuclear envelope and caused enough harm to stimulate the origin of the envelope. I am unconvinced, for several reasons, the first two already expressed by Forterre, who refereed his paper [376]. First, there is no evidence that the mitochondrion evolved before the nuclear envelope; Jékely himself earlier noted, as I had previously, that the recent conclusion that the ancestral eukaryote already had mitochondria has been misinterpreted as evidence for mitochondria being enslaved by a eukaryotic host before the origin of phagotrophy, when it is nothing of the kind. This is not a strong objection because I suspect that they evolved roughly simultaneously. Secondly, supposing that the transfer of genes could be tolerated with impunity for a period and then became a compelling force for evolving the envelope is problematic. This explanation is like the spliceosomal intron one in putting the damage first with a rescue later: the cart before the horse. If transfer were so strongly disadvantageous, it would be more likely either that cells with transferred genes would lose out in competition or that host genes and transferred genes were simply modified to reduce the problem, as I proposed earlier (we know that substantial modification took place, though whether the reason I gave is the best explanation is hard to test). However, there is little doubt that sometimes cells do suffer non-lethal harmful effects that are later phenotypically corrected in roundabout ways, so this style of reasoning is not intrinsically unsound; but its plausibility hinges on difficult judgements of selective disadvantages and advantages, and the likelihood of different modes of correction and weighing all these against other likely factors that are hard to evaluate. Thirdly, the benefits of differential concentration and nucleocytoplasmic distribution would have been the net benefit of excluding hundreds of different proteins from the nucleus and the positive import of hundreds of others. If avoiding ribosomal chimaerism were one of them, it was probably of such minor importance that the nucleus would have evolved as it did without it. Singling out exclusion of one class of proteins as markedly more critical than others is very arbitrary. In any case, my earlier suggestion of a modification of host ribosomal proteins and

RNA (helping to causing their long-known much greater divergence from prokaryotic ancestors than for mitochondrial ribosomes) [3,377] would be a much simpler and more direct way of reducing the problems of ribosomal chimaeras.

Nonetheless, Jékely's [376] simulation studies of concentration gradients between localised chromatin and cytoplasm in a cell without full nucleocytoplasmic compartmentation and transport are extremely important as showing how in principle proteins that can bind to chromatin can be localised and influence the local concentration of interactors/product in ways beneficial to the cell. Thus even though I am sceptical of a dominant role being assigned to the avoidance of ribosomal chimaerism, the principles he demonstrates can be applied to all kinds of molecules whose concentration collectively could have provided a strong selective force at all stages of nuclear envelope evolution. It is highly likely that many molecules were initially concentrated in chromatin regions by binding directly or indirectly to chromatin or by being generated by enzymes that were thus bound. The origin of the pore complex and nucleocytoplasmic exchange was so complex that many components must have evolved initially independently and been combined in successive selectively advantageous stages. As a general principle, compartmentation can explain the origin of the nucleus irrespective of whether it slightly preceded, slightly followed, or was contemporaneous with mitochondrial enslavement. That makes it a more general explanation that unlike Jékely's is insensitive to the precise order of events; his interpretation is a special case of the general principle.

The primacy of the Precursor and Mutation; Selection is a Secondary Metaphor

Many theories of eukaryote origins and other megaevolutionary events make the mistake of assuming that selection is the primary force in evolution. It is not. Selection is simply a consequence of a mutation and the environment in which it is found; it is not a third force, or even a force at all. In a stable environment, a novel mutation increases reproductive success, decreases it or is neutral. The relative increase over generations of individuals bearing a novel beneficial mutation is a direct consequence of the phenotype of that mutation; its most important phenotypic property is that it increases reproductive success. Selection is just a metaphorical name given to the mathematical fact that genotypes that inherently increase that success necessarily spread at the expense of those reduce it. Thus major innovation comes only from within the organism by mutation and is not imposed from outside by the environment or a designer or even 'natural selection'. The environment was irrelevant except in a very general permissive way for the origin of eukaryotes. The possibility of an organism getting its food by eating another existed even before the origin of life. There is no reason to think that the environment prevented phagotrophy and the origin of the nucleus for the roughly 2.8 billion years that elapsed between the origin of the first bacterial cell and the origin of the nucleus. The possibility of syntrophy existed for billions of years without making eukaryotes, and the possibility of being an aerobic eukaryote heterotroph existed for about 1.6 Gy before the first one evolved. Eukaryogenesis was so long delayed because of the exceeding difficulty of evolving phagocytosis in a eubacterial cell enclosed by a rigid corset of murein peptidoglycan, and because replacement of murein by a more flexible glycoprotein coat that allowed the evolution of phagocytosis was itself so difficult and unlikely that it did not happen till 2.8 billion years after life began [6]. Thus the nature of existing precursor organisms limits evolution. Most changes are harmful and eliminated by selection. Selection for efficient function in bacteria prevented the origin of the nucleus for nearly three billion years. No selective force could make a normal eubacterium or even a normal archaebacterium evolve a nucleus. There would be no advantage in giving up its traditional attachment of DNA to the cytoplasmic membrane and DNA segregation by membrane motors.

Only the forcible disruption of prokaryotic cell organization by novel forces - actomyosin, mediating phagocytosis - could have triggered such a radical change. No DNA mutation ever directly made endomembranes. Actomyosin provides the real physical force that powers our muscles, enables speech, writing, eukaryotic cell division, and phagocytosis, thereby making endomembranes. The first proto-endomembranes are easily understood as incidental physical consequences of phagocytosis [3,4,11]. Membrane heredity was as important as DNA heredity in stabilising them as distinct genetic membranes [85]. The origin of eukaryotes involved no metabolic innovations, unlike much bacterial evolution. At its core were structural innovations in the cell skeleton and its association with membranes, catalysts and genes. Only phagotrophy first and intracellular coevolution theories provide a clear logical explanation of the origin of the nucleus and of the way in which endomembranes, cytoskeleton, and eukaryotic chromosomes are so mutually interdependent in their functions. They must have coevolved. Enslavement of a proteobacterium to make a mitochondrion provided no novel metabolism. It merely made phagotrophy more efficient in an already aerobic host by improved division of labour by compartmentalisation [3]. Metabolic/bioenergetic theories of eukaryogenesis fail to explain why or how eukaryotes evolved. Prokaryotic host or prokaryotic cell fusion theories are incompatible with cell biology and have explained nothing of significance, yet may remain popular among those who avoid the fundamental issue of radical cellular transformation and how to explain it plausibly without assumptions that would kill the cell.

Novel Adaptive Zones, New Body Plans, and Megaevolution

Eukaryote origins has all the hallmarks of what Simpson [10,113] called megaevolution; the origin of major new body plans that distinguish higher taxa like phyla, classes and orders. He argued that such evolution occurred by the normal processes of mutation and selection but was exceptionally fast and radical in its consequences. He showed that the fossil record indicated that body plan innovation invariably occupied a very short time compared with the subsequent history of an innovatory new body plan. Almost invariably the most important adaptive radiations of a new major type also occur relatively soon after its origin with subsequent megaevolutionary stasis. New inventions, like arthropod cuticle and jointed limbs, land plant cuticle and vascular systems, the vertebrate jaw or tetrapod limbs, create what Simpson called a new adaptive zone (in plainer language, a novel way of life). Within it an immense variety of organisms able to fill related but distinct niches can relatively easily evolve by making minor variants of their shared body plan. New body plans always develop by gradual piecemeal modifications of pre-existing ones, but gradual does not mean slow! Typically they do so very suddenly, because transitions from one major adaptive zone to another are difficult because of the sharply conflicting selection pressures that apply to them. Unless the right organism is present in the right place at the right time and experiences a suitable succession of lucky chances, it will not happen - especially if the properties selectively advantageous in the new zone are disadvantageous in the old one.

Megaevolution, quantum evolution, and mosaic evolution are distinct but complementary concepts. Quantum evolution can occur independently of megaevolution. It is not necessarily associated with a novel body plan or higher taxon - it could just affect one molecule that suddenly found a novel function, e.g. following a gene duplication. It refers only to the rate being abnormally high - way beyond the normal range. Mosaic evolution simply refers to a basic truth about organisms: they are not homogeneous, so different parts can evolve in different ways at different times and vastly

CONCLUSION: PHAGOTROPHY, THE NOVEL ADAPTIVE ZONE THAT MADE THE EUKARYOTIC CELL

Now that we are reasonably certain that the ancestral eukaryote was a phagotrophic protozoan [9], not a non-phagotrophic photosynthetic alga or osmotrophic fungus, as on some past now firmly rejected theories, it is beyond serious question that eukaryogenesis involved the origin of phagotrophy [91]. De Duve [131], who discovered lysosomes and peroxisomes, and Stanier [2], apparently independently, first emphasised the key role of phagotrophy in explaining the origins of the endomembrane system. Stanier [2] proposed that phagocytosis was also the stimulus for the evolution of larger cells, microtubules and cilia. Then my coevolutionary theory [11] argued that compartmentation by endomembranes and the cytoskeleton must have coevolved and together fundamentally changed the chromosomes and gene expression machinery, because DNA segregation, transcription and ribosome biogenesis are tied up with cell structure and compartmentation. It proposed that loss of the bacterial peptidoglycan wall and subsequent origin of actomyosin were the primary stimuli for the origin of phagocytosis and the nucleus [11]. Thus the genetic peculiarities of eukaryotes were indirectly also consequences of the origin of phagotrophy in a wall-free prokaryote [4,5,11,163,379,380]. No compelling evidence or arguments have yet been found against this thesis, yet the idea that the novel genetic systems of eukaryotes are a consequence, not a cause, of their feeding behaviour and novel cell structure has not been popular, perhaps because of a widespread misconception that DNA makes cells. It does not; cells make DNA, which is an inert informational repository. Proteins and RNA are the catalysts and effectors; lipids, proteins, and glycans the builders. Trophic behaviour is the architect within the ecological landscape. DNA is not primary, but is important. We must explain its reasonably efficient segregation to daughter cells across the prokaryote/eukaryote divide.

Recently De Duve [367] reappraised the evidence for eukaryotic origins in the light of modern molecular cell biology and now accepts that peroxisomes were early offshoots of the endomembrane system [3,11] and that theories that he and I once espoused of a separate symbiotic origin for them are entirely superfluous. According to the coevolutionary theory of the eukaryote cell, cytoskeleton, endomembranes, peroxisomes, cilia and genetic systems are part of a unified whole, within which food came first and sex was an afterthought [3]. The autogenous origin of all these structures was far more radical than the minor tinkering when the resulting phagotroph converted an engulfed but undigested proteobacterium into a mitochondrion - another afterthought [20]; ultrastructurally and functionally mitochondria are still easily recognisable as slightly modified negibacteria. Intracellular coevolution is strongly exemplified between the endomembrane system and cytoskeleton; and in the way that internalization of membrane-attached DNA revolutionised DNA segregation [70]; in how the origin of centromeres led to other eukaryotic chromosomal properties through mutation pressure [70]; in the way the origin of motor ATPases provided selective forces for nuclear origins; and the origins of coat proteins by duplications both diversified endomembranes into ER, Golgi, endosomes [4] and yielded NPCs and ciliary transport particles [374] necessary for the simultaneous origin of cilia from astral microtubules attached orthogonally to the cell surface by transitional fibres [4].

It is thus hardly surprising that except for mitochondria piecemeal discussions of the origin of only one eukaryotic component largely failed. Symbiogenetic theories, especially, have been a 40-year distraction from the core problems of how a bacterium was transformed into a eukaryote.

Competing Interests

The author declares that he has no competing interests.

Appendix 1

Key innovations in the origin of the nucleus and eukaryotic cell cycle

1. Internalization of DNA attachment sites as protoNE/rough ER [3,4,11,380]

2. Cell division by actomyosin not FtsZ [3,4,6,11,70,240]

3. Chromatin condensation cycle: histone phosphorylation, methylation, acetylation; heterochromatin [18,170]

4. Mcm replication licensing system controlled by cyclins [244,251]

5. FtsZ triplication to make tubulins; (γ for centrosome) and α and β for microtubules fixing it to cell surface [3,4,6]

6. Kinesin to separate centromeres via antiparallel microtubules [6]

7. Centromeres/kinetochores (CenpA from core histone) for attaching DNA to microtubules [3,4,6] (note that as discussed in the text CenpA might be a property of neokaryotes only, but even Euglenozoa have some kind of kinetochores)

8. Dynein for moving cargo towards the minus end of microtubules and related midasin for ribosome export [6]

9. meiosis and synaptonemal complex [3,63,70,332,333,381]

10. telomerases and telomeres [3-5,70]

11. post-transcriptional gene silencing [177,199,201,203,382]

12. proteinaceous interphase nuclear matrix with bound DNA-topoisomerase II and its ability to reorganize as mitotic chromosome cores [163]

13. nuclear lamina [18,163]

14. nuclear pore complexes (NPCs) [14-17,85]

15. nucleolus and more complex rRNA processing (e.g. 5.8S rRNA) [3]

16. Ran GTP/GDP cycle for directionality of NE export/import [14,17,107]

17. karyopherins [14,17]

18. ribosome subunit export machinery [14,17] (note that most proteins involved have not been detected in Euglenozoa so they might use a simpler system than neokaryotes)

19. mRNA capping and export machinery [14,17,70]

20. polyA transcription termination system [70]

21. 26S proteasomes [27]

22. ubiquitin system [329,331,383,384]

23. sumoylation of nuclear and other proteins [318-323]

24. cell cycle resetting by anaphase proteolysis [6,13,18]

25. nuclear envelope fusion and syngamy [332]

26. spliceosomes and spliceosomal introns [4,70,81,211,385]

27. nonsense-mediated mRNA decay [314]

Innovations are listed in likely order of occurrence, but some (e.g. 3-6, 17-19, 22-23) were likely to have been simultaneous. Some were likely to have been rapid consequences of others (e.g. 9 of 1-7; 16-20 of 13,14; 26 & 27 of 14; 27 of 26). References indicate where their evolution or molecular basis is reviewed in more detail.

Reviewers' Comments

Anthony Poole

This is a fascinating and important paper on a complicated topic. Overall, I rate Cavalier-Smith's take on the evolutionary origin of the nucleus as the most thoroughly considered attempt published to date, and there is little doubt in my mind that the main conclusions he makes are better supported than those of competing models. I see no significant problems with the theory as presented, and agree with the central points.

There are two general take-home messages here. The first, eloquently summarised in the final paragraph of the paper, is that symbiogenetic theories have been a '40-year distraction'. Cavalier-Smith's forceful and cogent attack on these views makes for enjoyable reading (though perhaps not if one is on the receiving end), and the arguments here amply demonstrate why the position advanced by Yutin and collegues in their recently published paper (Biology Direct 2009, 4:9) is specious at best. The second general message is that, to have any hope of explaining the origin of the nucleus, it cannot be considered in isolation from other key eukaryotic cellular features (as has all too often been the case).

The only real gripe I have with this paper concerns its proportions. It is like a sandwich made with bread sliced too thickly, and meat sliced too thinly. The Introduction is an interesting though long and, at times, meandering read, and the Discussion is nine tenths diatribe (on why everyone else is utterly wrong on almost everything). Let me make it clear that I don't have any major concerns about the content of these sections, but the really exciting material (the 'Results' section) is concertinaed into a few brief pages. To give some perspective, there is about 10 pages in the Discussion devoted to the slating of other published theories (I found this interesting reading but note that many of these critiques have been made elsewhere, either by Cavalier-Smith or by others) and around the same proportion of the 58 pages of text in this manuscript is dedicated to developing the novel ideas. This has two effects. One is that the novel material is deeply buried, and as a reader one has to work hard to extract the author's insights from such a lengthy composition. The second is that, on account of the brevity of the 'Results' one has to spend a lot of time figuring out exactly what Cavalier-Smith means. I am probably not alone in acknowledging that Cavalier-Smith has a broader knowledge of the subject matter than I, but it makes it hard to follow the reasoning when a number of points are assumed to be common knowledge and stated without some sort of background. While one might say I should have done my homework, I think much would be gained from acknowledging that the topic of this paper is of interest to a broad range of evolutionary and cell biologists, and providing the requisite detail would therefore be of value. In some cases the arguments are indeed well-supported and developed, but for other points there may be a greater degree of speculation, and the difficulty is that without a fair recollection of Cavalier-Smith's extensive canon, it may be hard to follow all the reasoning. This is a pity because, some points might therefore be mistaken as superficial speculation when they are in fact well-supported.

Authors response: *Only one bread slice (the introduction) was substantially thicker than the meat. That is because it provides background to help digest the meat. The referee requests more background for certain topics in the results section; the problem with such a broad field is that readers differ greatly in which parts of the background they really need; some may want more on some points and less on others than provided. I have amplified only parts of the introduction for which at least one referee seemed to need more information. Cutting some or inserting more might please some readers but irritate others: each can skim what to them is obvious and linger over what seems novel. I also amplified the meat section on the evolution of nucleocytoplasmic transport to clarify points raised by the next referee, moved the general section on megaevolution to the discussion so readers reach the meat faster, and made the results discussion of the origins of mitosis (now mentioned in the title), meiosis and heterochromatin more thorough with more background even though this needs some revisiting of arguments and evidence discussed previously. A diatribe is defined as bitter and abusive; my discussion has much robust but carefully reasoned criticism of arguments, including some of each referee, but is neither bitter nor abusive so is not a diatribe. I would welcome similarly robust, reasoned and discriminating criticisms of my ideas, much preferable to their being ignored or uncritically accepted. I thank all referees for their comments; more such dialogue is needed in this field.*

What follows are a few specific comments or questions concerning ideas presented:

p7, "The eubacterial ancestor of neomura could not have been a negibacterium with two bounding membranes, but was a posibacterium with a single surface membrane, like neomura; probably an actinobacterium, but possibly an early intermediate between Endobacteria and Actinobacteria".

Here I wish to request a clarification. This statement builds on two points. One is that archaea and eukaryotes are sister groups. The other is that the neomura (archaea + eukaryotes) evolved from within the bacteria. The view that neomura evolved from bacteria is based on an argument Cavalier-Smith has made before (see in particular Biology Direct 2006, 1:19), and is where the statement 'probably an actinobacterium' comes from. This gives the impression that neomura (archaea + eukaryotes, and ignoring any genetic contribution from mitochondria) evolved from the actinobacterial crown. I would like to ask whether the author's statement should be taken to mean this is a possibility, or whether actinobacterial-neomuran affinity refers to a split before the most recent common ancestor of actinobacteria (i.e. a stem actinobacterium).

Author's reply: *I thought there was no compelling evidence either way, so stated the point thus, and simply referred to my previous discussion of the uncertainties* [13]. *The radical changes in so many genes that caused the neomuran revolution make it exceedingly difficult to answer this question using gene sequence trees - more difficult than establishing the branching order of the eubacterial phyla (itself largely unresolved by sequence trees alone). However, reasons now specified in the introduction indicate that neomura are most likely sisters to crown actinobacteria, as shown in the more resolved revision of Fig.* Fig.1.1. *Insofar as the posibacterial ancestor of neomura must already have evolved phosphatidylinositol, sterols, and 20S proteasomes, it was a stem actinobacterium rather than an endobacterium (though it may well have had endospores as do the most divergent actinobacteria). Phosphatidylinositol can be considered a synapomorphy of actinobacteria that clearly differentiates them from Endobacteria, which unlike the other two has not been secondarily lost within the group. If, as previously suggested* [12] *that ancestor was a thermophile (**not** hyperthermophilic), it might also have had a GC-rich genome.*

I also think it is worth me pointing out for those who might disagree with the bacterial origin of neomura that one does not need to accept all the steps preceding the archaeal-eukaryote split in figure figure11 - for the purposes of following the novel arguments developed here the reader should concentrate on the three stages marked with stars.

Author's reply: *That is correct. The nature of the mechanistic and selective forces that led to the nucleus is logically independent of the phylogenetic origin of neomura (seemingly not realised by the third referee). However it is likely that destabilization of eubacterial structure assumed by the overall* *theory during the origin of neomura almost immediately prior to the origin of the eukaryote cell was of key evolutionary importance in facilitating the radical changes during the origin of phagotrophy. A close association in time between this putative destabilization and the divergent origins of both eukaryotes and archaebacteria and the probable rapidity of the transitions is a major part of the explanation of the absence of extant intermediates between the three domains; it helps explain why they are so phenotypically distinct. Moreover, readers also need to take on board the palaeontological evidence that eubacteria are much older than eukaryotes, combined with the phylogenetic evidence that archaebacteria are sisters rather than ancestral to eukaryotes and the absence of any compelling evidence that archaebacteria are as old as eubacteria because this is important for understanding the phylogenetic origin of the individual components of the eukaryotic host, especially those that are absent from archaebacteria. It is important to appreciate the major role that unsubstantiated assumptions, fashion, and the very name archaebacteria, rather than critical reasoning, has played in the prevailing view (in my view wrong) of the equal age of neomura and eubacteria.*

p23 'and possibly triggering the Neoproteozoic snowball earth' and 'in the only really major explosive adaptive radiation prior to the Cambrian explosion' and (on p25) 'Every gene probably duplicated many times in just a few days'

One factor that makes this paper frustrating to read is all the distraction. The above comments do not really seem necessary for presenting the central ideas. The first is a speculative aside that Cavalier-Smith has given elsewhere and does nothing more than derail the reader's concentration by throwing in an unnecessary spanner. The second is a qualitative statement of the throwaway kind, better suited to a narrative piece, and is again distracting. The statement from p25 is again terribly speculative, and, to my mind these types of statement (there are many more) detract from the strengths of the paper.

Author's Response: *these are fair comments, reflecting our differing opinions of how one should present ideas; however, it would have been better to have said that such brief, less central statements may detract from some readers' perceptions of the strength of the paper. They do not reduce the actual strength of the central arguments, which stand on their own merits. I retain them to alert readers to these possibilities, the last one as a deliberate irritant to stimulate readers to consider that key changes* *can be much faster than often assumed. Because they are peripheral and/or speculative it is not useful to amplify them here.*

I like the discussion of Simpson's writings, but think the author might get to the last paragraph, and the last sentence of the last paragraph ('These are just descriptive terms; they do not invoke special mechanisms beyond mutation and selection'), a little earlier. I found that, having an idea of what the author was actually saying made the second reading of the paper much easier; the first time through I was heavily distracted by all the strong statements about microevolution, e.g. p4, '.dramatic innovations like the origin of the nucleus that are incomprehensible by just extrapolating normal microevolutionary changes'. I think that one does have to be careful with this sort of blanket statement since 'microevolutionary processes' has in the past been very broadly defined. Suffice it to say, it is clear with careful reading that Cavalier-Smith is not invoking special ad hoc mechanisms to account for the origin of the nucleus, so this is perhaps a case of smoke without fire, though with statements such as 'the origin of eukaryotes was an intrinsically abnormal event, not one understandable simply by extrapolating the trivial tinkering that occupies most evolution. To understand it we must invoke something quite exceptional' (p25), it is easy to think this is the case.

This melodramatic wordsmithery is followed by a brief list of four important items (note to the author: only three are listed): preadaptation to phagotrophy, 'novel selective forces' that phagotrophy brings (these are left unexplained), and disruption to cell division and DNA segregation as a consequence of phagotrophy. This at least tempers the hyperbole immediately preceding it. What follows that is an interesting discussion of preadaptations that the ancestor of eukaryotes must have possessed: a single membrane, loss of the murein cell wall and several cellular attributes (large cell and genome sizes, diverse lipids, a secretome and a facultatively aerobic/anaerobic metabolism). Cavalier-Smith argues that Actinobacteria appear to carry these preadaptations. In this view, the lineage leading to archaea secondarily lost these preadaptations.

The points concerning preadaptation are essential for Cavalier-Smith to place this work within the wider context of his own ideas on the neomuran revolution, and are interspersed with slicing critiques of others' proposals for the cellular nature of the host that engulfed the mitochondrial ancestor (i.e. an archaeal host or a mycoplasma). What is most important to my mind is that identifying these preadaptations provides important cell biological insight into the origin of the eukaryote cell, regardless of the specifics of the actinobacterial-eukaryote affinity.

Author's response: *These perceptive comments highlight the difficulty of getting over the idea of radical differences in scale, style, and speed of megaevolution compared with day to day microevolution without exaggerating or underplaying the importance of the descriptive distinction, and without being misunderstood because of past exaggerations or mistaken claims for radically different mechanisms or contradictions that do not exist that make many reasonably wary of such distinctions. It is pleasing that nonetheless you understood the essence of what I tried to say. I have now placed the last paragraph as the second paragraph with some modifications that may help other readers understand my message more easily. Emphasizing the dramatic contrast between megaevolutionary events and most evolution is not hyperbole; it is an essential truth. It is the extravagant exaggeration by some students of microevolution to the effect that 'all evolution is just changes in allele frequency' that is truly hyperbolic and wrong. The origin of eukaryotes is a prime example where the origins of hundreds of new genes and protein domains and of dozens of radically new cellular structures was far more important than changes in the frequency of alleles of existing genes, and thus involves considerations entirely outside the scope of standard population genetics, which does not consider cell structure or unique events at all. The more general parts of this discussion are now moved to the discussion.*

I have three questions for the author concerning the evolution of chromosomes from a circular to linear form (p33):

1. What do you make of the multiple secondary emergences of linear chromosomes among bacteria [386]? The constraints you raise ('the need for replication forks to converge on a single terminus and signal completion', p33) do not seem to apply to these cases.

Author's response: *These examples show that some bacteria can evolve linear chromosomes if they solve the end replication problem, which can be done in several different ways (no bacteria use mechanisms homologous to telomeres; the spirochaete Borrellia, like some viruses, uses the hairpin mechanism I originally proposed for eukaryotes* [221], *and the actinobacterium Streptomyces uses terminal proteins like some plasmids or viruses). Thus the constraints against bacterial linear chromosomes are not absolute, but that does not mean that no constraints other than the end replication problem exist (the same must apply to mitochondria, where linear* *genomes have also evolved multiply; apparently plastid genomes never did). It would be interesting to study the control of replication termination and its coordination with segregation and division in each case of secondary origin of linear chromosomes. A thorough understanding of this should show either that the putative constraint I invoked is irrelevant or that these bacteria found novel ways around it or for some reason can better tolerate linear chromosomes. I doubt whether they would reveal an advantage for linearity, which I suspect is a functionless evolutionary accident driven by mutation pressure, but do not exclude the possibility of an advantage.*

2. The discussion presented regarding the consequences of phagocytosis includes the idea that, prior to this, there were strong constraints on the number of origins of replication. The relaxation of these constraints ('only a single replicon per chromosome so replication termination could directly signal division to occur directly between the only two daughter replicon termini') is thus something that is proposed to postdate the archaeal-eukaryote split. Sulfolobus species are known to have more than one origin of replication [387,388] - do you have an opinion as to why multiple origins are found in some archaea? - this observation doesn't obviously support your statement that additional origins would be 'stringently removed by purifying selection for >2Gy'.

Author's response: *Sulfolobus is an interesting exception to the general rule. In so far as I consider that Sulfolobus evolved well after eukaryotes, this important exception does not invalidate my statement from which you omitted the key word 'previously': my actual statement 'previously stringently removed by purifying selection for >2 Gy' remains true. As this exception is found in an archaebacterium, which unlike eubacteria have Mcm-related DNA synthesis initiation proteins (Cdc6 proteins) related to those of eukaryotes, it is clear that replication initiation was significantly changed during the neomuran revolution (I argue as a coevolutionary result of the origin of core histones). These changes might have preadapted both eukaryotes and archaebacteria to evolve multiple replicon origins, provided there is a selective advantage for this. In the case of eukaryotes I suggested that this selective advantage was to ensure that their much larger genomes were replicated by their much slower replication forks in a small fraction of the cell cycle. Though Sulfolobus has a rather small genome, its replication forks move an order of magnitude more slowly than in most prokaryotes* [387]. *With three origins replication takes 40% of the cell cycle; with only one it would take 120%, clearly impossible; without multiple origins the cell cycle would be markedly longer, a severe selective* *disadvantage. Thus the exceptional slowness of Sulfolobus replication forks imposed a novel selective force for multiple origins, as in eukaryotes (I would argue probably independently, though the mechanisms of its synchronous initiation of multiple origin control need elucidating in detail to test this). Sulfolobus has asynchronous replicon termination, which raises the same issue of coordination with division as linear chromosomes: understanding this better would also help test these ideas. It seems that in general in archaebacteria there is a long gap after replication termination before division and no archaebacteria are known to use the trick of successive initiations at single origins (at shorter intervals than total fork transit time) whereby Escherichia coli and other enterobacteria speed up DNA replication and allow cell cycles shorter than replication time. The absence of such mechanisms combined with slow forks would favour multiple origins. In these and some other respects* [389,390] *there are closer resemblances in cell cycle controls of archaebacteria with eukaryotes than with eubacteria; possibly when archaebacterial cell cycles are better understood, additional preadaptations for what are currently thought of as typically eukaryotic mechanisms will become apparent. If they do, such cell features would have arisen in the ancestral neomuran - and thus be neomuran not eukaryotic innovations. In Sulfolobus multiple origins clearly arose by gene duplication (two of the three are linked to cdc6 genes)* [387], *as I originally suggested for eukaryotes* [4]. *Why replication should be so exceptionally slow in Sulfolobus could be related to its exceptional environment (80C and pH 3) that may have led to unusually stable chromatin structure, likely to delay strand separation by DNA helicases. Overall this exception illuminates the rule and supports my interpretation of the reasons for it, which explain both the rule and this exception.*

3. Thermophily has previously been proposed to be a selective pressure favouring the emergence of circular chromosomes [391]. Earlier in the manuscript (p6), you state that the ancestral archaebacterium was subjected to massive gene loss (including genes such as MreB) during 'secondary adaptation to hyperthermophily'. Given the statement that the ancestral archaeon was not hyperthermophilic, it would be nice to hear the author's view on the proposal that Cenarchaeum symbiosum (and related mesophilic 'Crenarchaeota') be classified as a distinct phylum (Thaumarchaeota - [232]), one possible implication from that work is of course that the ancestral archaeon was not hyperthermophilic. I haven't discussed this with Forterre yet, but an ancestrally mesophilic archaeon could potentially kill

the argument for circular chromosomes as a direct thermoadaptation (since Cenarchaeum has a circular genome).

Author's response: *You misinterpreted the significance of 'secondary' in my sentence. I did not state that the ancestral archaebacterium was not hyperthermophilic. As explained in detail before* [12] *I have consistently argued since 1987 that it was hyperthermophilic and that hyperthermophily was the key innovation involved in the origin of the special archaebacterial lipids. Secondary means only that the ancestral neomuran and its eubacterial ancestors were not hyperthermophilic. Thus in the context of the whole tree of life ancestral archaebacterial hyperthermophily is secondary. I have never favoured Foterre's argument for thermophily as an explanation of the circularity of the chromosome as I have never accepted that archaebacteria are as old as eubacteria, which also mostly have circular chromosomes and are mostly non-thermophiles, which means that circularity evolved prior to the last common ancestor of all prokaryotes, for which there is no reason to assume hyperthermophily or even thermophily. Forterre's idea of the origin of prokaryotes by reductive evolution of eukaryotes is mechanistically totally implausible (nobody has ever suggested plausible intermediate stages or how you could undo the eukaryotic endomembrane system or cytoskeleton, still less how the first cell could have evolved it) and refuted by the fossil record, so his 'explanation' of circular chromosomes was a non-starter.*

Your question about Cenarchaeum has a phylogenetic and a taxonomic aspect. Taxonomically I oppose making it a distinct phylum, just as I oppose cren- and euryarchaeotes as distinct phyla, for their phenotypic disparity is too small to justify a rank above subphylum. There have been too many unmerited designations of new phyla among bacteria based just on sequence trees rather than phenotype. Another unmerited phylum ranking is the hyperthermophilic Korarchaeota [392]. *It would be more appropriate to make thaumarchaeotes and korarchaeotes new crenarchaeote orders, possibly classes. Ranking is a subtle thing and bacteriologists underuse the intermediate ranks of class and subphylum compared with order and phylum (technically division), yielding unbalanced and less useful classifications.*

Phylogenetically there are two contradictory positions for Cenarchaeum: as sister to hyperthermophilic crenarchaeotes as on the 2-gene rRNA tree of Brochier-Armanet et al. [232] *and both protein trees of* [393] *and as sister to all other archaebacteria as in the 53 ribosomal protein tree of* [232]. *It was unwise to assume as they did that the latter is the correct topology, as Cenarchaeum is represented only by one lineage and its position on the tree could reflect systematic bias, not true history. Indeed the EF2 tree and one based on 35 ribosomal plus 3 RNA polymerase proteins give strong support to Cenarchaeum being sister to hyperthermophilic crenarchaeotes. Furthermore the 16S-23S rRNA tree including Korarachaeum gave unequivocal support to it and Cenarchaeum both being the deepest branching members of the crenarchaeote clade. The latter trees based on a greater diversity of different proteins and including also a korarchaeote (unlike* [232]) *are more likely to be correct than the ribosomal protein tree of* [232]. *Interestingly the 38-gene tree places Cenarchaeum and 'Korarchaeum' as sisters with good support and this clade as sister to the traditional thermophilic crenarchaeotes with 100% support. If this is correct then your question would be irrelevant, as the conclusion that the cenancestral archaebacterium was hyperthermophilic would remain, as it would be most parsimonious to suppose that Cenarchaeum, like several euryarchaeote lineages, was secondarily mesophilic (it could also be sensible to group Cenarchaeum and Korarchaeum in one new class). I agree with Brochier-Armanet et al. that the characters shared by Cenarchaeum and euryarchaeotes are likely to be ancestral to archaebacteria and were probably lost by hyperthermophilic crenarchaeotes, but if such characters are indeed ancestral one cannot use their presence in Cenarchaeum as a reason for preferring one topology over the other. If Cenarchaeum is sister to all other archaebacteria, which seems to me unlikely in the light of* [393] *this would make their ancestral state harder to deduce; Brochier-Armanet et al. candidly admit that one could not confidently decide whether the cenancestor was hyperthermophilic or mesophilic just from the topology* [232]; *that is because a hyperthermophilic ancestry would involve only one more reversion to mesophily than the several we are confident of in euryarchaeotes; thus it is not so unlikely that we can reasonably reject it purely for parsimony. Even in the unlikely eventuality that Cenarchaeum is truly the deepest branching archaebacterium it would remain more likely that the cenancestor was hyperthermophilic, as that more satisfactorily explains the origin of their lipids and reverse DNA gyrase.*

Forterre's explanation of the origin of reverse gyrase as a chimaera of two eubacterial proteins strongly shows that eubacteria are indeed ancestral to archaebacteria and neomura, as all the most compelling evidence congruently indicates. The trees and genome data of [393] *confirm that erecting Korarchaeota as a separate phylum based on bad and misleading 16S rRNA trees (an all too common practice for bacteria) was premature and unwise, and together with the genome data for Cenarchaeum make the same point for it; neither is more radically different from other crenarchaeotes than the very diverse euryarchaeotes are from each other; moreover they show that the differences between crenarchaeotes and euryarchaeotes have been exaggerated; many now clearly stem from secondary losses in the cenancestor of the traditional hyperthermophilic crenarchaeotes, just as I argued* [12]. *It is increasingly clear that archaebacteria were ancestrally hyperthermophilic, evolved from eubacteria, that both subphyla generated secondary mesophiles, and that reductive evolution by multiple gene losses has been far more important in their origin and diversification than is yet widely accepted. But there is absolutely no evidence for the generation of prokaryotes by the loss of eukaryotic organelles. I would rank Thaumarchaeota and Crenarchaeota each as classes, not phyla.*

p39: 'Because selfish DNA, especially transposons and endogenous retroviruses, spread much more slowly in sexual than in clonal populations, a sexual protoeukaryote would have suddenly been inundated by such genetic parasites.' There is an error here: presumably this sentence was supposed to read 'spread much more slowly in clonal than in sexual populations'. Here I think it would be an oversight not to cite the seminal paper by Hickey [280]. I am not aware of 'spread' in clonal systems, though there may be 'persistence' of some types of mobile element. Arkhipova & Meselson [394] showed loss of retroelements during the evolution of asexuality in Bdelloid rotifers, which makes sense given these elements propagate through replicative transposition. It seems that DNA transposons persist, and, given that they can transpose by conservative mechanisms, this may well be a case persistence rather than slow spread. It is noteworthy that bacterial transposons are primarily conservative, but can occasionally transpose by replicative means - overall there doesn't seem to be much evidence for proliferation (at least judging by prevalence in prokaryote genomes).

Author's response: *It was an inversion error, now corrected. I inserted reference to Hickey* [280] *and also to my paper that was probably first to argue that non-infectious selfish DNA cannot spread easily in clonal populations* [279], *and which directly stimulated Hickey to show that it can easily in sexual ones. I originally cited neither because this is old established history, just as one does not cite Darwin every time one mentions evolution or Watson and Crick every time one mentions a DNA double helix. With respect to bdelloids, they contain many more transposons than previously thought* [395], *indicating that transposons can spread even in clonal asexuals by lateral gene transfer (e.g. by viral infection), for which there is also good evidence in bdelloids* [396] *(or that bdelloids are not strictly* *asexual). This does not alter the fact that evolving sex added an extra easy way for transposons to spread.*

Gáspár Jékely

Max Planck Institute for Developmental Biology, Tübingen, Germany

This paper presents a critical review of various models on the evolution of the nucleus along with a long argument that intracellular coevolution is the key to understanding eukaryote origins, as well as a scenario on endomembrane and nucleus evolution. I find this a very detailed and insightful synthesis. Below I comments on certain parts of the paper, which relate to some cell biological aspects of the author's model. I focus on those parts where the author's scenarios disagree with some of the scenarios I had proposed on the origin of secretory membranes, predation and the nucleus. Hopefully this discussion will help to improve our models and also help to recognise the merits and weaknesses of the somewhat contrasting scenarios.

Comments about the small GTPase tree

You challenge some of my earlier suggestions that were based on the phylogeny of eukaryotic small GTPases [65]. You write, that "It is evident from other work that the Arf branch is the longest among the eukaryote paralogues and that the small GTPase tree is essentially unresolved at its base [83]; therefore rooting of the tree with the far longer eubacterial outgroup could have wrongly attracted it to the base." Ref. 83 does not show that Arf is the longest branch, and in the tree shown in this paper only Ran is unresolved. The small GTPase tree is also resolved at its base in other analyses (see also my Bayesian tree in [ref. [67]]). Besides, I didn't claim that Arf came first, but that the first split separated Sar1/Arf/SR? from the rest. Sar1/Arf/SR? also forms a distinct group in [ref. [87], which paper suggests independent origin for this branch and the Ras/Rab/Rac/Ran branch. Looking back at my paper [65], I realised that you probably misinterpret the numbers at the nodes of the tree in my Fig. 1[65], which solely indicated the nodes, to simplify discussion about them, and not branching orders.

You also write, that "It is as unwise as for other paralogue trees to assume that the rooting is accurate." Rooting with the closest eukaryotic paralogs, the trimeric G-protein alpha subunits, also gives the same position for the root [67].

Further down you write: "The worst argument, however, was that the Arf-1 branch is involved in secretion and the Rab one in phagocytosis, and to combine these two bad arguments to conclude that exocytosis evolved before phagocytosis [65]." I disagree. It can be safely deduced that the Sar1/Arf/SR? branch ancestrally regulated secretory membranes. Sar1 and SR? still exclusively do this, and Arf1 has indispensable functions in Golgi. The ancestral function is not so clear for the other branch. Later you write, that "almost certainly endocytosis and exocytosis coevolved, recruiting from a common pool of enzymes when assembling their toolkit." I agree with this, but I still think that putting whole-cell phagocytosis first is cell biologically unrealistic. There had to be some preadaptation also in the membrane trafficking system to allow complete cell engulfment.

You also write, that "Thus was born the primary divergence between ER outwards secretory traffic and plasma membrane inwards endocytic traffic seen in GTPase [65] and SNARE [89] trees." Funnily, this is the same over-simplification that you just criticised above as my worst argument (in ref [65]).

Author's reply: *We both agree that 'endocytosis and exocytosis coevolved, recruiting from a common pool of enzymes when assembling their toolkit' and also that it is not easy to infer whether the Rab branch was originally involved in phagocytosis or exocytosis. That being so, it is not safe to infer what was the ancestral role in the common ancestor of both branches. The difference between us may not be that huge. I agree that there must have been much preadaptation in several respects not least with respect to surface membrane properties. I also agree that on the grand scale of things secretion preceded phagocytosis as it is done by all prokaryotes whereas phagocytosis is derived. However secretion and exocytosis are not the same. My 'oversimplification' and your 'worst argument' are not the same - both for this reason and also because my statement about SNAREs is obviously not the same as yours about Rabs. You argued from the Rab tree that exocytosis preceded phagocytosis. I argued that secretion (which ancestrally did not involve exocytosis) preceded both and that primitive phagocytosis marginally preceded exocytosis, but that well-developed exocytosis and phagocytosis must have coevolved; even my statement about both trees mentioned only primary divergence and did not involve a claim that exocytosis came first. With respect to 'putting whole cell phagocytosis first' my recent discussion of the origin of the endomembrane system argues that predation was the primary driving force for eukaryogenesis but that some novel features of the endomembrane system evolved during an earlier phase of extracellular digestion prior to phagocytosis itself, but that (as argued earlier) cell ingestion was what first generated internal membranes* *and thus membranes that could for the first time undergo exocytosis* [27]. *I refer the reader to that paper for details.*

Comments about the nature of the first endomembranes

You also disagree with my secretory membranes-first scenario, although the difference between our scenarios may not be so great as it seems from this criticism, and I think that our models will eventually converge. You step back a little in this paper, but in your recent paper [91] you postulated an early phase of membrane evolution, even discussing the possibility of some tubulation before phagotrophy, as I suggested in [65]. In ref [91] you also speculate on the primacy of ER-linked functions, such as surface secretion of membrane-anchored digestive enzymes and ERAD. In my scenario these evolved on the early membrane tubules, somewhat before total cell engulfment and were preadaptations to allow regulated secretion during phagotrophy and possibly membrane splitting/fusion. The prey might not have been exclusively cellular, as you also write [91], it could have been digested nutrients or viruses and these could have been digested/absorbed by a tubular endomembrane network (and I agree that ERAD is a very plausible mechanism [91]). I would argue that the phagocytosis machinery, which is based on actin, Rac and pseudopodia, evolved independently, although always functionally linked (e.g. via digestive enzyme targeting) to the secretory membranes, which is rather based on microtubule motor, tubules and vesicles. Your figures are still not updated to include these tubules, which would emanate from the prey vesicle in your scheme, or directly from the PM, before total engulfment, according to my scheme.

You also write, that "So we should not ask 'did phagocytosis or exocytosis evolve first?': both evolved together, with phagotrophy being the entirely novel selective advantage, as De Duve [90] and Stanier [2] first argued." Actually Stanier in ref [2] wrote 'endocytosis', and De Duve expressed views, which are more similar to the enzyme secreting, food capturing tubules I described in ref [65], rather than to whole-cell engulfment. According to De Duve [90] endomembranes evolved by "infolding of the cell membrane, allowing the formation of internalized extracellular pockets into which captured food and secreted enzymes were trapped together ... If this hypothetical reconstruction is correct, then the decisive event may have been simply the progressive spreading, with concomitant infolding, of a membrane already adapted for secretion and absorption." Assuming that the food was not cellular, it is easier to imagine such (tubular) invaginations, generated by primitive membrane bending coats and molecular motors pulling on membranes [65]. These membranes could have had both secretory and absorptive function. In these membrane invaginations and networks fusion and fission could evolve, all preadaptations for total cell engulfment. Otherwise I cannot imagine what your prekaryote did when it first engulfed a whole bacterium and topologically separated the phagosome from the plasmamembrane. How did it re-fuse the vacuole if no membrane fusion machinery was present? How did the cell do it on a regular basis to earn a living? I would like to emphasize, that my model also posits that eukaryogenesis was driven by the evolution of predation, but I incorporate some intermediate steps before total cell engulfment, that also make the scenario cell biologically more plausible. The selective logic and a possible social scenario for the evolution of predation is described in (Biol Direct. 2007 Jan 19;2:3.).

Author's reply: *Our thinking is certainly much more similar than either is to some deeply contrasting views that seem to ignore cell biology, especially in that we both see membrane budding and fusion as the most crucial innovation in eukaryogenesis and give a central role to novel selective forces and mechanisms stimulated by predation. You also rightly guessed that the present paper and figures were prepared long before my recent paper that proposes a predatory and partially extracellular digestion phase prior to ingestive phagocytosis* [27]. *That modification to my earlier ideas was made because, like you, I think that some degree of preadaptation is essential in explaining complex innovation and because my proposal also neatly explains many aspects of the evolution of the ERAD machinery that was unknown when I first thought about the origin of phagocytosis. I imagined that some refusion of food vacuoles with the plasma membrane could have occurred accidentally because of an inherent tendency of membranes to refuse, but grant that this would probably be inefficient and that an evolution of phagocytosis without a simultaneous improvement to exocytosis could hardly have occurred; I pointed that out long ago* [11], *and thus have never thought that phagocytosis evolved in detail prior to exocytosis. Thus without a very rapid improvement of refusion by the origin of specific membrane fusion proteins, there would have been a severe bottleneck. This would have given an extremely strong selective advantage to the evolution of exocytosis, stronger I think than in your scenario. The intermediate where I now postulate both were perfected is assumed to have 'made its living' initially by extracellular breakup the cellular prey and import of the proteins, so the transition to engulfing whole cells would make absorption of proteins more efficient and thus not be as radical a change in way of 'earning a* *living' as your question implies; complete internalization would also allow the better exploitation of energy rich lipids. In a chicken-and-egg problem like this it is more likely that exocytosis and endocytosis evolved together rather than either developed alone. I did not complicate the present figures by introducing tubulation because this paper focuses on the later origin of the nucleus, the necessarily prior origin of endomembranes having been discussed in detail in* [27].

I am not against membrane tubulation per se as an intermediate stage; the difference between us is that I invoked it to improve the efficiency of absorption, as did de Duve, whereas you did so to increase the efficiency of secretion at the cell surface, which is less plausible to me. If as you now say such invaginations 'could have had both secretory and absorptive function', we are indeed coming closer; what I objected to was singling out secretion as their primary one. Until at least some membrane budding to internalize membrane had occurred, there would have been no internal membranes that could be externalized by exocytosis or from which exocytotic vesicles could bud and later fuse with the cell surface. Thus exocytosis must have followed **some** *membrane internalization; such internalization would be most simply initiated by a rudimentary form of endocytosis/phagocytosis. If so, the rudiments of phagocytosis at least briefly preceded the rudiments of exocytosis. With respect to history, you misinterpret Stanier's use of 'endocytosis'; at the time he wrote the term included both phagocytosis and pinocytosis (see p. 9 of his paper); it was specifically invented to embrace both; its much more recent use to refer mainly to receptor mediated micropinocytosis in contrast to phagocytosis is a complete distortion of its original meaning. As you probably know, Stanier went on to say (p. 27): 'the capacity for endocytosis would have conferred on its early possessors a new biological means for obtaining nutrients; predation on other cells.' prior to discussing the impact of this on cell structural complexification. There is no doubt that he had in mind mainly phagocytosis rather than pinocytosis as the driving force, just as I did when proposing actin as the key molecular invention enabling phagocytosis and eukaryogenesis* [10].

Comment about the constriction of the NPC cylinder

About the early evolution of the NPC you write, that "Later this wide cylinder, allowing nucleocytoplasmic exchange by passive diffusion (Fig,5a), was constricted to exclude ribosomes from the nucleus" A simple constriction of the proto-NPC cannot exclude ribosomes and other factors from the nucleus. For exclusion FG-repeats are needed. A constriction will only result in more limited diffusion, but not (size) selectivity. Ribosomes cannot be excluded this way, because they are assembled in the nucleus. What such a diffusion barrier could achieve is increased RanGTP levels in the nucleus and a sharper RanGTP gradient (as shown in ref [139]). Everything then follows from this, as nuclear transport evolves. An increase in localized RanGTP was probably the primary reason for constricting the proto-NE and NPCs.

Author's reply: *there is a misunderstanding here; the Figure citation relating to 'constriction' is 5b, which does have the FG repeats. I thought it was clear that in 5a where there are no FG repeats the ribosomes had free passage but in 5b when they are added they do not. To prevent others misinterpreting my meaning I changed 'constricted' to 'constricted by inserting the inner FG repeat ring'. I do not really understand the statement about ribosomes not being excluded because they are assembled in the nucleus; precursor ribosomal subunits are indeed thus assembled, but complete ribosomes with messenger are larger and generally only assembled in the cytoplasm and thus generally are excluded.*

You also write, that "possibly their first cargo was inner-membrane DNA-binding proteins [18], that would otherwise be impeded by the novel integral membrane Nups, unless from the start NPCs opened to let them cross the NPC NE domain" It is not clear that NPCs have to open to allow the crossing of transmembrane cargo. There are clear gaps between the NE and the NPC that may allow the passage of transmembrane proteins (Cell 1992 69:1133-41, Nature 2007 449:611-5, Science, 2008 322:1369-73).

Author's reply: *an interesting comment. However there is evidence that some integral membrane proteins are imported by karyopherins* [397]. *I suspect that my and your proposals about which were the first things to be transported may both be oversimplified guesswork, and that we may never be able to reconstruct with confidence the order in detail.*

Comment about the selective forces during the evolution of the nucleus

I am unconvinced by the kinetic argument for phase 2 of the evolution of the nucleus (the first steps in the evolution of selective transport). In ribosome assembly the rate-limiting step is rRNA transcription and maturation, and is not limited by the diffusion of ribosomal proteins, which is very fast. So even if the cell can concentrate its ribosomal proteins by active nuclear transport, it will not make more ribosomes. The anuclear cell can simply make the amount of ribosomal proteins needed, which will quickly diffuse to the site of ribosome assembly as the rRNA is made. Similarly, during nucleosome assembly the rate limiting step is replication, and concentrating histones doesn't help. So no cost is spared, but it is costly to drive transport by RanGTP. So evolving nuclear transport in order to concentrate ribosomal proteins and histones around DNA is not a good explanation. In my model [ref [139]] the evolution of nuclear transport is not simply about absolute, but relative concentrations (host ribosomal proteins relative to mitochondrial ones).

Author's reply: *As explained previously* [4,5] *but not reiterated here, compartmentation advantages centrally involve the lower cost of maintaining a high concentration of key molecule if they are excluded from the bulk of the cell. By largely excluding free histone, other nuclear proteins and ribosomal proteins from the cytoplasm, roughly 10-fold less of each protein is needed per cell cycle to maintain a given nuclear concentration than if they were evenly spread through the cell. That is a large economy. Thus it is not just a question of rate limitation as you assume. If replication were the rate-limiting step for nucleosome assembly why would there be so many different nucleosome assembly chaperones?*

Comments on the criticism of the ribosome chimerism model

You also criticize my model on the origin of the nucleus [139]. You write, that "First, there is no evidence that the mitochondrion evolved before the nuclear envelope" This I also acknowledge in the paper. However, you also note, that "We now see that the origins of mitochondria, nuclei, and cilia were virtually simultaneous. As each was a complex series of processes it is most likely that they overlapped in time, so disputing which was first may be meaningless." So, simply, we don't know the order. My theory is based on the assumption that the mitochondrium came slightly earlier, than the nucleus, but clearly after phagotrophy evolved. You also write, that "Secondly, it is not very logical to suppose that the transfer of genes could be tolerated with impunity for a period and then became a compelling force for evolving the envelope." In my theory the nucleus evolve to correct the bad effects of mutation pressure. It is the same logic as the one you apply to explain the origin of NMD: "Thus like splicing itself, and many genomic oddities like RNA editing [105], nonsense-mediated decay is a consequence of selfish DNA evolution; correcting the bad effects of mutation pressure, not positively adaptive." I say the same about the gene transfer ratchet of mitochondrial ribosomal proteins (slightly deleterious mutations fixed by drift), and when the harm is done, the problem is fixed by compensatory advantageous mutations as the nucleus evolves. Since there are 24 individual transfers, the harm is done in small steps, and compensation can also evolve in small steps.

You also write, that "If transfer were so strongly disadvantageous, it would be more likely either that cells with transferred genes would lose out in competition" It was probably not more strongly disadvantageous, than the shearing forces you postulate. This was collateral damage, caused by the evolution of other cellular features, which were either selected for or spread by drift, and then had to be fixed. Further on you write, that "Thirdly, just as for the intron harm theory criticised above, the early stages in nuclear envelope assembly could not have helped solve the problem" This is not a valid criticism either. Figure Figure22 and and33 in the paper [139] clearly show how the first stages could have reduced the problem. Actually, I can even give an explanation for the origin of the RanGTP gradient around chromatin, which can create a compositionally distinct region of the cytoplasm, which is not explained by invoking only shearing forces, as you do. Later you write: «Fourthly, in the final stages of evolution of nuclear import - increased selectivity - the benefit would have been the net benefit of excluding hundreds of different proteins from the nucleus and the positive import of hundreds of others.» I agree, and I also mention [139] that in the final stages many transport processes and cargos had to coevolve.

Author's reply: *I am not against the principle of phenotypic correction of unavoidable mutation pressure being important in cell evolution, having invoked it several times myself, e.g. in the origin of spliceosomal introns, RNA editing, chloroplast DNA minicircles, and even with respect to ribosomal evolution as a consequence of the chimaera problem* [3,211,398,399]. *The difference between us is primarily over the likely selective disadvantages and advantages of the steps discussed. I apologize for not having properly appreciated your simulation studies, which I agree refute my original third objection to your ideas. I have therefore deleted it, making the fourth objection the third. I have also revised this objection to cast it in the form of strong support for your general thinking of how a degree of compartmentation could have been achieved prior to complete pores combined with scepticism over the special importance of avoiding chimaerism. We are both agreed in the central importance of compartmentation and coevolution for the origin of the nucleus, but I think it likely that there were many simultaneous benefits of compartmentation.*

Minor comments:

The conservation of EJC-dependent NMD in plants has also been demonstrated experimentally (EMBO J 2008 27:1585-95).

The structural similarity of COPII and NPC has recently been demonstrated (Science, 2008 322:1369-73).

Author's reply: *references added.*

The section "Origins of nuclear protein modification by sumoylation" does not really fit into the Discussion. Sumoylation is not mentioned in the main text. It would be better to include it into the Results section.

Author's reply: *moved.*

Eugene Koonin

'Intracellular coevolution and the origin of the cell nucleus and sex'

This is a very far-reaching, extensive discussion of a paramount problem in evolutionary biology, at least, with regard to the evolution of eukaryotes, the origin of the nucleus and sex.

The article is quite long but is overall an excellent read. Furthermore, I am very sympathetic with the succinct conclusion of the abstract on the importance of studying coevolution of different eukaryotic organelles for understanding eukaryogenesis. Yes, I think such a systemic approach is indeed key.

I am afraid, however, that this is where I have to stop with my comments. The text of the article, interesting as it undoubtedly is, does not appear to be an objective discussion of the problem, but rather a one-sided narrative that I am not inclined to analyze and criticize in detail. The show stopper to me is that the "painted picture" is based on several major assumptions that are accepted here as unquestionable but that I find either highly controversial or outright implausible. The highly controversial assumptions are the archaezoan nature of the proto-eukaryotes that was the host of the mitochondrial endosymbiont, that is, the assertion that this organism was a bona fide phagotroph that possessed the principal eukaryotic features such as the endomembrane system, the cytoskeleton, and the nucleus itself; and the bikont-unikont phylogeny of eukaryotes.

Author's response: *The core subject of this paper is not the origin of mitochondria, treated in detail elsewhere* [20], *but new proposals on the physical mechanism and selective forces causing the origin of the nucleus. These are entirely independent of whether or not the eukaryote tree is rooted between unikonts and bikonts, as the original version of this paper assumed, or not. Indeed, new evidence suggests that that rooting was mistaken* [9,55] *and that the root is instead within Eozoa, specifically between excavates and Euglenozoa (or possibly even within Euglenozoa) as* *explained elsewhere* [9]. *But this does not in any way affect the mechanistic and selective arguments proposed here, which are independent of the precise rooting. As I stated, no reasonable alternative alters the conclusion 'that the last common ancestor of all eukaryotes was a phagotrophic protozoan with nucleus, at least one centriole and cilium, facultatively aerobic mitochondria, sex (meiosis and syngamy) and dormant cyst with cell wall of chitin and/or cellulose, and peroxisomes'.*

The referee inaccurately states that I asserted that the host for the origin of the mitochondrion was a fully developed eukaryote with a nucleus. I did not. I said it was either such a well-developed eukaryote or else a somewhat earlier 'intermediate stage' that already had evolved 'rudiments of phagocytosis and endomembranes'. My arguments as to the mechanistic cause of the origin of the nucleus depend (as do most scenarios for its origin) on the prior evolution of the endomembrane system, but are independent of whether mitochondria were acquired after the origin of the nucleus (as in the disproved archezoan hypothesis [400]*), before the origin of the nucleus (as Martin and Koonin postulated* [21]*)or temporally overlapping with the origin of the nucleus. My positive arguments for the selective forces that favoured the origin of the nucleus in successive stages are also independent of the relative timing of nuclear and mitochondrial origins. My negative arguments against the prior acquisition of mitochondria and of introns derived from them as being a plausible selective force for the initiation of the origin of the nucleus are also independent of the historical fact of whether the mitochondrion came before, after, or during the origin of the nucleus. Thus even were it true that the mitochondrion came first, I would not consider it plausible that the origin of spliceosomal introns was the primary selective cause of the origin of the nucleus. It is also misleading to refer to my present views as favouring an archezoan host, as for over a decade I have rejected my earlier hypothesis that there are extant primitively mitochondrial eukaryotes (Archezoa) as all putative Archezoa turned out to be secondary anaerobes with relict degenerate mitochondria.*

To refer to the idea that the host for mitochondrial enslavement was a bona fide phagotroph as 'highly controversial' is misleading and tendentious. Unless you place the root within non-phagotrophic Plantae (which the single symbiogenetic origin of chloroplasts forbids) or Fungi (which much evidence contradicts), we can safely conclude that the last common ancestor of eukaryotes was a bona fide aerobic and sexual protozoan phagotroph with all standard eukaryotic organelles. That its earlier ancestor which engulfed the ancestor of mitochondria was also a phagotroph obviously cannot be stated with equal confidence. Nonetheless, as no free-living bacteria (=prokaryotes) have ever been shown to internalize a foreign cell and maintain it as a symbiont, the alternative assumption [21] *that the host that thus internally enslaved an proteobacterium was a prokaryote is mechanistically implausible and much less justifiable than to suppose that the host was a phagotrophic prekaryote or protoeukaryote that regularly engulfed prey cells. Phagocytosis provides the easiest and most widespread mechanism whereby symbiotic cells enter host cells. Intracellular symbiosis is so widespread in eukaryotes mainly (not entirely) because of the existence of phagocytosis; billions of cells daily are thus taken up by free-living eukaryotes. Conversely there are no examples of free-living prokaryotes able to engulf other cells or harbour intracellular symbionts (the only known example of intracellular symbiosis within a prokaryote is a parasite of eukaryotes); thus there is no evidence that a free-living prokaryote did so even once during 3.5 billion years. It is therefore immensely more likely that mitochondria entered eukaryotes after the origin of at least a rudimentary phagocytosis. Furthermore there is no phylogenetic evidence that mitochondria were taken up before the origin of the nucleus. To assume that they were is phylogenetically gratuitous and to assume that the host was a non-phagotrophic bacterium is mechanistically extremely implausible. Therefore it is not scientifically sound to invoke the formal possibility that mitochondria came before phagocytosis in evolutionary explanations; unless new evidence or compelling arguments to the contrary are adduced, it is more parsimonious and more likely that phagocytosis preceded mitochondrial acquisition.*

Evolutionary explanations should distinguish between mere formal possibilities and genuinely likely events based on known properties of organisms. A remote formal possibility is not a valid explanation of an historical fact if a simpler one based on known phenomena exists. Therefore it is the alternative invocation of an unknown, unprecedented mechanism for engulfment by a free-living prokaryotic host (e.g. by [21]*) that is unwarranted and 'highly controversial', not the idea that at least a rudimentary version of phagocytosis had probably evolved prior to the origin of mitochondria, which is mechanistically much sounder and phylogenetically entirely acceptable.*

Because there are no phagotrophic prokaryotes it is unlikely that phagotrophy would have long preceded the origin of the nucleus (if such organisms existed for millions of years there is no reason why they should all now be extinct as the niches would still be available), and because there appear to be no primitively amitochondrial eukaryotes (if such organisms existed for millions of years there is no reason why they should all now be extinct as the niches would still be available), it is equally unlikely that the acquisition of mitochondria long preceded that of the origin of nuclei. Therefore for some years, in contrast to my earlier, now disproved, archezoan hypothesis [60,400,401], *I have considered that the origins of phagocytosis, mitochondria, and nuclei were most likely essentially contemporaneous* [3,27]*and that we are unlikely ever to be able to order the two latter with certainty in time. All these events may have occupied well under a million years* [27].

An implausible one, as far as I can judge, is the "neomuran" origin or archaea (archaebacteria, under the terminology used here). All these assumptions are accepted as statements of fact, largely, on the force of previous publications, without much elaboration. My understanding, however, is that the arguments in those publications were seriously flawed, so I cannot accept the assumptions.

Author's response: *I wish you had said what you think the flaws are so that we could see if your objections have any weight. As you agree with me that archaebacteria are holophyletic sisters to eukaryotes* [24], *I do not see how you can logically object to my deducing that characters shared by both groups were present in their common ancestor, whereas characters unique to each probably arose only in their individual last common ancestors. The latter means that with respect to lipids the ancestor of neomura had acyl esters and was thus eubacterial not archaebacterial in nature. From your previous writings, I assume that you do not like the idea that eubacteria are older than both eukaryotes and archaebacteria, but you ought to consider both the palaeontological and transition analysis arguments for the greater antiquity of eubacteria and if you disagree for definite reasons should attempt to refute them. I refer readers to the latest discussion by Valas and Bourne* [41] *on the position of the root of the tree of life, which concludes that available evidence most strongly supports my conclusion, and that the post-1977 mindset of so many microbiologists of assuming the antiquity of archaebacteria without proper evaluation of the evidence is probably mistaken.*

I did not present this view dogmatically but did so only briefly 'relying on previous arguments'. As the arguments in my previous papers on this topic all remain valid and have not been refuted, and as the logic of my explanation for the rooting of the tree (both mechanisms and selective forces) is valid independently of the rooting of the overall tree of life, about which I had few novel arguments, I simply cited those in my earlier papers. As the referee criticises my doing that and as the root position is important for another aspect of the origin of the nucleus (the phylogenetic origin of each key protein), I have now introduced two even more compelling new arguments from lipid evolution for the posibacterial ancestry of neomura and summarise the overall logic below (readers familiar with this can skip the next two paragraphs):

First is the phylogenetic argument that archaebacteria are holophyletic sisters to eukaryotes, not their paraphyletic ancestors. I originally based that argument on (1) the view that the unique lipids of archaebacteria are a shared derived character of archaebacteria alone and that the acyl ester lipids of eukaryotes are an ancestral eubacterial character vertically inherited by the host from an ultimately eubacterial ancestor, plus (2) the palaeontological fact that eubacteria are much older than eukaryotes but no unambiguous fossil evidence exists that archaebacteria are any older than eukaryotes, together with (3) the view that there is no mechanistically plausible way that the far simpler prokaryotic cells could be secondarily derived from eukaryotes, which was congruent with the fossil evidence that prokaryotes, specifically eubacteria, are older [4]. *This original argument was made somewhat less decisive by the demonstration that all extant archezoa are secondarily derived from aerobic mitochondriate ancestors, as this meant that the origin of the nucleus and mitochondrion must have roughly coincided, raising the formal possibility that the eukaryotic acyl ester lipids were derived by lipid replacement from the mitochondrial rather than the host ancestor* [31]. *Elsewhere I explained why that formal possibility is mechanistically and selectively extremely implausible* [3]. *At that time I introduced a new argument that three gene splits found in archaebacteria alone, which are unlikely to have been reversed in the last common ancestor of eukaryotes, are additional evidence for the holophyly of archaebacteria* [12]. *Two new arguments introduced here are that Martin's idea of host lipid replacement by the mitochondrion* [31] *cannot explain the origin of phosphatidylinositol or sterols; phosphatidylinositol was crucial for eukaryogenesis* [27] *but is totally absent from both archaebacteria and proteobacteria but is universally present in actinobacteria, the likely ancestors of neomura; and the enzymes for making sterols are very widespread in actinobacteria and were not acquired by LGT from eukaryotes. Thus neomura must either have evolved from actinobacteria or be their sisters as suggested here. The latest trees from the referee›s own group based on 355 genes argue much more strongly for the holophyly of archaebacteria than did any previous trees* [24]. *Thus, despite conflicting/ ambiguous but unconvincing evidence from some gene trees* [46], *evidence for archaebacterial holophyly is stronger than ever.*

If archaebacteria are indeed holophyletic, one cannot justifiably infer that the last common ancestor of archaebacteria and eukaryotes had any of the characters that are unique to archaebacteria or any that are unique to eukaryotes. Instead it is most parsimonious to suppose that this common ancestor had all the properties shared by archaebacteria and eukaryotes, plus all those shared by archaebacteria and eubacteria alone (i.e. prokaryotic genome and cell structure), plus all those shared by eukaryotes and the phylogenetically older eubacteria (i.e. acyl ester lipids, and dozens of genes ancestrally absent in archaebacteria such as Hsp70). In other words this last common ancestor of eukaryotes and archaebacteria (collectively neomura) was a not an archaebacterium, because on the most parsimonious assumptions it had acyl ester lipids like a eubacterium and prokaryotic cell structure. It resembled a posibacterial eubacterium because it had a single membrane not a double bounding envelope as do all other eubacteria (negibacteria). But it would be equally incorrect to call it a eubacterium as it would have had N-linked glycoproteins on its surface and presumably had already lost the peptidoglycan wall. Thus the last common ancestor of neomura was prokaryotic, but neither an archaebacterium nor a posibacterial eubacterium, but a missing link with some properties of each that was most probably derived from a posibacterium prior to the origin of either eukaryotic or archaebacteria-specific characters. That is why it is misleading to use modern names of extant groups to label such inferred intermediates; better call it an ancestral neomuran. The ancestor of the neomuran clade prior to the replacement of the peptidoglycan by glycoprotein and the origin of histones was a eubacterium, specifically a stem actinobacterium that still retained some key endobacterial properties that persist in neomura.

As the referee does not give any reasons why these arguments for a posibacterial/neomuran origin of archaebacteria are 'implausible' or 'deeply flawed' or cite any from the literature (I am aware of none that have been clearly laid out) I cannot do more to defend them than I did in previous publications. The referee does not mention any better explanation of the ancestry of Archaebacteria than that given previously [4,12]. *The widespread assumption that they came from a mythical 'progenote' devoid of specific properties* [37], *which he presumably shares, is no more informative or better based than the idea of special creation.*

One important discovery from the referee's research [402] *illuminates the nature of the cenancestral neomuran further. They have shown that crenarchaeote and korarchaeote archaebacteria possess actin-like proteins that are more similar to actin and actin-related proteins (Arp 2 and 3) than are eubacterial MreB and ParM proteins. This implies that certain key changes that differentiate actin from MreB (and related eubacterial proteins ParM and MamK) occurred in the neomuran ancestor prior to the divergence of eukaryotes and archaebacteria. It is reasonable to suppose that these were associated with the loss of the peptidoglycan wall and the development instead of an improved internal cell skeleton for osmotic protection and mechanical stability. This finding does not alter the fact that the gene duplications that generated actin and the related Arp2/3 which nucleate its polymerization and mediate branching took place* ***after*** *the pre-eukaryotic lineage diverged from archaebacteria. This is consistent with the argument that this major step in actin evolution was associated with the origin of phagocytosis* [6,27]. *Thus archaebacteria lack this key feature of the actin skeleton; a three dimensionally cross-linked actin gel remains unique to eukaryotes. It was misleading to write of the archaeal ancestry of actins* [402] *as the tree for their actin-related proteins provides no evidence that crenarchaeotes are ancestral to eukaryotes; instead they appear as their sisters. Thus the novel features shared by actin/Arp2/3 and crenarchaeote actin-related protein (notably large related insertions near the C-terminal) are most parsimoniously interpreted as having evolved in the ancestral neomuran. Their absence in euryarchaeotes can be attributed to secondary loss. Very likely during the neomuran burst of gene duplication of MreB/actin-related proteins a number of paralogues was generated; one would have been ancestral to actin/Arp2/3 and crenarchaeote actin and a more conservative one to the methanogenic euryarchaeote MreB. Additional paralogue loss within crenarchaeotes can explain the findings of Yutin et al.* [207]*more simply than lateral transfers and protein replacements that they invoke between crenarchaeotes and korarchaeotes.*

Their assumption of lateral gene transfer of MreB from eubacteria to methanogenic archaebacteria is also entirely unnecessary if one accepts the rooting of the tree of life within eubacteria [4,6,12,13,40,47,48,50,381]; *vertical transmission from posibacteria to archaebacteria alone would suffice. Their assumption of lateral transfer to Thermoplasma is a bit more plausible because such transfer may be inherently more likely for a plasmid encoded paralogue like ParM than for MreB, but is also probably unnecessary. It is not obvious that its placement on the tree within the ParM/StpA clade is correct; I suspect it is a long branch artefact (LBA): a strong divergence in Thermoplasma actin-like protein caused by the loss of the eukaryote cell wall and ensuing selection for a better internal skeleton would have made it so divergent from its ancestral type (whether MreB or the crenarchaeote/actin paralogue) that it might not group correctly with them. It is evident from the tree that ParM evolves much faster than MreB. The alignment indicates that Thermoplasma protein is very idiosyncratic compared with all others in respect of both indels and sequence - just the sort of protein likely to be misplaced on trees; I do not think one can reliably distinguish between 3-5 possibilities: (1) that it is really sister to the crenarchaeote protein, but misplaced on the tree by LBA, (2) that it is really a ParM protein (descended either vertically from the neomuran ancestor or by LGT from another eubacterium) or (4) that it is an independent derivative of an MreB ancestor (descended either vertically or by LGT, but if the latter not necessarily from a eubacterium). Nonetheless the first possibility seems simplest.*

Portrayal of the last common ancestor of eukaryotes by the referee's group [402] *as a cell with a branched actin skeleton and mitochondrion but no nucleus or cilium was wrong. No matter where one roots the eukaryotic tree the last common eukaryotic ancestor had both nucleus and mitochondrion. Provided that one places the root somewhere between the five supergroups shown it is inescapable that it also had a cilium; that figure confused the last common ancestor that we can rigorously infer cladistically with a purely hypothetical intermediate between prokaryotes and eukaryotes. As Jékely pointed out, the implication that Rhizaria may lack phagocytosis is entirely erroneous; the group has two phyla: Retaria (Foraminifera and Radiozoa, in which every single one of the over 10,000 species feeds by phagocytosis) and Cercozoa in which all but a handful of the thousands of free-living species feed by phagocytosis. As referees Gribaldo and Jékely correctly argued, the conclusion that phagocytosis was absent from the cenancestral eukaryote and evolved several times independently in different eukaryotic supergroups is entirely unjustified. The authors conceded that the actin branching and control machinery was present in the last common ancestor of eukaryotes and thus accept that molecules of this core machinery like Arp2/3 which have not been detected in diplomonads must either have been lost or evolved dramatically in sequence beyond bioinformatic detection in a diplomonad ancestor. It is perfectly reasonable to argue that the same is true of other near universal eukaryotic proteins, e.g. Jékely also pointed out that the assumption that Ras was derived from mitochondria was not justified. Thus Yutin et al.* [402]*provided no new convincing evidence for their assumption that mitochondria arose before the nucleus, nor any reason for their calling the hypothetical primitively phagotrophic intermediate an archaebacterium rather than an intermediate pre-eukaryotic derivative of the ancestral neomuran. The first referee of the present paper shares my scepticism of their basic thesis, calling it 'specious at best'.*

Once again, this is interesting reading that reveals remarkable erudition of the author, a lot of interesting literature is cited, and the reader will benefit from a variety of exposition to an important and exciting research field, and possibly, from some of the ideas proposed in the manuscript. However, for reasons outlined above, I find that this manuscript does not call for serious discussion, so I offer none such.

Author's response: *A pity you did not give any reasons for your disagreement. It is important that those many who do not accept the phylogenetically older character of eubacteria compared with both eukaryotes and archaebacteria seriously consider (and not just ignore) the now quite compelling and varied evidence for this, which rests on a combination of palaeontological and phylogenetic evidence, both critically interpreted* [6,12,13,41]. *To establish an historical fact like the position of the root, we need arguments and evidence, not unjustified assumptions. Your assertion when reviewing another paper on the position of the root of the universal tree* [41] *that the sheer magnitude of the differences between archaebacteria and eubacteria is sufficient to place the root between neomura and eubacteria and to ‹close› the debate on the position of the root and that you do not need to go into ‹the minute details› of the evidence was phylogenetically unsound and not a good scientific attitude. Thank you, however, for being open-minded enough to recommend publication despite your disagreement. Please try to open your mind still further to the arguments and evidence* [4,6,12,13,23,41,403] *that refute this idea stemming from the extremely weak speculations of Woese and Fox* [404]. *Alternatively, please explicitly explain any errors in the reasoning concerning the roughly 20 polarizations from eubacteria to neomura* [12] *and 13 additional ones within eubacteria* [6,13]; *i.e. over 30 independent arguments that are mutually consistent evidence for the root being in Eobacteria. No similarly coherent integration of palaeontology and cell and molecular evolution has ever been presented in support of their now pervasive but almost certainly* *erroneous assumption (it is nothing more) that archaebacteria are as ancient as eubacteria.*

ACKNOWLEDGEMENTS

I thank NERC for grant and fellowship support.

REFERENCES

1. Cavalier-Smith T. A revised six-kingdom system of life. *Biol Rev Camb Philos Soc.* 1998;73(3):203–266. doi: 10.1017/S0006323198005167.
2. Stanier RY. In: *Organization and Control in Prokaryotic and Eukaryotic Cells (Society for General Microbiology Symposium no 20)* Charles HP, Knight JG, editor. Cambridge: Cambridge University Press; 1970. Some aspects of the biology of cells and their possible evolutionary significance; pp. 1–38.
3. Cavalier-Smith T. The phagotrophic origin of eukaryotes and phylogenetic classification of Protozoa. *Int J Syst Evol Microbiol.* 2002;52(Pt 2):297–354.
4. Cavalier-Smith T. The origin of eukaryotic and archaebacterial cells. *Ann N Y Acad Sci.* 1987;503:17–54. doi: 10.1111/j.1749-6632.1987.tb40596.x.
5. Cavalier-Smith T. Origin of the cell nucleus. *BioEssays.* 1988;9(2-3):72–78. doi: 10.1002/bies.950090209.
6. Cavalier-Smith T. Cell evolution and Earth history: stasis and revolution. *Philos Trans R Soc Lond B Biol Sci.* 2006;361(1470):969–1006. doi: 10.1098/rstb.2006.1842.
7. Richards TA, Cavalier-Smith T. Myosin domain evolution and the primary divergence of eukaryotes. *Nature.* 2005;436:1113–1118. doi: 10.1038/nature03949.
8. Stechmann A, Cavalier Smith T. The root of the eukaryote tree pinpointed. *Curr Biol.* 2003;13(17):R665–R666. doi: 10.1016/S0960-9822(03)00602-X.
9. Cavalier-Smith T. Kingdoms Protozoa and Chromista and the eozoan root of eukaryotes. *Biol Lett.* 2009. in press . [PubMed]
10. Simpson GC. *The Major Features of Evolution.* New York: Columbia University Press; 1953.
11. Cavalier-Smith T. The origin of nuclei and of eukaryotic cells. *Nature.* 1975;256:463–468. doi: 10.1038/256463a0.
12. Cavalier-Smith T. The neomuran origin of archaebacteria, the negibacterial root of the universal tree and bacterial megaclassification. *Int J Syst Evol Microbiol.* 2002;52(Pt 1):7–76.
13. Cavalier-Smith T. Rooting the tree of life by transition analysis. *Biology Direct.* 2006;1:19. doi: 10.1186/1745-6150-1-19.

14. Bapteste E, Charlebois RL, MacLeod D, Brochier C. The two tempos of nuclear pore complex evolution: highly adapting proteins in an ancient frozen structure. *Genome Biol.* 2005;6:R85. doi: 10.1186/gb-2005-6-10-r85.
15. Devos D, Dokudovskaya S, Alber F, Williams R, Chait BT, Sali A, Rout MP. Components of coated vesicles and nuclear pore complexes share a common molecular architecture. *PLoS Biol.* 2004;2(12):e380. doi: 10.1371/journal.pbio.0020380.
16. Devos D, Dokudovskaya S, Williams R, Alber F, Eswar N, Chait BT, Rout MP, Sali A. Simple fold composition and modular architecture of the nuclear pore complex. *Proc Natl Acad Sci USA.* 2006;103:2172–2177. doi: 10.1073/pnas.0506345103.
17. Mans BJ, Anantharaman V, Aravind L, Koonin EV. Comparative genomics, evolution and origins of the nuclear envelope and nuclear pore complex. *Cell Cycle.* 2004;3:1612–1637.
18. Cavalier-Smith T. Economy, speed and size matter: evolutionary forces driving nuclear genome miniaturization and expansion. *Ann Bot.* 2005;95:147–175. doi: 10.1093/aob/mci010.
19. King MC, Lusk CP, Blobel G. Karyopherin-mediated import of integral inner nuclear membrane proteins. *Nature.* 2006;442:1003–1007. doi: 10.1038/nature05075.
20. Cavalier-Smith T. Origin of mitochondria by intracellular enslavement of a photosynthetic purple bacterium. *Proc Biol Sci.* . 2006;273(1596):1943–1952. doi: 10.1098/rspb.2006.3531.
21. Martin W, Koonin EV. Introns and the origin of nucleus-cytosol compartmentalization. *Nature.* 2006;440:41–45. doi: 10.1038/nature04531.
22. Cavalier-Smith T. In: *Origin of Mitochondria and Hydrogenosomes.* Martin WF, Müller M, editor. Berlin: Springer-Verlag; 2007. The chimaeric origin of mitochondria: photosynthetic cell enslavement, gene-transfer pressure, and compartmentation efficiency; pp. 161–199. full_text.
23. Cavalier-Smith T. Deep phylogeny, ancestral groups, and the four ages of life. *Philos Trans R Soc Lond B Biol Sci.* . 2010;365(1537):111–132. doi: 10.1098/rstb.2009.0161.
24. Yutin N, Makarova KS, Mekhedov SL, Wolf YI, Koonin EV. The deep archaeal roots of eukaryotes. *Mol Biol Evol.* 2008;25(8):1619–1630.

doi: 10.1093/molbev/msn108.

25. Rogozin IB, Makarova KS, Pavlov YI, Koonin EV. A highly conserved family of inactivated archaeal B family DNA polymerases. *Biol Direct.* 2008;3(1):32. doi: 10.1186/1745-6150-3-32.
26. Martin W, Müller M. The hydrogen hypothesis for the first eukaryote. *Nature.* 1998;392:37–44. doi: 10.1038/32096.
27. Cavalier-Smith T. Predation and eukaryote cell origins: a coevolutionary perspective. *Int J Biochem Cell Biol.* 2009;41:307–322. doi: 10.1016/j.biocel.2008.10.002.
28. Boucher Y, Kamekura M, Doolittle WF. Origins and evolution of isoprenoid lipid biosynthesis in archaea. *Mol Microbiol.* 2004;52(2):515–527. doi: 10.1111/j.1365-2958.2004.03992.x.
29. Peretó J, López-García P, Moreira D. Ancestral lipid biosynthesis and early membrane evolution. *Trends Biochem Sci.* 2004;29:469–477. doi: 10.1016/j.tibs.2004.07.002.
30. Van Valen LM, Maiorana VC. The archaebacteria and eukaryotic origins. *Nature.* 1980;287:248–250. doi: 10.1038/287248a0.
31. Martin W. A briefly argued case that mitochondria and plastids are descendants of endosymbionts, but that the nuclear compartment is not. *Philos Trans R Soc Lond B Biol Sci.* 1999;266:1387–1395. doi: 10.1098/rspb.1999.0792. [CrossRef]
32. Margulis L. *Origin of eukaryotic cells.* New Haven: Yale University Press; 1970.
33. Margulis L. *Symbiosis in cell evolution.* San Francisco: Freeman; 1981.
34. Desmond E, Gribaldo S. Phylogenomics of sterol synthesis: insights into the origin, evolution and diversity of a key eukaryotic feature. *Genome Biol Evol.* 2009;2009:364. doi: 10.1093/gbe/evp036.
35. Summons RE, Bradley AS, Jahnke LL, Waldbauer JR. Steroids, triterpenoids and molecular oxygen. *Philos Trans R Soc Lond B Biol Sci.* . 2006;361(1470):951–968. doi: 10.1098/rstb.2006.1837.
36. Martin W, Russell MJ. On the origins of cells: a hypothesis for the evolutionary transitions from abiotic geochemistry to chemoautotrophic prokaryotes, and from prokaryotes to nucleated cells. *Philos Trans R Soc Lond B Biol Sci.* 2003;358(1429):59–85. doi: 10.1098/rstb.2002.1183.
37. Woese CR, Fox GE. The concept of cellular evolution. *J Mol Evol.* 1977;10:1–6. doi: 10.1007/BF01796132.

38. Gao B, Sugiman-Marangos S, Junop MS, Gupta RS. Structural and phylogenetic analysis of a conserved actinobacteria-specific protein (ASP1; SCO1997) from *Streptomyces coelicolor*. *BMC Struct Biol.* 2009;9:40. doi: 10.1186/1472-6807-9-40.
39. Santos CL, Vieira J, Tavares F, Benson DR, Tisa LS, Berry AM, Moradas-Ferreira P, Normand P. On the nature of fur evolution: a phylogenetic approach in Actinobacteria. *BMC Evol Biol.* 2008;8:185. doi: 10.1186/1471-2148-8-185.
40. Skophammer RG, Servin JA, Herbold CW, Lake JA. Evidence for a gram-positive, eubacterial root of the tree of life. *Mol Biol Evol.* 2007;24(8):1761–1768. doi: 10.1093/molbev/msm096.
41. Valas RE, Bourne PE. Structural analysis of polarizing indels: an emerging consensus on the root of the tree of life. *Biol Direct.* 2009;4:30. doi: 10.1186/1745-6150-4-30.
42. Servin JA, Herbold CW, Skophammer RG, Lake JA. Evidence excluding the root of the tree of life from the actinobacteria. *Mol Biol Evol.* 2008;25(1):1–4. doi: 10.1093/molbev/msm249.
43. Forterre P, Gadelle D. Phylogenomics of DNA topoisomerases: their origin and putative roles in the emergence of modern organisms. *Nucleic Acids Res.* 2009;37(3):679–692. doi: 10.1093/nar/gkp032.
44. Forterre P, Gribaldo S, Gadelle D, Serre MC. Origin and evolution of DNA topoisomerases. *Biochimie.* 2007;89(4):427–446. doi: 10.1016/j.biochi.2006.12.009.
45. Cox CJ, Foster PG, Hirt RP, Harris SR, Embley TM. The archaebacterial origin of eukaryotes. *Proc Natl Acad Sci USA.* 2008;105(51):20356–20361. doi: 10.1073/pnas.0810647105.
46. Foster PG, Cox CJ, Embley TM. The primary divisions of life: a phylogenomic approach employing composition-heterogeneous methods. *Philos Trans R Soc Lond B Biol Sci.* 2009;364(1527):2197–2207. doi: 10.1098/rstb.2009.0034.
47. Lake JA, Skophammer RG, Herbold CW, Servin JA. Genome beginnings: rooting the tree of life. *Philos Trans R Soc Lond B Biol Sci.* 2009;364(1527):2177–2185. doi: 10.1098/rstb.2009.0035.
48. Lake JA, Servin JA, Herbold CW, Skophammer RG. Evidence for a new root of the tree of life. *Syst Biol.* 2008;57(6):835–843. doi: 10.1080/10635150802555933.
49. Lake JA, Herbold CW, Rivera MC, Servin JA, Skophammer

RG. Rooting the tree of life using nonubiquitous genes. *Mol Biol Evol.* 2007;24(1):130–136. doi: 10.1093/molbev/msl140.

50. Skophammer RG, Herbold CW, Rivera MC, Servin JA, Lake JA. Evidence that the root of the tree of life is not within the Archaea. *Mol Biol Evol.* 2006;23(9):1648–1651. doi: 10.1093/molbev/msl046.
51. Lake JA, Rivera MC. Deriving the genomic tree of life in the presence of horizontal gene transfer: conditioned reconstruction. *Mol Biol Evol.* 2004;21(4):681–690. doi: 10.1093/molbev/msh061.
52. Kurland CG, Collins LJ, Penny D. Genomics and the irreducible nature of eukaryote cells. *Science.* 2006;312(5776):1011–1014. doi: 10.1126/science.1121674.
53. Hampl V, Silberman JD, Stechmann A, Diaz-Trivino S, Johnson PJ, Roger AJ. Genetic evidence for a mitochondriate ancestry in the 'amitochondriate' flagellate *Trimastix pyriformis*. *PLoS One.* 2008;3(1):e1383. doi: 10.1371/journal.pone.0001383.
54. Dagley MJ, Dolezal P, Likic VA, Smid O, Purcell AW, Buchanan SK, Tachezy J, Lithgow T. The protein import channel in the outer mitosomal membrane of *Giardia intestinalis*. *Mol Biol Evol.* 2009;26(9):1941–1947. doi: 10.1093/molbev/msp117.
55. Roger A, Simpson AGB. Evolution: revisiting the root of the eukaryotic tree. *Curr Biol.* 2009;19:R165–167. doi: 10.1016/j.cub.2008.12.032.
56. Kim E, Simpson AG, Graham LE. Evolutionary relationships of apusomonads inferred from taxon-rich analyses of 6 nuclear encoded genes. *Mol Biol Evol.* 2006;23(12):2455–2166. doi: 10.1093/molbev/msl120.
57. Stechmann A, Cavalier-Smith T. Rooting the eukaryote tree by using a derived gene fusion. *Science.* 2002;297(5578):89–91. doi: 10.1126/science.1071196.
58. Cavalier-Smith T. In: *The Flagellates.* Green JC, Leadbeater BSC, editor. London: Taylor and Francis; 2000. Flagellate megaevolution: the basis for eukaryote diversification; pp. 361–390.
59. Clayton C, Shapira M. Post-transcriptional regulation of gene expression in trypanosomes and leishmanias. *Mol Biochem Parasitol.* 2007;156(2):93–101. doi: 10.1016/j.molbiopara.2007.07.007.
60. Cavalier-Smith T. The simultaneous symbiotic origin of mitochondria, chloroplasts, and microbodies. *Ann N Y Acad Sci.* 1987;503:55–71.

doi: 10.1111/j.1749-6632.1987.tb40597.x.

61. Cavalier-Smith T. Genomic reduction and evolution of novel genetic membranes and protein-targeting machinery in eukaryote-eukaryote chimaeras (meta-algae) *Philos Trans R Soc Lond B Biol Sci.* . 2003;358(1429):109–133. doi: 10.1098/rstb.2002.1194.
62. Cavalier-Smith T. In: *Organelles, genomes and eukaryote phylogeny.* Hirt RP, Horner DS, editor. London: CRC Press; 2004. Chromalveolate diversity and cell megaevolution: interplay of membranes, genomes and cytoskeleton; pp. 75–108.
63. Cavalier-Smith T. Origins of the machinery of recombination and sex. *Heredity.* 2002;88(2):125–141. doi: 10.1038/sj.hdy.6800034.
64. Gibson W, Peacock L, Ferris V, Williams K, Bailey M. The use of yellow fluorescent hybrids to indicate mating in *Trypanosoma brucei. Parasit Vectors.* 2008;1(1):4. doi: 10.1186/1756-3305-1-4.
65. Gibson W, Peacock L, Ferris V, Williams K, Bailey M. Analysis of a cross between green and red fluorescent trypanosomes. *Biochem Soc Trans.* 2006;34(Pt 4):557–559.
66. Conte I, Labriola C, Cazzulo JJ, Docampo R, Parodi AJ. The interplay between folding-facilitating mechanisms in *Trypanosoma cruzi* endoplasmic reticulum. *Mol Biol Cell.* 2003;14(9):3529–3540. doi: 10.1091/mbc.E03-04-0228.
67. Banerjee S, Vishwanath P, Cui J, Kelleher DJ, Gilmore R, Robbins PW, Samuelson J. The evolution of N-glycan-dependent endoplasmic reticulum quality control factors for glycoprotein folding and degradation. *Proc Natl Acad Sci USA.* 2007;104(28):11676–11681. doi: 10.1073/pnas.0704862104.
68. Mignot J-P. Contribution à l'étude cytologique de *Scytomonas pusilla* (Stein) (Flagellé euglénien) *Bull Biol Fr Belg.* 1961;95:665–678.
69. Hampl V, Hug L, Leigh JW, Dacks JB, Lang BF, Simpson AG, Roger AJ. Phylogenomic analyses support the monophyly of Excavata and resolve relationships among eukaryotic "supergroups". *Proc Natl Acad Sci USA.* 2009;106(10):3859–3864. doi: 10.1073/pnas.0807880106.
70. Cavalier-Smith T. In: *Molecular and cellular aspects of microbial evolution.* Carlile MJ, Collins JF, Moseley BEB, editor. Cambridge: Cambridge University Press; 1981. The origin and early evolution of the eukaryotic cell; pp. 33–84.

71. Strnad P, Leidel S, Vinogradova T, Euteneuer U, Khodjakov A, Gonczy P. Regulated HsSAS-6 levels ensure formation of a single procentriole per centriole during the centrosome duplication cycle. *Dev Cell.* 2007;13(2):203–213. doi: 10.1016/j.devcel.2007.07.004.
72. Salisbury JL. A mechanistic view on the evolutionary origin for centrin-based control of centriole duplication. *J Cell Physiol.* 2007;213(2):420–428. doi: 10.1002/jcp.21226.
73. Haren L, Remy MH, Bazin I, Callebaut I, Wright M, Merdes A. NEDD1-dependent recruitment of the gamma-tubulin ring complex to the centrosome is necessary for centriole duplication and spindle assembly. *J Cell Biol.* 2006;172(4):505–515. doi: 10.1083/jcb.200510028.
74. Bettencourt-Dias M, Rodrigues-Martins A, Carpenter L, Riparbelli M, Lehmann L, Gatt MK, Carmo N, Balloux F, Callaini G, Glover DM. SAK/PLK4 is required for centriole duplication and flagella development. *Curr Biol.* 2005;15(24):2199–2207. doi: 10.1016/j.cub.2005.11.042.
75. Zou C, Li J, Bai Y, Gunning WT, Wazer DE, Band V, Gao Q. Centrobin: a novel daughter centriole-associated protein that is required for centriole duplication. *J Cell Biol.* 2005;171(3):437–345. doi: 10.1083/jcb.200506185.
76. Habedanck R, Stierhof YD, Wilkinson CJ, Nigg EA. The Polo kinase Plk4 functions in centriole duplication. *Nat Cell Biol.* 2005;7(11):1140–1146. doi: 10.1038/ncb1320.
77. de Lichtenberg U, Jensen TS, Brunak S, Bork P, Jensen LJ. Evolution of cell cycle control: same molecular machines, different regulation. *Cell Cycle.* 2007;6(15):1819–1825.
78. Chang P, Giddings TH Jr, Winey M, Stearns T. Epsilon-tubulin is required for centriole duplication and microtubule organization. *Nat Cell Biol.* 2003;5(1):71–76. doi: 10.1038/ncb900.
79. Mereschkovsky C. Theorie der Zwei Plasmaarte als Grundlage der Symbiogenesis, einer neue Lehre von der Entstehung der Organismen. *Biol Zentralbl.* 1910;30:278–303.
80. Rotte C, Martin W. Related Articles L: Does endo-symbiosis explain the origin of the nucleus? *Nat Cell Biol.* 2001;3:E173–174. doi: 10.1038/35087104.
81. Cavalier-Smith T. Intron phylogeny: a new hypothesis. *Trends*

Genet. 1991;7(5):145–148. doi: 10.1016/0168-9525(91)90377-3.

82. Zhang YJ, Tian HF, Wen JF. The evolution of YidC/Oxa/Alb3 family in the three domains of life: a phylogenomic analysis. *BMC Evol Biol.* 2009;9:137. doi: 10.1186/1471-2148-9-137.
83. Cavalier-Smith T. Membrane heredity and early chloroplast evolution. *Trends Plant Sci.* 2000;5(4):174–182. doi: 10.1016/S1360-1385(00)01598-3.
84. Cavalier-Smith T. Obcells as proto-organisms: membrane heredity, lithophosphorylation, and the origins of the genetic code, the first cells, and photosynthesis. *J Mol Evol.* 2001;53(4-5):555–595. doi: 10.1007/s002390010245.
85. Cavalier-Smith T. In: *Organelles, genomes and eukaryote phylogeny.* Hirt RP, Horner DS, editor. London: Taylor & Francis; 2004. The membranome and membrane heredity in development and evolution; pp. 335–351.
86. Cavalier-Smith T. Chloroplast evolution: secondary symbiogenesis and multiple losses. *Curr Biol.* 2002;12(2):R62–64. doi: 10.1016/S0960-9822(01)00675-3.
87. Cavalier-Smith T. Principles of protein and lipid targeting in secondary symbiogenesis: euglenoid, dinoflagellate, and sporozoan plastid origins and the eukaryotic family tree. *J Euk Microbiol.* 1999;46:347–366. doi: 10.1111/j.1550-7408.1999.tb04614.x.
88. Cavalier-Smith T. In: *Biodiversity and Evolution.* Arai R, Kato M, Doi Y, editor. Tokyo: The National Science Museum Foundation; 1995. Membrane heredity, symbiogenesis, and the multiple origins of algae; pp. 75–114.
89. Leipe DD, Wolf M, Koonin EV, Aravind L. Classification and evolution of P-loop GTPases and related ATPases. *J Mol Biol.* 2002;317:41–72. doi: 10.1006/jmbi.2001.5378.
90. Odronitz F, Kollmar M. Drawing the tree of eukaryotic life based on the analysis of 2,269 manually annotated myosins from 328 species. *Genome Biol.* 2007;8(9):R196. doi: 10.1186/gb-2007-8-9-r196.
91. Wickstead B, Gull K. A "holistic" kinesin phylogeny reveals new kinesin families and predicts protein functions. *Mol Biol Cell.* 2006;17(4):1734–1743. doi: 10.1091/mbc.E05-11-1090.
92. Garbarino JE, Gibbons IR. Expression and genomic analysis of

midasin, a novel and highly conserved AAA protein distantly related to dynein. *BMC Genomics.* 2002;3(1):18. doi: 10.1186/1471-2164-3-18.
93. Wilkes DE, Watson HE, Mitchell DR, Asai DJ. Twenty-five dyneins in *Tetrahymena* : A re-examination of the multidynein hypothesis. *Cell Motil Cytoskeleton.* 2008;65(4):342–351. doi: 10.1002/cm.20264.
94. Wickstead B, Gull K. Dyneins across eukaryotes: a comparative genomic analysis. *Traffic.* 2007;8(12):1708–1721. doi: 10.1111/j.1600-0854.2007.00646.x.
95. Lee YJ, Szumlanski A, Nielsen E, Yang Z. Rho-GTPase-dependent filamentous actin dynamics coordinate vesicle targeting and exocytosis during tip growth. *J Cell Biol.* 2008;181(7):1155–1168. doi: 10.1083/jcb.200801086.
96. Doreian BW, Fulop TG, Smith CB. Myosin II activation and actin reorganization regulate the mode of quantal exocytosis in mouse adrenal chromaffin cells. *J Neurosci.* 2008;28(17):4470–4478. doi: 10.1523/JNEUROSCI.0008-08.2008.
97. Sokac AM, Bement WM. Kiss-and-coat and compartment mixing: coupling exocytosis to signal generation and local actin assembly. *Mol Biol Cell.* 2006;17(4):1495–1502. doi: 10.1091/mbc.E05-10-0908.
98. Jerdeva GV, Wu K, Yarber FA, Rhodes CJ, Kalman D, Schechter JE, Hamm-Alvarez SF. Actin and non-muscle myosin II facilitate apical exocytosis of tear proteins in rabbit lacrimal acinar epithelial cells. *J Cell Sci.* 2005;118(Pt 20):4797–4812. doi: 10.1242/jcs.02573.
99. Bader MF, Doussau F, Chasserot-Golaz S, Vitale N, Gasman S. Coupling actin and membrane dynamics during calcium-regulated exocytosis: a role for Rho and ARF GTPases. *Biochim Biophys Acta.* 2004;1742(1-3):37–49. doi: 10.1016/j.bbamcr.2004.09.028.
100. Vats P, Rothfield L. Duplication and segregation of the actin (MreB) cytoskeleton during the prokaryotic cell cycle. *Proc Natl Acad Sci USA.* 2007;104(45):17795–17800. doi: 10.1073/pnas.0708739104.
101. Divakaruni AV, Baida C, White CL, Gober JW. The cell shape proteins MreB and MreC control cell morphogenesis by positioning cell wall synthetic complexes. *Mol Microbiol.* 2007;66(1):174–188. doi: 10.1111/j.1365-2958.2007.05910.x.
102. Mazza P, Noens EE, Schirner K, Grantcharova N, Mommaas AM, Koerten HK, Muth G, Flardh K, van Wezel GP, Wohlleben W. MreB of *Streptomyces coelicolor* is not essential for vegetative

growth but is required for the integrity of aerial hyphae and spores. *Mol Microbiol.* 2006;60(4):838–852. doi: 10.1111/j.1365-2958.2006.05134.x.

103. Kruse T, Blagoev B, Lobner-Olesen A, Wachi M, Sasaki K, Iwai N, Mann M, Gerdes K. Actin homolog MreB and RNA polymerase interact and are both required for chromosome segregation in *Escherichia coli. Genes Dev.* 2006;20(1):113–124. doi: 10.1101/gad.366606.
104. Shih YL, Kawagishi I, Rothfield L. The MreB and Min cytoskeletal-like systems play independent roles in prokaryotic polar differentiation. *Mol Microbiol.* 2005;58(4):917–928. doi: 10.1111/j.1365-2958.2005.04841.x.
105. Kruse T, Gerdes K. Bacterial DNA segregation by the actin-like MreB protein. *Trends Cell Biol.* 2005;15(7):343–345. doi: 10.1016/j.tcb.2005.05.002.
106. Makarova KS, Wolf YI, Mekhedov SL, Mirkin BG, Koonin EV. Ancestral paralogs and pseudoparalogs and their role in the emergence of the eukaryotic cell. *Nucleic Acids Res.* 2005;33:4626–4638. doi: 10.1093/nar/gki775.
107. Jékely G. Small GTPases and the evolution of the eukaryotic cell. *BioEssays.* 2003;25:1129–1138. doi: 10.1002/bies.10353.
108. Raasi S, Wolf DH. Ubiquitin receptors and ERAD: a network of pathways to the proteasome. *Semin Cell Dev Biol.* 2007;18(6):780–791. doi: 10.1016/j.semcdb.2007.09.008.
109. Przewloka MR, Glover DM. The kinetochore and the centromere: a working long distance relationship. *Annu Rev Genet.* 2009;43:439–465. doi: 10.1146/annurev-genet-102108-134310.
110. Triemer RE, Farmer MA. In: *The biology of free-living heterotrophic flagellates.* Patterson DJ, Larsen J, editor. Oxford: Clarendon Press; 1991. The ultrastructural organization of heterotrophic euglenids and its evolutionary implications; pp. 185–204.
111. Jékely G. Origin of eukaryotic endomembranes: a critical evaluation of different model scenarios. *Adv Exp Med Biol.* 2007;607:38–51. full_text.
112. Erickson HP. Evolution of the cytoskeleton. *BioEssays.* 2007;29(7):668–677. doi: 10.1002/bies.20601.
113. Simpson GC. *Tempo and Mode in Evolution.* New York: Columbia University Press; 1944.

114. De Beer G. Archaeopteryx and evolution. *Adv Sci.* 1954;42:160–170.

115. Esser C, Ahmadinejad N, Wiegand C, Rotte C, Sebastiani F, Gelius-Dietrich G, Henze K, Kretschmann E, Richly E, Leister D, Bryant D, Steel MA, Lockhart PJ, Penny D, Martin W. A genome phylogeny for mitochondria among alpha-proteobacteria and a predominantly eubacterial ancestry of yeast nuclear genes. *Mol Biol Evol.* 2004;21(9):1643–1660. doi: 10.1093/molbev/msh160.

116. Cavalier-Smith T. In: *Origin of Plastids: Symbiogenesis, Prochlorophytes and the Origins of Chloroplasts.* Lewin RA, editor. New York: Chapman & Hall; 1993. The origin, losses and gains of chloroplasts; pp. 291–348.

117. Ryall K, Harper JT, Keeling PJ. Plastid-derived Type II fatty acid biosynthetic enzymes in chromists. *Gene.* 2003;313:139–148. doi: 10.1016/S0378-1119(03)00671-1.

118. Mendell JE, Clements KD, Choat JH, Angert ER. Extreme polyploidy in a large bacterium. *Proc Natl Acad Sci USA.* 2008;105(18):6730–6734. doi: 10.1073/pnas.0707522105.

119. Hu B, Yang G, Zhao W, Zhang Y, Zhao J. MreB is important for cell shape but not for chromosome segregation of the filamentous cyanobacterium *Anabaena* sp. PCC 7120. *Mol Microbiol.* 2007;63(6):1640–1652. doi: 10.1111/j.1365-2958.2007.05618.x.

120. Margulis L, Dolan M, Whiteside JH. "Imperfections and oddities" in the origin of the nucleus. *Paleobiology.* 2005;31:175–191. doi: 10.1666/0094-8373(2005)031[0175:IAOITO]2.0.CO;2. [CrossRef]

121. Lynch M, Conery JS. The origins of genome complexity. *Science.* 2003;302:1401–1404. doi: 10.1126/science.1089370.

122. López-García P, Moreira D. Selective forces for the origin of the eukaryotic nucleus. *BioEssays.* 2006;28:525–533. doi: 10.1002/bies.20413.

123. Cavalier-Smith T. In: *Endocytobiology: Endosymbiosis and Cell biology, a Synthesis of Recent Research.* Schwemmler W, Schenk HEA, editor. Berlin: de Gruyter; 1980. Cell compartmentation and the origin of eukaryote membranous organelles; pp. 893–916.

124. Jiang SY, Ramachandran S. Comparative and evolutionary analysis of genes encoding small GTPases and their activating proteins in

eukaryotic genomes. *Physiol Genomics.* 2006;24(3):235–251.

125. Siriputthaiwan P, Jauneau A, Herbert C, Garcin D, Dumas B. Functional analysis of CLPT1, a Rab/GTPase required for protein secretion and pathogenesis in the plant fungal pathogen *Colletotrichum lindemuthianum. J Cell Sci.* 2005;118(Pt 2):323–329. doi: 10.1242/jcs.01616.
126. Lipatova Z, Tokarev AA, Jin Y, Mulholland J, Weisman L, Segev N. Direct interaction between a myosin V motor and the Rab GTPases Ypt31/32 is required for polarized secretion. *Mol Biol Cell.* 2008;19:4177–4187. doi: 10.1091/mbc.E08-02-0220.
127. Braun V, Deschamps C, Raposo G, Benaroch P, Benmerah A, Chavrier P, Niedergang F. AP-1 and ARF1 control endosomal dynamics at sites of FcR mediated phagocytosis. *Mol Biol Cell.* 2007;18(12):4921–4931. doi: 10.1091/mbc.E07-04-0392.
128. Dong JH, Wen JF, Tian HF. Homologs of eukaryotic Ras superfamily proteins in prokaryotes and their novel phylogenetic correlation with their eukaryotic analogs. *Gene.* 2007;396(1):116–124. doi: 10.1016/j.gene.2007.03.001.
129. Kloepper TH, Kienle CN, Fasshauer D. An elaborate classification of SNARE proteins sheds light on the conservation of the eukaryotic endomembrane system. *Mol Biol Cell.* 2007;18(9):3463–3471. doi: 10.1091/mbc.E07-03-0193.
130. Dacks JB, Doolittle WF. Molecular and phylogenetic characterization of syntaxin genes from parasitic protozoa. *Mol Biochem Parasitol.* 2004;136:123–136. doi: 10.1016/j.molbiopara.2004.02.014.
131. De Duve C, Wattiaux R. Functions of lysosomes. *Ann Rev Physiol.* 1966;28:435–492.doi:10.1146/annurev.ph.28.030166.002251.
132. Gruber S, Errington J. Recruitment of condensin to replication origin regions by ParB/SpoOJ promotes chromosome segregation in *B. subtilis. Cell.* 2009;137(4):685–696. doi: 10.1016/j.cell.2009.02.035.
133. Sullivan NL, Marquis KA, Rudner DZ. Recruitment of SMC by ParB-parS organizes the origin region and promotes efficient chromosome segregation. *Cell.* 2009;137(4):697–707. doi: 10.1016/j.cell.2009.04.044.
134. Vaughan S, Wickstead B, Gull K, Addinall SG. Molecular evolution of FtsZ protein sequences encoded within the genomes of archaea, bacteria, and eukaryota. *J Mol Evol.* 2004;58(1):19–29. doi: 10.1007/

s00239-003-2523-5.

135. Chen Y, Erickson HP. In vitro assembly studies of FtsZ/tubulin-like proteins (TubZ) from *Bacillus* plasmids: evidence for a capping mechanism. *J Biol Chem.* 2008;283(13):8102–8109. doi: 10.1074/jbc.M709163200.
136. Löwe J, Amos LA. Evolution of cytomotive filaments: the cytoskeleton from prokaryotes to eukaryotes. *Int J Biochem Cell Biol.* 2009;41(2):323–329. doi: 10.1016/j.biocel.2008.08.010.
137. Larsen RA, Cusumano C, Fujioka A, Lim-Fong G, Patterson P, Pogliano J. Treadmilling of a prokaryotic tubulin-like protein, TubZ, required for plasmid stability in *Bacillus thuringiensis*. *Genes Dev.* 2007;21(11):1340–1352. doi: 10.1101/gad.1546107.
138. Nasmyth K, Haering CH. The structure and function of SMC and kleisin complexes. *Annu Rev Biochem.* 2005;74:595–648. doi: 10.1146/annurev.biochem.74.082803.133219.
139. Peters JM, Tedeschi A, Schmitz J. The cohesin complex and its roles in chromosome biology. *Genes Dev.* 2008;22(22):3089–3114. doi: 10.1101/gad.1724308.
140. Nasmyth K, Haering CH. Cohesin: its roles and mechanisms. *Annu Rev Genet.* 2009;43:525–558. doi: 10.1146/annurev-genet-102108-134233.
141. Cavalier-Smith T. Bacterial DNA segregation: its motors and positional control. *J Theor Biol.* 1987;127(3):361–372. doi: 10.1016/S0022-5193(87)80112-1.
142. Graumann PL, Knust T. Dynamics of the bacterial SMC complex and SMC-like proteins involved in DNA repair. *Chromosome Res.* 2009;17(2):265–275. doi: 10.1007/s10577-008-9014-x.
143. Cobbe N, Heck MM. The evolution of SMC proteins: phylogenetic analysis and structural implications. *Mol Biol Evol.* 2004;21(2):332–347. doi: 10.1093/molbev/msh023.
144. Gluenz E, Sharma R, Carrington M, Gull K. Functional characterization of cohesin subunit SCC1 in *Trypanosoma brucei* and dissection of mutant phenotypes in two life cycle stages. *Mol Microbiol.* 2008;69(3):666–680. doi: 10.1111/j.1365-2958.2008.06320.x.
145. Leonard CJ, Aravind L, Koonin EV. Novel families of putative protein kinases in bacteria and archaea: evolution of the "eukaryotic" protein kinase superfamily. *Genome Res.* 1998;8(10):1038–1047.
146. Fiuza M, Canova MJ, Zanella-Cleon I, Becchi M, Cozzone AJ, Mateos

LM, Kremer L, Gil JA, Molle V. From the characterization of the four serine/threonine protein kinases (PknA/B/G/L) of *Corynebacterium glutamicum* toward the role of PknA and PknB in cell division. *J Biol Chem.* 2008;283(26):18099–180112. doi: 10.1074/jbc.M802615200.

147. Krupa A, Srinivasan N. Diversity in domain architectures of Ser/Thr kinases and their homologues in prokaryotes. *BMC Genomics.* 2005;6:129. doi: 10.1186/1471-2164-6-129.

148. de Graffenried CL, Ho HH, Warren G. Polo-like kinase is required for Golgi and bilobe biogenesis in *Trypanosoma brucei. J Cell Biol.* 2008;181(3):431–438. doi: 10.1083/jcb.200708082.

149. Bessat M, Ersfeld K. Functional characterization of cohesin SMC3 and separase and their roles in the segregation of large and minichromosomes in *Trypanosoma brucei. Mol Microbiol.* 2009;71(6):1371–1385. doi: 10.1111/j.1365-2958.2009.06611.x.

150. Brown JR, Koretke KK, Birkeland ML, Sanseau P, Patrick DR. Evolutionary relationships of Aurora kinases: implications for model organism studies and the development of anti-cancer drugs. *BMC Evol Biol.* 2004;4(1):39. doi: 10.1186/1471-2148-4-39.

151. Tu X, Kumar P, Li Z, Wang CC. An aurora kinase homologue is involved in regulating both mitosis and cytokinesis in *Trypanosoma brucei. J Biol Chem.* 2006;281(14):9677–9687. doi: 10.1074/jbc.M511504200.

152. Kumar P, Wang CC. Dissociation of cytokinesis initiation from mitotic control in a eukaryote. *Eukaryot Cell.* 2006;5(1):92–102. doi: 10.1128/EC.5.1.92-102.2006.

153. Hammarton TC, Kramer S, Tetley L, Boshart M, Mottram JC. *Trypanosoma brucei* Polo-like kinase is essential for basal body duplication, kDNA segregation and cytokinesis. *Mol Microbiol.* 2007;65(5):1229–1248. doi: 10.1111/j.1365-2958.2007.05866.x.

154. Umeyama T, Wang CC. Polo-like kinase is expressed in S/G2/M phase and associated with the flagellum attachment zone in both procyclic and bloodstream forms of *Trypanosoma brucei. Eukaryot Cell.* 2008;7(9):1582–1590. doi: 10.1128/EC.00150-08.

155. Moreira-Leite FF, Sherwin T, Kohl L, Gull K. A trypanosome structure involved in transmitting cytoplasmic information during cell division. *Science.* 2001;294(5542):610–612. doi: 10.1126/science.1063775.

156. Selvapandiyan A, Kumar P, Morris JC, Salisbury JL, Wang CC, Nakhasi HL. Centrin1 Is required for organelle segregation and cytokinesis in *Trypanosoma brucei. Mol Biol Cell.* 2007;18(9):3290–301. doi: 10.1091/mbc.E07-01-0022.
157. Vaughan S, Gull K. The structural mechanics of cell division in *Trypanosoma brucei. Biochem Soc Trans.* 2008;36(Pt 3):421–424. doi: 10.1042/BST0360421.
158. Cavalier-Smith T, Lewis R, Chao EE, Oates B, Bass D. Morphology and phylogeny of *Sainouron acronematica* sp. n. and the ultrastructural unity of Cercozoa. *Protist.* 2008;159:591–620. doi: 10.1016/j.protis.2008.04.002.
159. Shi J, Franklin JB, Yelinek JT, Ebersberger I, Warren G, He CY. Centrin4 coordinates cell and nuclear division in *T. brucei. J Cell Sci.* 2008;121(Pt 18):3062–3070. doi: 10.1242/jcs.030643.
160. Gogendeau D, Klotz C, Arnaiz O, Malinowska A, Dadlez M, de Loubresse NG, Ruiz F, Koll F, Beisson J. Functional diversification of centrins and cell morphological complexity. *J Cell Sci.* 2008;121(Pt 1):65–74. doi: 10.1242/jcs.019414.
161. Zhou Y, Yang W, Kirberger M, Lee HW, Ayalasomayajula G, Yang JJ. Prediction of EF-hand calcium-binding proteins and analysis of bacterial EF-hand proteins. *Proteins.* 2006;65(3):643–655. doi: 10.1002/prot.21139.
162. Mahajan B, Selvapandiyan A, Gerald NJ, Majam V, Zheng H, Wickramarachchi T, Tiwari J, Fujioka H, Moch JK, Kumar N, Aravind L, Nakhasi HL, Kumar S. Centrins, cell cycle regulation proteins in human malaria parasite Plasmodium falciparum. *J Biol Chem.* 2008;283(46):31871–31883. doi: 10.1074/jbc.M800028200.
163. Cavalier-Smith T. In: *The nuclear envelope and matrix.* Maul GG, editor. New York: Alan R. Liss; 1982. The evolution of the nuclear matrix and envelope; pp. 307–318.
164. Cavalier-Smith T. In: *Evolutionary biology of the Fungi.* Rayner ADM, Brasier CM, Moore D, editor. Cambridge University Press; 1987. The origin of Fungi and pseudofungi; pp. 339–353.
165. Cavalier-Smith T. In: *The Mycota.* Part A. McLaughlin DJ, McLaughlin EJ, Lemke P, editor. VII. Berlin: Springer-Verlag; 2000. What are Fungi? pp. 3–37.
166. Cavalier-Smith T. Basal body and flagellar development during

the vegetative cell cycle and the sexual cycle of *Chlamydomonas reinhardii*. *J Cell Sci.* 1974;16(3):529–556.

167. Bornens M. Organelle positioning and cell polarity. *Nat Rev Mol Cell Biol.* 2008;9(11):874–886. doi: 10.1038/nrm2524.
168. Haeckel E. *Generelle Morphologie der Organismen.* Berlin: Reimer; 1866.
169. Cavalier-Smith T. Nuclear volume control by nucleoskeletal DNA, selection for cell volume and cell growth rate, and the solution of the DNA C-value paradox. *J Cell Sci.* 1978;34:247–278.
170. Cavalier-Smith T. Skeletal DNA and the evolution of genome size. *Annu Rev Biophys Bioeng.* 1982;11:273–302. doi: 10.1146/annurev.bb.11.060182.001421.
171. Krauss SW, Chen C, Penman S, Heald R. Nuclear actin and protein 4.1: essential interactions during nuclear assembly in vitro. *Proc Natl Acad Sci USA.* 2003;100(19):10752–10757. doi: 10.1073/pnas.1934680100.
172. Gieni RS, Hendzel MJ. Actin dynamics and functions in the interphase nucleus: moving toward an understanding of nuclear polymeric actin. *Biochem Cell Biol.* 2009;87(1):283–306. doi: 10.1139/O08-133.
173. Kitayama K, Kamo M, Oma Y, Matsuda R, Uchida T, Ikura T, Tashiro S, Ohyama T, Winsor B, Harata M. The human actin-related protein hArp5: nucleo-cytoplasmic shuttling and involvement in DNA repair. *Exp Cell Res.* 2009;315(2):206–2017. doi: 10.1016/j.yexcr.2008.10.028.
174. Fransz P, Soppe W, Schubert I. Heterochromatin in interphase nuclei of *Arabidopsis thaliana*. *Chromosome Res.* 2003;11(3):227–240. doi: 10.1023/A:1022835825899.
175. Cavalier-Smith T. In: *The evolution of genome size.* Cavalier-Smith T, editor. Chichester: Wiley; 1985. Eukaryote gene numbers, non-coding DNA and genome size; pp. 69–103.
176. Cavalier-Smith T. In: *Symposium on the evolution of terrestrial vertebrates Selected Symposia and Monographs V ZI, 4.* Ghiara G, et al, editor. Modena: Muchi; 1991. Coevolution of vertebrate genome, cell and nuclear sizes; pp. 51–88.
177. Carthew RW, Sontheimer EJ. Origins and mechanisms of miRNAs and siRNAs. *Cell.* 2009;136(4):642–655. doi: 10.1016/j.cell.2009.01.035.
178. Lu J, Gilbert DM. Cell cycle regulated transcription of heterochromatin in mammals vs. fission yeast: functional conservation or coincidence? *Cell Cycle.* 2008;7(13):1907–1910.

179. Kim SM, Dubey DD, Huberman JA. Early-replicating heterochromatin. *Genes Dev.* 2003;17(3):330–335. doi: 10.1101/gad.1046203.
180. Djupedal I, Ekwall K. Epigenetics: heterochromatin meets RNAi. *Cell Res.* 2009;19(3):282–295. doi: 10.1038/cr.2009.13.
181. Volpe TA, Kidner C, Hall IM, Teng G, Grewal SI, Martienssen RA. Regulation of heterochromatic silencing and histone H3 lysine-9 methylation by RNAi. *Science.* 2002;297(5588):1833–1837. doi: 10.1126/science.1074973.
182. Cam HP, Chen ES, Grewal SI. Transcriptional scaffolds for heterochromatin assembly. *Cell.* 2009;136(4):610–614. doi: 10.1016/j.cell.2009.02.004.
183. Chen ES, Zhang K, Nicolas E, Cam HP, Zofall M, Grewal SI. Cell cycle control of centromeric repeat transcription and heterochromatin assembly. *Nature.* 2008;451(7179):734–737. doi: 10.1038/nature06561.
184. Cam HP, Sugiyama T, Chen ES, Chen X, FitzGerald PC, Grewal SI. Comprehensive analysis of heterochromatin- and RNAi-mediated epigenetic control of the fission yeast genome. *Nat Genet.* 2005;37(8):809–819. doi: 10.1038/ng1602.
185. Cam H, Grewal SI. RNA interference and epigenetic control of heterochromatin assembly in fission yeast. *Cold Spring Harb Symp Quant Biol.* 2004;69:419–427. doi: 10.1101/sqb.2004.69.419.
186. Figueiredo LM, Cross GA, Janzen CJ. Epigenetic regulation in African trypanosomes: a new kid on the block. *Nat Rev Microbiol.* 2009;7(7):504–513. doi: 10.1038/nrmicro2149.
187. Mizuguchi G, Xiao H, Wisniewski J, Smith MM, Wu C. Nonhistone Scm3 and histones CenH3-H4 assemble the core of centromere-specific nucleosomes. *Cell.* 2007;129(6):1153–1164. doi: 10.1016/j.cell.2007.04.026.
188. Sandman K, Reeve JN. Archaeal chromatin proteins: different structures but common function? *Curr Opin Microbiol.* 2005;8(6):656–661. doi: 10.1016/j.mib.2005.10.007.
189. Marsh VL, McGeoch AT, Bell SD. Influence of chromatin and single strand binding proteins on the activity of an archaeal MCM. *J Mol Biol.* 2006;357(5):1345–1350. doi: 10.1016/j.jmb.2006.01.074.
190. Aravind L, Iyer LM, Anantharaman V. The two faces of Alba: the

evolutionary connection between proteins participating in chromatin structure and RNA metabolism. *Genome Biol.* 2003;4(10):R64. doi: 10.1186/gb-2003-4-10-r64.

191. Farazi TA, Juranek SA, Tuschl T. The growing catalog of small RNAs and their association with distinct Argonaute/Piwi family members. *Development.* 2008;135(7):1201–1214. doi: 10.1242/dev.005629.
192. Drider D, Condon C. The continuing story of endoribonuclease III. *J Mol Microbiol Biotechnol.* 2004;8(4):195–200. doi: 10.1159/000086700.
193. Wang Y, Juranek S, Li H, Sheng G, Tuschl T, Patel DJ. Structure of an argonaute silencing complex with a seed-containing guide DNA and target RNA duplex. *Nature.* 2008;456(7224):921–926. doi: 10.1038/nature07666.
194. Wang Y, Sheng G, Juranek S, Tuschl T, Patel DJ. Structure of the guide-strand-containing argonaute silencing complex. *Nature.* 2008;456(7219):209–213. doi: 10.1038/nature07315.
195. Dennis PP, Omer A. Small non-coding RNAs in Archaea. *Curr Opin Microbiol.* 2005;8(6):685–694. doi: 10.1016/j.mib.2005.10.013.
196. Reichow SL, Hamma T, Ferre-D'Amare AR, Varani G. The structure and function of small nucleolar ribonucleoproteins. *Nucleic Acids Res.* 2007;35(5):1452–1464. doi: 10.1093/nar/gkl1172.
197. Reichow SL, Varani G. Nop10 is a conserved H/ACA snoRNP molecular adaptor. *Biochemistry.* 2008;47(23):6148–6156. doi: 10.1021/bi800418p.
198. Comella P, Pontvianne F, Lahmy S, Vignols F, Barbezier N, Debures A, Jobet E, Brugidou E, Echeverria M, Saez-Vasquez J. Characterization of a ribonuclease III-like protein required for cleavage of the pre-rRNA in the 3'ETS in *Arabidopsis*. *Nucleic Acids Res.* 2008;36(4):1163–1175. doi: 10.1093/nar/gkm1130.
199. Jin L, Kryukov K, Suzuki Y, Imanishi T, Ikeo K, Gojobori T. The evolutionary study of small RNA-directed gene silencing pathways by investigating RNase III enzymes. *Gene.* 2009;435(1-2):1–8. doi: 10.1016/j.gene.2008.12.022.
200. Macrae IJ, Zhou K, Li F, Repic A, Brooks AN, Cande WZ, Adams PD, Doudna JA. Structural basis for double-stranded RNA processing by Dicer. *Science.* 2006;311(5758):195–198. doi: 10.1126/science.1121638.

201. Cerutti H, Casas-Mollano JA. On the origin and functions of RNA-mediated silencing: from protists to man. *Curr Genet.* 2006;50(2):81–99. doi: 10.1007/s00294-006-0078-x.

202. Kagansky A, Folco HD, Almeida R, Pidoux AL, Boukaba A, Simmer F, Urano T, Hamilton GL, Allshire RC. Synthetic heterochromatin bypasses RNAi and centromeric repeats to establish functional centromeres. *Science.* 2009;324(5935):1716–1719. doi: 10.1126/science.1172026.

203. Kawahara T, Siegel TN, Ingram AK, Alsford S, Cross GA, Horn D. Two essential MYST-family proteins display distinct roles in histone H4K10 acetylation and telomeric silencing in trypanosomes. *Mol Microbiol.* 2008;69(4):1054–1068. doi: 10.1111/j.1365-2958.2008.06346.x.

204. Durand-Dubief M, Absalon S, Menzer L, Ngwabyt S, Ersfeld K, Bastin P. The Argonaute protein TbAGO1 contributes to large and mini-chromosome segregation and is required for control of RIME retroposons and RHS pseudogene-associated transcripts. *Mol Biochem Parasitol.* 2007;156(2):144–153. doi: 10.1016/j.molbiopara.2007.07.016.

205. Shi H, Chamond N, Djikeng A, Tschudi C, Ullu E. RNA interference in *Trypanosoma brucei* : the role of the amino-terminal RGG domain and the polyribosome association of Argonaute1. *J Biol Chem.* 2009;284(52):36511–36520. doi: 10.1074/jbc.M109.073072.

206. Shi H, Ullu E, Tschudi C. Function of the trypanosome Argonaute 1 protein in RNA interference requires the N-terminal RGG domain and arginine 735 in the Piwi domain. *J Biol Chem.* 2004;279(48):49889–49893. doi: 10.1074/jbc.M409280200.

207. Patrick KL, Shi H, Kolev NG, Ersfeld K, Tschudi C, Ullu E. Distinct and overlapping roles for two Dicer-like proteins in the RNA interference pathways of the ancient eukaryote *Trypanosoma brucei*. *Proc Natl Acad Sci USA.* 2009;106(42):17933–17938. doi: 10.1073/pnas.0907766106.

208. Murphy D, Dancis B, Brown JR. The evolution of core proteins involved in microRNA biogenesis. *BMC Evol Biol.* 2008;8:92. doi: 10.1186/1471-2148-8-92.

209. Archer S, Queiroz R, Stewart M, Clayton C. Trypanosomes as a model to investigate mRNA decay pathways. *Methods Enzymol.* 2008;448:359–377. full_text.

210. Ivens AC, Peacock CS, Worthey EA, Murphy L, Aggarwal G, Berriman M, Sisk E, Rajandream MA, Adlem E, Aert R. The genome of the kinetoplastid parasite, *Leishmania* major. *Science.* 2005;309(5733):436–442. doi: 10.1126/science.1112680.
211. Cavalier-Smith T. In: *The Eukaryotic Genome.* Broda P, Oliver SG, Sims P, editor. Cambridge University Press; 1993. Evolution of the eukaryotic genome; pp. 333–385.
212. Talbert PB, Henikoff S. Chromatin-based transcriptional punctuation. *Genes Dev.* 2009;23(9):1037–1041. doi: 10.1101/gad.1806409.
213. Siegel TN, Hekstra DR, Kemp LE, Figueiredo LM, Lowell JE, Fenyo D, Wang X, Dewell S, Cross GA. Four histone variants mark the boundaries of polycistronic transcription units in *Trypanosoma brucei. Genes Dev.* 2009;23(9):1063–1076. doi: 10.1101/gad.1790409.
214. Harrison BR, Yazgan O, Krebs JE. Life without RNAi: noncoding RNAs and their functions in *Saccharomyces cerevisiae. Biochem Cell Biol.* 2009;87(5):767–779. doi: 10.1139/O09-043.
215. Drinnenberg IA, Weinberg DE, Xie KT, Mower JP, Wolfe KH, Fink GR, Bartel DP. RNAi in budding yeast. *Science.* 2009;326(5952):544–550. doi: 10.1126/science.1176945.
216. Saraiya AA, Wang CC. snoRNA, a novel precursor of microRNA in *Giardia lamblia. PLoS Pathog.* 2008;4(11):e1000224. doi: 10.1371/journal.ppat.1000224.
217. Marz M, Stadler PF. Comparative analysis of eukaryotic U3 snoRNA. *RNA Biol.* 2009;6(5):503–507. doi: 10.4161/rna.6.5.9607.
218. Luo J, Teng M, Zhang GP, Lun ZR, Zhou H, Qu LH. Evaluating the evolution of *G. lamblia* based on the small nucleolar RNAs identified from Archaea and unicellular eukaryotes. *Parasitol Res.* 2009;104(6):1543–1546. doi: 10.1007/s00436-009-1403-3.
219. Doniger T, Michaeli S, Unger R. Families of H/ACA ncRNA molecules in trypanosomatids. *RNA Biol.* 2009;6(4):370–374. doi: 10.4161/rna.6.4.9270.
220. Russell AG, Schnare MN, Gray MW. A large collection of compact box C/D snoRNAs and their isoforms in *Euglena gracilis* : structural functional and evolutionary insights. *J Mol Biol.* 2006;367:1548–1565. doi: 10.1016/j.jmb.2006.01.063.
221. Cavalier-Smith T. Palindromic base sequences and replication of

eukaryote chromosome ends. *Nature.* 1974;250(5466):467–470. doi: 10.1038/250467a0.

222. Cavalier-Smith T. In: *Endocytobiology II.* Schwemmler W, Schenk HEA, editor. Berlin, de Gruyter; 1983. Genetic symbionts and the origin of split genes and linear chromosomes; pp. 29–45.
223. Greenwood J, Cooper JP. Trapping Rap1 at the telomere to prevent chromosome end fusions. *Embo J.* 2009;28(21):3277–3278. doi: 10.1038/emboj.2009.299.
224. Horn D. Antigenic variation: extending the reach of telomeric silencing. *Curr Biol.* 2009;19(12):R496–498. doi: 10.1016/j.cub.2009.04.049.
225. Yang X, Figueiredo LM, Espinal A, Okubo E, Li B. RAP1 is essential for silencing telomeric variant surface glycoprotein genes in *Trypanosoma brucei. Cell.* 2009;137(1):99–109. doi: 10.1016/j.cell.2009.01.037.
226. Adams DW, Errington J. Bacterial cell division: assembly, maintenance and disassembly of the Z ring. *Nat Rev Microbiol.* 2009;7(9):642–653. doi: 10.1038/nrmicro2198.
227. Lutkenhaus J. FtsN--trigger for septation. *J Bacteriol.* 2009;191(24):7381–7382. doi: 10.1128/JB.01100-09.
228. McDonald B, Martin-Serrano J. No strings attached: the ESCRT machinery in viral budding and cytokinesis. *J Cell Sci.* 2009;122(Pt 13):2167–2177. doi: 10.1242/jcs.028308.
229. Lindas AC, Karlsson EA, Lindgren MT, Ettema TJ, Bernander R. A unique cell division machinery in the Archaea. *Proc Natl Acad Sci USA.* 2008;105(48):18942–18946. doi: 10.1073/pnas.0809467105.
230. Ettema TJ, Bernander R. Cell division and the ESCRT complex: A surprise from the archaea. *Commun Integr Biol.* 2009;2(2):86–88.
231. Ellen AF, Albers SV, Huibers W, Pitcher A, Hobel CF, Schwarz H, Folea M, Schouten S, Boekema EJ, Poolman B, Driessen AJ. Proteomic analysis of secreted membrane vesicles of archaeal *Sulfolobus* species reveals the presence of endosome sorting complex components. *Extremophiles.* 2009;13(1):67–79. doi: 10.1007/s00792-008-0199-x.
232. Brochier-Armanet C, Boussau B, Gribaldo S, Forterre P. Mesophilic Crenarchaeota: proposal for a third archaeal phylum, the Thaumarchaeota. *Nat Rev Microbiol.* 2008;6(3):245–252. doi: 10.1038/nrmicro1852.

233. Miyagishima SY, Kuwayama H, Urushihara H, Nakanishi H. Evolutionary linkage between eukaryotic cytokinesis and chloroplast division by dynamin proteins. *Proc Natl Acad Sci USA.* 2008;105(39):15202–15207. doi: 10.1073/pnas.0802412105.
234. Sherratt DJ, Soballe B, Barre FX, Filipe S, Lau I, Massey T, Yates J. Recombination and chromosome segregation. *Philos Trans R Soc Lond B Biol Sci.* 2004;359(1441):61–69. doi: 10.1098/rstb.2003.1365.
235. Grainge I, Bregu M, Vazquez M, Sivanathan V, Ip SC, Sherratt DJ. Unlinking chromosome catenanes in vivo by site-specific recombination. *Embo J.* 2007;26(19):4228–4238. doi: 10.1038/sj.emboj.7601849.
236. Graham JE, Sivanathan V, Sherratt DJ, Arciszewska LK. FtsK translocation on DNA stops at XerCD-dif. *Nucleic Acids Res.* 2009. [PubMed]
237. Pickett-Heaps JD. Evolution of mitosis and the eukaryote condition. *BioSystems.* 1974;6:37–45. doi: 10.1016/0303-2647(74)90009-4.
238. Cavalier-Smith T. The evolutionary origin and phylogeny of microtubules, mitotic spindles and eukaryote flagella. *BioSystems.* 1978;10(1-2):93–114. doi: 10.1016/0303-2647(78)90033-3.
239. Cavalier-Smith T. In: *Prokaryotic and Eukaryotic Flagella 35th Symposium of the Society of Experimental Biology.* Amos WB, Duckett JG, editor. Cambridge University Press; 1982. The evolutionary origin and phylogeny of eukaryote flagella; pp. 465–493.
240. Cavalier-Smith T. In: *The Origin and Evolution of the Cell.* Hartman H, Matsuno K, editor. Singapore: World Scientific Publishers; 1992. Origin of the cytoskeleton; pp. 79–106.
241. Inoue I, Ino R, Nishimura A. New model for assembly dynamics of bacterial tubulin in relation to the stages of DNA replication. *Genes Cells.* 2009;14(3):435–544. doi: 10.1111/j.1365-2443.2009.01280.x.
242. Ferrari S. Protein kinases controlling the onset of mitosis. *Cell Mol Life Sci.* 2006;63(7-8):781–795. doi: 10.1007/s00018-005-5515-3.
243. Nakamura A, Arai H, Fujita N. Centrosomal Aki1 and cohesin function in separase-regulated centriole disengagement. *J Cell Biol.* 2009;187(5):607–614. doi: 10.1083/jcb.200906019.
244. Nasmyth K. Evolution of the cell cycle. *Philos Trans R Soc Lond B Biol Sci.* 1995;349(1329):271–281. doi: 10.1098/rstb.1995.0113.

245. Maiorano D, Lutzmann M, Mechali M. MCM proteins and DNA replication. *Curr Opin Cell Biol.* 2006;18(2):130–136. doi: 10.1016/j.ceb.2006.02.006.

246. Duncker BP, Chesnokov IN, McConkey BJ. The origin recognition complex protein family. *Genome Biol.* 2009;10(3):214. doi: 10.1186/gb-2009-10-3-214.

247. Wirth KG, Wutz G, Kudo NR, Desdouets C, Zetterberg A, Taghybeeglu S, Seznec J, Ducos GM, Ricci R, Firnberg N, Peters JM, Nasmyth K. Separase: a universal trigger for sister chromatid disjunction but not chromosome cycle progression. *J Cell Biol.* 2006;172(6):847–860. doi: 10.1083/jcb.200506119.

248. Gruber S, Arumugam P, Katou Y, Kuglitsch D, Helmhart W, Shirahige K, Nasmyth K. Evidence that loading of cohesin onto chromosomes involves opening of its SMC hinge. *Cell.* 2006;127(3):523–537. doi: 10.1016/j.cell.2006.08.048.

249. Haering CH, Farcas AM, Arumugam P, Metson J, Nasmyth K. The cohesin ring concatenates sister DNA molecules. *Nature.* 2008;454(7202):297–301. doi: 10.1038/nature07098.

250. Godoy PD, Nogueira-Junior LA, Paes LS, Cornejo A, Martins RM, Silber AM, Schenkman S, Elias MC. Trypanosome prereplication machinery contains a single functional orc1/cdc6 protein, which is typical of archaea. *Eukaryot Cell.* 2009;8(10):1592–1603. doi: 10.1128/EC.00161-09.

251. Krylov DM, Nasmyth K, Koonin EV. Evolution of eukaryotic cell cycle regulation: stepwise addition of regulatory kinases and late advent of the CDKs. *Curr Biol.* 2003;13(2):173–177. doi: 10.1016/S0960-9822(03)00008-3.

252. Bayliss R, Littlewood T, Strawn LA, Wente SR, Stewart M. GLFG and FxFG nucleoporins bind to overlapping sites on importin-beta. *J Biol Chem.* 2002;277(52):50597–55606. doi: 10.1074/jbc.M209037200.

253. Pyhtila B, Rexach M. A gradient of affinity for the karyopherin Kap95p along the yeast nuclear pore complex. *J Biol Chem.* 2003;278(43):42699–42709. doi: 10.1074/jbc.M307135200.

254. Brohawn SG, Leksa NC, Spear ED, Rajashankar KR, Schwartz TU. Structural evidence for common ancestry of the nuclear pore complex and vesicle coats. *Science.* 2008;322(5906):1369–1373. doi: 10.1126/science.1165886.

255. Sorokin AV, Kim ER, Ovchinnikov LP. Nucleocytoplasmic transport of proteins. *Biochemistry (Mosc)* 2007;72(13):1439–1457. doi: 10.1134/S0006297907130032.

256. Bayliss R, Leung SW, Baker RP, Quimby BB, Corbett AH, Stewart M. Structural basis for the interaction between NTF2 and nucleoporin FxFG repeats. *Embo J.* 2002;21(12):2843–2853. doi: 10.1093/emboj/cdf305.

257. Isgro TA, Schulten K. Association of nuclear pore FG-repeat domains to NTF2 import and export complexes. *J Mol Biol.* 2007;366(1):330–345. doi: 10.1016/j.jmb.2006.11.048.

258. Kalab P, Heald R. The RanGTP gradient - a GPS for the mitotic spindle. *J Cell Sci.* 2008;121(Pt 10):1577–1586. doi: 10.1242/jcs.005959.

259. Jékely G. Origin of the nucleus and Ran-dependent transport to safeguard ribosome biogenesis in a chimeric cell. *Biol Direct.* 2008;3:31. doi: 10.1186/1745-6150-3-31.

260. Jékely G. Small GTPases and the evolution of the eukaryotic cell. *BioEssays.* 2003;25:1129–1138. doi: 10.1002/bies.10353.

261. Bentley M, Liang Y, Mullen K, Xu D, Sztul E, Hay JC. SNARE status regulates tether recruitment and function in homotypic COPII vesicle fusion. *J Biol Chem.* 2006;281(50):38825–38833. doi: 10.1074/jbc.M606044200.

262. Xu D, Hay JC. Reconstitution of COPII vesicle fusion to generate a pre-Golgi intermediate compartment. *J Cell Biol.* 2004;167(6):997–1003. doi: 10.1083/jcb.200408135.

263. Oakley BR, Dodge JD. Kinetochores associated with the nuclear envelope in the mitosis of a dinoflagellate. *J Cell Biol.* 1974;63(1):322–325. doi: 10.1083/jcb.63.1.322.

264. Goode D. Evolution of mitosis in protozoa: the association of chromosomes, nuclear envelope, kinetochores and microtubules. *Biosystems.* 1975;7(3-4):318–325. doi: 10.1016/0303-2647(75)90009-X.

265. Oakley BR, Dodge JD. Mitosis and cytokinesis in the dinoflagellate *Amphidinium carterae. Cytobios.* 1976;17(65):35–46.

266. Costas E, Goyanes V. Architecture and evolution of dinoflagellate chromosomes: an enigmatic origin. *Cytogenet Genome Res.* 2005;109(1-3):268–275. doi: 10.1159/000082409.

267. Cavalier-Smith T, Chao EE. Protalveolate phylogeny and systematics

and the origins of Sporozoa and dinoflagellates (phylum Myzozoa nom. nov.) *Eur J Protistol.* 2004;40(3):185–212. doi: 10.1016/j.ejop.2004.01.002. [CrossRef]

268. Strom AC, Weis K. Importin-beta-like nuclear transport receptors. *Genome Biol.* 2001;2(6):REVIEWS3008. doi: 10.1186/gb-2001-2-6-reviews3008.
269. Hutten S, Kehlenbach RH. CRM1-mediated nuclear export: to the pore and beyond. *Trends Cell Biol.* 2007;17(4):193–201. doi: 10.1016/j.tcb.2007.02.003.
270. Torosantucci L, De Luca M, Guarguaglini G, Lavia P, Degrassi F. Localized RanGTP accumulation promotes microtubule nucleation at kinetochores in somatic mammalian cells. *Mol Biol Cell.* 2008;19(5):1873–1882. doi: 10.1091/mbc.E07-10-1050.
271. Neuber A, Franke J, Wittstruck A, Schlenstedt G, Sommer T, Stade K. Nuclear export receptor Xpo1/Crm1 is physically and functionally linked to the spindle pole body in budding yeast. *Mol Cell Biol.* 2008;28(17):5348–5358. doi: 10.1128/MCB.02043-07.
272. Jäkel S, Mingot JM, Schwarzmaier P, Hartmann E, Gorlich D. Importins fulfil a dual function as nuclear import receptors and cytoplasmic chaperones for exposed basic domains. *Embo J.* 2002;21(3):377–386. doi: 10.1093/emboj/21.3.377.
273. Bäuerle M, Doenecke D, Albig W. The requirement of H1 histones for a heterodimeric nuclear import receptor. *J Biol Chem.* 2002;277(36):32480–32489. doi: 10.1074/jbc.M202765200.
274. Trotta CR, Lund E, Kahan L, Johnson AW, Dahlberg JE. Coordinated nuclear export of 60 S ribosomal subunits and NMD3 in vertebrates. *Embo J.* 2003;22(11):2841–2851. doi: 10.1093/emboj/cdg249.
275. Timney BL, Tetenbaum-Novatt J, Agate DS, Williams R, Zhang W, Chait BT, Rout MP. Simple kinetic relationships and nonspecific competition govern nuclear import rates in vivo. *J Cell Biol.* 2006;175(4):579–593. doi: 10.1083/jcb.200608141.
276. Chen IJ, Wang IA, Tai LR, Lin A. The role of expansion segment of human ribosomal protein L35 in nuclear entry, translation activity, and endoplasmic reticulum docking. *Biochem Cell Biol.* 2008;86(3):271–277. doi: 10.1139/O08-032.
277. Hardeland U, Hurt E. Coordinated nuclear import of RNA polymerase

III subunits. *Traffic.* 2006;7(4):465–473. doi: 10.1111/j.1600-0854.2006.00399.x.

278. Soutoglou E, Demeny MA, Scheer E, Fienga G, Sassone-Corsi P, Tora L. The nuclear import of TAF10 is regulated by one of its three histone fold domain-containing interaction partners. *Mol Cell Biol.* 2005;25(10):4092–4104. doi: 10.1128/MCB.25.10.4092-4104.2005.
279. Cavalier-Smith T. How selfish is DNA? *Nature.* 1980;285(5767):617–618. doi: 10.1038/285617a0.
280. Hickey DA. Selfish DNA: a sexually-transmitted nuclear parasite. *Genetics.* 1982;101(3-4):519–531.
281. Fridkin A, Penkner A, Jantsch V, Gruenbaum Y. SUN-domain and KASH-domain proteins during development, meiosis and disease. *Cell Mol Life Sci.* 2009;66(9):1518–1533. doi: 10.1007/s00018-008-8713-y.
282. Tzur YB, Wilson KL, Gruenbaum Y. SUN-domain proteins: 'Velcro' that links the nucleoskeleton to the cytoskeleton. *Nat Rev Mol Cell Biol.* 2006;7(10):782–798. doi: 10.1038/nrm2003.
283. Östlund C, Folker ES, Choi JC, Gomes ER, Gundersen GG, Worman HJ. Dynamics and molecular interactions of linker of nucleoskeleton and cytoskeleton (LINC) complex proteins. *J Cell Sci.* 2009;122(Pt 22):4099–4108. doi: 10.1242/jcs.057075.
284. Minn IL, Rolls MM, Hanna-Rose W, Malone CJ. SUN-1 and ZYG-12, mediators of centrosome-nucleus attachment, are a functional SUN/KASH pair in *Caenorhabditis elegans*. *Mol Biol Cell.* 2009;20(21):4586–4595. doi: 10.1091/mbc.E08-10-1034.
285. Xiong H, Rivero F, Euteneuer U, Mondal S, Mana-Capelli S, Larochelle D, Vogel A, Gassen B, Noegel AA. *Dictyostelium* Sun-1 connects the centrosome to chromatin and ensures genome stability. *Traffic.* 2008;9(5):708–724. doi: 10.1111/j.1600-0854.2008.00721.x.
286. Burke B, Roux KJ. Nuclei take a position: managing nuclear location. *Dev Cell.* 2009;17(5):587–597. doi: 10.1016/j.devcel.2009.10.018.
287. Penkner A, Tang L, Novatchkova M, Ladurner M, Fridkin A, Gruenbaum Y, Schweizer D, Loidl J, Jantsch V. The nuclear envelope protein Matefin/SUN-1 is required for homologous pairing in *C. elegans* meiosis. *Dev Cell.* 2007;12(6):873–885. doi: 10.1016/j.devcel.2007.05.004.

288. Sato A, Isaac B, Phillips CM, Rillo R, Carlton PM, Wynne DJ, Kasad RA, Dernburg AF. Cytoskeletal forces span the nuclear envelope to coordinate meiotic chromosome pairing and synapsis. *Cell.* 2009;139(5):907–919. doi: 10.1016/j.cell.2009.10.039.
289. King MC, Drivas TG, Blobel G. A network of nuclear envelope membrane proteins linking centromeres to microtubules. *Cell.* 2008;134(3):427–438. doi: 10.1016/j.cell.2008.06.022.
290. Hiraoka Y, Dernburg AF. The SUN rises on meiotic chromosome dynamics. *Dev Cell.* 2009;17(5):598–605. doi: 10.1016/j.devcel.2009.10.014.
291. Güttinger S, Laurell E, Kutay U. Orchestrating nuclear envelope disassembly and reassembly during mitosis. *Nat Rev Mol Cell Biol.* 2009;10(3):178–191. doi: 10.1038/nrm2641.
292. Karpov SA, Mylnikov AP. Preliminary observations on the ultrastructure of mitosis in choanoflagellates. *Eur J Protistol.* 1993;29:19–23.
293. Cavalier-Smith T. In: *Evolutionary relationships among Protozoa.* Coombs GH, Vickerman K, Sleigh MA, Warren A, editor. London: Chapman and Hall; 1998. Neomonada and the origin of animals and fungi; pp. 375–407.
294. Raikov IB. *The protozoan nucleus: morphology and evolution.* Vienna: Springer-Verlag; 1982.
295. Smirnov AV, Nassonova ES, Chao EE, Cavalier-Smith T. Phylogeny, evolution and taxonomy of vannellid amoebae. *Protist.* 2007;158:295–324. doi: 10.1016/j.protis.2007.04.004.
296. Bhaud Y, Guillebault D, Lennon J, Defacque H, Soyer-Gobillard MO, Moreau H. Morphology and behaviour of dinoflagellate chromosomes during the cell cycle and mitosis. *J Cell Sci.* 2000;113(Pt 7):1231–1239.
297. Levi-Setti R, Gavrilov KL, Rizzo PJ. Divalent cation distribution in dinoflagellate chromosomes imaged by high-resolution ion probe mass spectrometry. *Eur J Cell Biol.* 2008;87(12):963–976. doi: 10.1016/j.ejcb.2008.06.002.
298. Chan YH, Wong JT. Concentration-dependent organization of DNA by the dinoflagellate histone-like protein HCc3. *Nucleic Acids Res.* 2007;35(8):2573–2583. doi: 10.1093/nar/gkm165.
299. Wong JT, New DC, Wong JC, Hung VK. Histone-like proteins of the dinoflagellate *Crypthecodinium cohnii* have homologies to bacterial DNA-binding proteins. *Eukaryot Cell.* 2003;2(3):646–650. doi:

10.1128/EC.2.3.646-650.2003.

300. Hackett JD, Scheetz TE, Yoon HS, Soares MB, Bonaldo MF, Casavant TL, Bhattacharya D. Insights into a dinoflagellate genome through expressed sequence tag analysis. *BMC Genomics.* 2005;6(1):80. doi: 10.1186/1471-2164-6-80.

301. Alverca E, Cuadrado A, Jouve N, Franca S, Moreno Diaz de la Espina S. Telomeric DNA localization on dinoflagellate chromosomes: structural and evolutionary implications. *Cytogenet Genome Res.* 2007;116(3):224–231. doi: 10.1159/000098191.

302. Alverca E, Franca S, Diaz de la Espina SM. Topology of splicing and snRNP biogenesis in dinoflagellate nuclei. *Biol Cell.* 2006;98(12):709–720. doi: 10.1042/BC20050083.

303. Cavalier-Smith T. In: *Unravelling the Algae.* Brodie J, Lewis J, editor. Boca Raton: CRC Press; 2007. Evolution and relationships of algae: major branches of the tree of life; pp. 21–55.

304. Chikashige Y, Tsutsumi C, Yamane M, Okamasa K, Haraguchi T, Hiraoka Y. Meiotic proteins bqt1 and bqt2 tether telomeres to form the bouquet arrangement of chromosomes. *Cell.* 2006;125(1):59–69. doi: 10.1016/j.cell.2006.01.048.

305. Chikashige Y, Yamane M, Okamasa K, Tsutsumi C, Kojidani T, Sato M, Haraguchi T, Hiraoka Y. Membrane proteins Bqt3 and -4 anchor telomeres to the nuclear envelope to ensure chromosomal bouquet formation. *J Cell Biol.* 2009;187(3):413–427. doi: 10.1083/jcb.200902122.

306. Cavalier-Smith T. r- and K-tactics in the evolution of protist developmental systems: cell and genome size, phenotype diversifying selection, and cell cycle patterns. *Biosystems.* 1980;12(1-2):43–59. doi: 10.1016/0303-2647(80)90037-4.

307. Cavalier-Smith T, Chao EE, Stechmann A, Oates B, Nikolaev S. Planomonadida ord. nov. (Apusozoa): ultrastructural affinity with *Micronuclearia podoventralis* and deep divergences within *Planomonas* gen. nov. *Protist.* 2008;159:535–562. doi: 10.1016/j.protis.2008.06.002.

308. Cavalier-Smith T, Chao EE. Phylogeny of Choanozoa, Apusozoa, and other Protozoa and early eukaryote megaevolution. *J Mol Evol.* 2003;56:540–563. doi: 10.1007/s00239-002-2424-z.

309. Cavalier-Smith T. In: *The evolution of genome size.* Cavalier-Smith

T, editor. Chichester: Wiley; 1985. Cell volume and the evolution of genome size; pp. 105–184.

310. Beaton MJ, Cavalier-Smith T. Eukaryotic non-coding DNA is functional: evidence from the differential scaling of cryptomonad genomes. *Proc Biol Sci.* . 1999;266(1433):2053–2059. doi: 10.1098/rspb.1999.0886.

311. Cavalier-Smith T, Beaton MJ. The skeletal function of non-genic nuclear DNA: new evidence from ancient cell chimaeras. *Genetica.* 1999;106(1-2):3–13. doi: 10.1023/A:1003701925110.

312. Cavalier-Smith T. Selfish DNA and the origin of introns. *Nature.* 1985;315:283–284. doi: 10.1038/315283b0.

313. Pendle AF, Clark GP, Boon R, Lewandowska D, Lam YW, Andersen J, Mann M, Lamond AI, Brown JW, Shaw PJ. Proteomic analysis of the *Arabidopsis* nucleolus suggests novel nucleolar functions. *Mol Biol Cell.* 2005;16(1):260–269. doi: 10.1091/mbc.E04-09-0791.

314. Maquat LE. Nonsense-mediated mRNA decay: splicing, translation and mRNP dynamics. *Nat Rev Mol Cell Biol.* . 2004;5(2):89–99. doi: 10.1038/nrm1310.

315. Anantharaman V, Aravind L. New connections in the prokaryotic toxin-antitoxin network: relationship with the eukaryotic nonsense-mediated RNA decay system. *Genome Biol.* 2003;4:R81. doi: 10.1186/gb-2003-4-12-r81.

316. Bercovich N, Levin MJ, Clayton C, Vazquez MP. Identification of core components of the exon junction complex in trypanosomes. *Mol Biochem Parasitol.* 2009;166(2):190–193. doi: 10.1016/j.molbiopara.2009.03.008.

317. Mair G, Shi H, Li H, Djikeng A, Aviles HO, Bishop JR, Falcone FH, Gavrilescu C, Montgomery JL, Santori MI, Stern LS, Wang Z, Ullu E, Tschudi C. A new twist in trypanosome RNA metabolism: cis-splicing of pre-mRNA. *Rna.* 2000;6(2):163–169. doi: 10.1017/S135583820099229X.

318. Matic I, Macek B, Hilger M, Walther TC, Mann M. Phosphorylation of SUMO-1 occurs in vivo and is conserved through evolution. *J Proteome Res.* 2008;7(9):4050–4057. doi: 10.1021/pr800368m.

319. Schimmel J, Larsen KM, Matic I, van Hagen M, Cox J, Mann M, Andersen JS, Vertegaal AC. The ubiquitin-proteasome system is a key component of the SUMO-2/3 cycle. *Mol Cell Proteomics.* 2008;7(11):2107–2122.

doi: 10.1074/mcp.M800025-MCP200.

320. Shiio Y, Eisenman RN. Histone sumoylation is associated with transcriptional repression. *Proc Natl Acad Sci USA.* 2003;100(23):13225–13230. doi: 10.1073/pnas.1735528100.

321. Nathan D, Ingvarsdottir K, Sterner DE, Bylebyl GR, Dokmanovic M, Dorsey JA, Whelan KA, Krsmanovic M, Lane WS, Meluh PB, Johnson ES, Berger SL. Histone sumoylation is a negative regulator in *Saccharomyces cerevisiae* and shows dynamic interplay with positive-acting histone modifications. *Genes Dev.* 2006;20(8):966–976. doi: 10.1101/gad.1404206.

322. Gao C, Ho CC, Reineke E, Lam M, Cheng X, Stanya KJ, Liu Y, Chakraborty S, Shih HM, Kao HY. Histone deacetylase 7 promotes PML sumoylation and is essential for PML nuclear body formation. *Mol Cell Biol.* 2008;28(18):5658–5667. doi: 10.1128/MCB.00874-08.

323. Zhao X, Blobel G. A SUMO ligase is part of a nuclear multiprotein complex that affects DNA repair and chromosomal organization. *Proc Natl Acad Sci USA.* 2005;102(13):4777–4782. doi: 10.1073/pnas.0500537102.

324. Xu XM, Rose A, Muthuswamy S, Jeong SY, Venkatakrishnan S, Zhao Q, Meier I. NUCLEAR PORE ANCHOR, the *Arabidopsis* homolog of Tpr/Mlp1/Mlp2/megator, is involved in mRNA export and SUMO homeostasis and affects diverse aspects of plant development. *Plant Cell.* 2007;19(5):1537–1548. doi: 10.1105/tpc.106.049239.

325. Saracco SA, Miller MJ, Kurepa J, Vierstra RD. Genetic analysis of SUMOylation in *Arabidopsis* : conjugation of SUMO1 and SUMO2 to nuclear proteins is essential. *Plant Physiol.* 2007;145(1):119–134. doi: 10.1104/pp.107.102285.

326. Shang Q, Xu C, Zhang J, Zhang X, Tu X. Solution structure of SUMO from *Trypanosoma brucei* and its interaction with Ubc9. *Proteins.* 2009;76(1):266–269. doi: 10.1002/prot.22409.

327. Burroughs AM, Balaji S, Iyer LM, Aravind L. A novel superfamily containing the beta-grasp fold involved in binding diverse soluble ligands. *Biol Direct.* 2007;2:4. doi: 10.1186/1745-6150-2-4.

328. Burroughs AM, Balaji S, Iyer LM, Aravind L. Small but versatile: the extraordinary functional and structural diversity of the beta-grasp fold. *Biol Direct.* 2007;2:18. doi: 10.1186/1745-6150-2-18.

329. Iyer LM, Burroughs AM, Aravind L. The prokaryotic antecedents of

the ubiquitin-signaling system and the early evolution of ubiquitin-like beta-grasp domains. *Genome Biol.* 2006;7(7):R60. doi: 10.1186/gb-2006-7-7-r60.

330. Pearce MJ, Mintseris J, Ferreyra J, Gygi SP, Darwin KH. Ubiquitin-like protein involved in the proteasome pathway of *Mycobacterium tuberculosis*. *Science.* 2008;322(5904):1104–1107. doi: 10.1126/science.1163885.
331. Iyer LM, Burroughs AM, Aravind L. Unraveling the biochemistry and provenance of pupylation: a prokaryotic analog of ubiquitination. *Biol Direct.* 2008;3:45. doi: 10.1186/1745-6150-3-45.
332. Cavalier-Smith T. Cell cycles, diplokaryosis, and the archezoan origin of sex. *Archiv f Protistenkunde.* 1995;145:189–207.
333. Cavalier-Smith T. *The Oxford Encyclopaedia of Evolution.* New York: Oxford University Press; 2002. Meiosis; pp. 700–708.
334. Cleveland LR. The origin and evolution of meiosis. *Science.* 1947;105(2724):287–289. doi: 10.1126/science.105.2724.287.
335. Kilmartin JV. Sfi1p has conserved centrin-binding sites and an essential function in budding yeast spindle pole body duplication. *J Cell Biol.* 2003;162(7):1211–1221. doi: 10.1083/jcb.200307064.
336. Pelletier L, Ozlu N, Hannak E, Cowan C, Habermann B, Ruer M, Muller-Reichert T, Hyman AA. The Caenorhabditis elegans centrosomal protein SPD-2 is required for both pericentriolar material recruitment and centriole duplication. *Curr Biol.* 2004;14(10):863–873. doi: 10.1016/j.cub.2004.04.012.
337. Nash B, Bowerman B. Uncoiling centriole duplication. *Nat Cell Biol.* 2004;6(7):573–575. doi: 10.1038/ncb0704-573.
338. Lange BM, Faragher AJ, March P, Gull K. Centriole duplication and maturation in animal cells. *Curr Top Dev Biol.* 2000;49:235–349. full_text.
339. Gerstein AC, McBride RM, Otto SP. Ploidy reduction in *Saccharomyces cerevisiae*. *Biol Lett.* 2008;4(1):91–94. doi: 10.1098/rsbl.2007.0476.
340. Keightley PD, Otto SP. Interference among deleterious mutations favours sex and recombination in finite populations. *Nature.* 2006;443:89–92. doi: 10.1038/nature05049.
341. Kondrashov AS. The asexual ploidy cycle and the origin of sex. *Nature.* 1994;370(6486):213–216. doi: 10.1038/370213a0.

342. Otto SP. The evolutionary enigma of sex. *Am Nat.* 2009;174(Suppl 1):S1–S14. doi: 10.1086/599084.
343. Hill JA, Otto SP. The role of pleiotropy in the maintenance of sex in yeast. *Genetics.* 2007;175(3):1419–1427. doi: 10.1534/genetics.106.059444.
344. Agrawal AF, Otto SP. Host-parasite coevolution and selection on sex through the effects of segregation. *Am Nat.* 2006;168(5):617–629. doi: 10.1086/508029.
345. Mandegar MA, Otto SP. Mitotic recombination counteracts the benefits of genetic segregation. *Proc Biol Sci.* 2007;274(1615):1301–1307. doi: 10.1098/rspb.2007.0056.
346. Cleveland LR. Brief accounts of the sexual cycle of the flagellates of *Cryptocercus*. *J Protozool.* 1956;3:161–180.
347. Hadany L, Otto SP. Condition-dependent sex and the rate of adaptation. *Am Nat.* 2009;174(Suppl 1):S71–78. doi: 10.1086/599086.
348. Cavalier-Smith T, Nikolaev S. The zooflagellates *Stephanopogon* and *Percolomonas* are a clade (class Percolatea: phylum Percolozoa) *J Eukaryot Microbiol.* 2008;55:501–509. doi: 10.1111/j.1550-7408.2008.00356.x.
349. Trigueros S, Tran T, Sorto N, Newmark J, Colloms SD, Sherratt DJ, Tolmasky ME. mwr Xer site-specific recombination is hypersensitive to DNA supercoiling. *Nucleic Acids Res.* 2009;37(11):3580–3587. doi: 10.1093/nar/gkp208.
350. Villasante A, Abad JP, Mendez-Lago M. Centromeres were derived from telomeres during the evolution of the eukaryotic chromosome. *Proc Natl Acad Sci USA.* 2007;104(25):10542–10547. doi: 10.1073/pnas.0703808104.
351. Tyler-Smith C, Floridia G. Many paths to the top of the mountain: diverse evolutionary solutions to centromere structure. *Cell.* 2000;102(1):5–8. doi: 10.1016/S0092-8674(00)00004-0.
352. Cavalier-Smith T. Electron microscopic evidence for chloroplast fusion in zygotes of *Chlamydomonas reinhardii*. *Nature.* 1970;228(269):333–335. doi: 10.1038/228333a0.
353. Tartakoff AM, Jaiswal P. Nuclear fusion and genome encounter during yeast zygote formation. *Mol Biol Cell.* 2009;20(12):2932–29342. doi: 10.1091/mbc.E08-12-1193.
354. Kiefel BR, Gilson PR, Beech PL. Diverse eukaryotes have retained

mitochondrial homologues of the bacterial division protein FtsZ. *Protist.* 2004;155:105–115. doi: 10.1078/1434461000168.

355. Kuroiwa T, Misumi O, Nishida K, Yagisawa F, Yoshida Y, Fujiwara T, Kuroiwa H. Vesicle, mitochondrial, and plastid division machineries with emphasis on dynamin and electron-dense rings. *Int Rev Cell Mol Biol.* 2008;271:97–152. full_text.

356. Rahaman A, Elde NC, Turkewitz AP. A dynamin-related protein required for nuclear remodeling in Tetrahymena. *Curr Biol.* 2008;18(16):1227–1233. doi: 10.1016/j.cub.2008.07.042.

357. Gerstein AC, Chun HJ, Grant A, Otto SP. Genomic convergence toward diploidy in *Saccharomyces cerevisiae. PLoS Genet.* 2006;2(9):e145. doi: 10.1371/journal.pgen.0020145.

358. Gerstein AC, Otto SP. Ploidy and the causes of genomic evolution. *J Hered.* 2009;100(5):571–581. doi: 10.1093/jhered/esp057.

359. Otto SP, Gerstein AC. The evolution of haploidy and diploidy. *Curr Biol.* 2008;18(24):R1121–1124. doi: 10.1016/j.cub.2008.09.039.

360. Moreira D, López-García P. Symbiosis between methanogenic archaea and delta-proteobacteria as the origin of eukaryotes: the syntrophic hypothesis. *J Mol Evol.* 1998;47(5):517–530. doi: 10.1007/PL00006408.

361. López-García P, Moreira D. Metabolic symbiosis at the origin of eukaryotes. *Trends Biochem Sci.* 1999;24(3):88–93. doi: 10.1016/S0968-0004(98)01342-5.

362. Rivera MC, Lake JA. The ring of life provides evidence for a genome fusion origin of eukaryotes. *Nature.* 2004;431(7005):152–155. doi: 10.1038/nature02848.

363. Gupta RS, Golding GB. The origin of the eukaryotic cell. *Trends Biochem Sci.* 1996;21(5):166–171.

364. Hartman H, Fedorov A. The origin of the eukaryotic cell: a genomic investigation. *Proc Natl Acad Sci USA.* 2002;99(3):1420–1425. doi: 10.1073/pnas.032658599.

365. Doolittle WF. You are what you eat: a gene transfer ratchet could account for bacterial genes in eukaryotic nuclear genomes. *Trends Genet.* . 1998;14(8):307–311. doi: 10.1016/S0168-9525(98)01494-2.

366. Sagan L. On the origin of mitosing cells. *J Theor Biol.* . 1967;14(3):255–274. doi: 10.1016/0022-5193(67)90079-3.

367. De Duve C. The origin of eukaryotes: a reappraisal. *Nat Rev Genet.* 2007;8:395–403. doi: 10.1038/nrg2071.
368. Lindsay MR, Webb RI, Strous M, Jetten MS, Butler MK, Forde RJ, Fuerst JA. Cell compartmentalisation in Planctomycetes: novel types of structural organisation for the bacterial cell. *Arch Microbiol.* 2001;175(6):413–429. doi: 10.1007/s002030100280.
369. Giezen M van der, Tovar J. Degenerate mitochondria. *EMBO Rep.* 2005;6(6):525–530. doi: 10.1038/sj.embor.7400440.
370. Staley JT, Bouzek H, Jenkins C. Eukaryotic signature proteins of *Prosthecobacter dejongeii* and *Gemmata* sp. Wa-1 as revealed by in silico analysis. *FEMS Microbiol Lett.* 2005;243(1):9–14. doi: 10.1016/j.femsle.2004.11.034.
371. Pilhofer M, Rosati G, Ludwig W, Schleifer KH, Petroni G. Coexistence of tubulins and FtsZ in different *Prosthecobacter* species. *Mol Biol Evol.* 2007;24(7):1439–1442. doi: 10.1093/molbev/msm069.
372. Pilhofer M, Bauer AP, Schrallhammer M, Richter L, Ludwig W, Schleifer KH, Petroni G. Characterization of bacterial operons consisting of two tubulins and a kinesin-like gene by the novel Two-Step Gene Walking method. *Nucleic Acids Res.* 2007;35(20):e135. doi: 10.1093/nar/gkm836.
373. Mitchell DR. The evolution of eukaryotic cilia and flagella as motile and sensory organelles. *Adv Exp Med Biol.* 2007;607:130–140. full_text.
374. Jékely G, Arendt D. Evolution of intraflagellar transport from coated vesicles and autogenous origin of the eukaryotic cilium. *Bioessays.* 2006;28(2):191–198. doi: 10.1002/bies.20369.
375. Bellare P, Small EC, Huang X, Wohlschlegel JA, Staley JP, Sontheimer EJ. A role for ubiquitin in the spliceosome assembly pathway. *Nat Struct Mol Biol.* 2008;15(5):444–451. doi: 10.1038/nsmb.1401.
376. Jékely G. Origin of the nucleus and Ran-dependent transport to safeguard ribosome biogenesis in a chimeric cell. *Biol Direct.* 2008;3:31. doi: 10.1186/1745-6150-3-31.
377. Cavalier-Smith T. Kingdom Protozoa and its 18 phyla. *Microbiol Rev.* 1993;57(4):953–994.
378. Haldane JBS. *The causes of evolution.* London: Longmans, Green and Co.; 1932.
379. Cavalier-Smith T. In: *Fundamentals of Medical Cell biology.* Bittar

GE, editor. I. Greenwich, Conn.: J.A.I. Press; 1991. The evolution of prokaryotic and eukaryotic cells; pp. 217–272.

380. Cavalier-Smith T. In: *Evolution of Life.* Osawa S, Honjo T, editor. Tokyo: Springer-Verlag; 1991. The evolution of cells; pp. 271–304.

381. Cavalier-Smith T. The origin of cells: a symbiosis between genes, catalysts, and membranes. *Cold Spring Harb Symp Quant Biol.* 1987;52:805–824.

382. Heard E, Colot V. Chromosome structural proteins and RNA-mediated epigenetic silencing. *Dev Cell.* 2008;14(6):813–814. doi: 10.1016/j.devcel.2008.05.013.

383. Larsen CN, Wang H. The ubiquitin superfamily: members, features, and phylogenies. *J Proteome Res.* 2002;1(5):411–419. doi: 10.1021/pr025522n.

384. Schwartz DC, Hochstrasser M. A superfamily of protein tags: ubiquitin, SUMO and related modifiers. *Trends Biochem Sci.* 2003;28(6):321–328. doi: 10.1016/S0968-0004(03)00113-0.

385. Cavalier-Smith T. Selfish DNA and the origin of introns. *Nature.* 1985;315(6017):283–284. doi: 10.1038/315283b0.

386. Hinnebusch J, Tilly K. Linear plasmids and chromosomes in bacteria. *Mol Microbiol.* 1993;10(5):917–922. doi: 10.1111/j.1365-2958.1993.tb00963.x.

387. Lundgren M, Andersson A, Chen L, Nilsson P, Bernander R. Three replication origins in *Sulfolobus* species: synchronous initiation of chromosome replication and asynchronous termination. *Proc Natl Acad Sci USA.* 2004;101(18):7046–7051. doi: 10.1073/pnas.0400656101.

388. Robinson NP, Dionne I, Lundgren M, Marsh VL, Bernander R, Bell SD. Identification of two origins of replication in the single chromosome of the archaeon *Sulfolobus solfataricus*. *Cell.* 2004;116(1):25–38. doi: 10.1016/S0092-8674(03)01034-1.

389. Lundgren M, Bernander R. Archaeal cell cycle progress. *Curr Opin Microbiol.* 2005;8(6):662–668. doi: 10.1016/j.mib.2005.10.008.

390. Lundgren M, Malandrin L, Eriksson S, Huber H, Bernander R. Cell cycle characteristics of Crenarchaeota: unity among diversity. *J Bacteriol.* 2008;190(15):5362–5367. doi: 10.1128/JB.00330-08.

391. Forterre P. Thermoreduction, a hypothesis for the origin of prokaryotes. *C R Acad Sci III.* 1995;318(4):415–422.

392. Barns SM, Delwiche CF, Palmer JD, Pace NR. Perspectives on

archaeal diversity, thermophily and monophyly from environmental rRNA sequences. *Proc Natl Acad Sci USA.* 1996;93(17):9188–9193. doi: 10.1073/pnas.93.17.9188.

393. Elkins JG, Podar M, Graham DE, Makarova KS, Wolf Y, Randau L, Hedlund BP, Brochier-Armanet C, Kunin V, Anderson I, Lapidus A, Goltsman E, Barry K, Koonin EV, Hugenholtz P, Kyrpides N, Wanner G, Richardson P, Keller M, Stetter KO. A korarchaeal genome reveals insights into the evolution of the Archaea. *Proc Natl Acad Sci USA.* 2008;105(23):8102–8107. doi: 10.1073/pnas.0801980105.
394. Arkhipova I, Meselson M. Transposable elements in sexual and ancient asexual taxa. *Proc Natl Acad Sci USA.* 2000;97(26):14473–14477. doi: 10.1073/pnas.97.26.14473.
395. Arkhipova IR, Meselson M. Diverse DNA transposons in rotifers of the class Bdelloidea. *Proc Natl Acad Sci USA.* 2005;102(33):11781–11786. doi: 10.1073/pnas.0505333102.
396. Gladyshev EA, Meselson M, Arkhipova IR. Massive horizontal gene transfer in bdelloid rotifers. *Science.* 2008;320(5880):1210–1213. doi: 10.1126/science.1156407.
397. King MC, Lusk CP, Blobel G. Karyopherin-mediated import of integral inner nuclear membrane proteins. *Nature.* 2006;442(7106):1003–1007. doi: 10.1038/nature05075.
398. Zhang Z, Cavalier-Smith T, Green BR. A family of selfish minicircular chromosomes with jumbled chloroplast gene fragments from a dinoflagellate. *Mol Biol Evol.* 2001;18(8):1558–1565.
399. Zhang Z, Cavalier-Smith T, Green BR. Evolution of dinoflagellate unigenic minicircles and the partially concerted divergence of their putative replicon origins. *Mol Biol Evol.* 2002;19(4):489–500.
400. Cavalier-Smith T. In: *Endocytobiology II.* Schwemmler W, Schenk HEA, editor. Berlin: de Gruyter; 1983. A 6-kingdom classification and a unified phylogeny; pp. 1027–1034.
401. Cavalier-Smith T. In: *Endocytobiology II.* Schwemmler W, Schenk HEA, editor. Berlin: de Gruyter; 1983. Endosymbiotic origin of the mitochondrial envelope; pp. 265–279.
402. Yutin N, Wolf MY, Wolf YI, Koonin EV. The origins of phagocytosis and eukaryogenesis. *Biol Direct.* 2009;4:9. doi: 10.1186/1745-6150-4-9.

403. Cavalier-Smith T. Concept of a bacterium still valid in prokaryote debate. *Nature.* 2007;446(7133):7257. doi: 10.1038/446257c.

404. Woese CR, Fox GE. Phylogenetic structure of the prokaryotic domain: the primary kingdoms. *Proc Natl Acad Sci USA.* 1977;74(11):5088–5090. doi: 10.1073/pnas.74.11.5088.

CHAPTER 5

CHANGES IN NUCLEAR SHAPE AND GENE EXPRESSION IN RESPONSE TO SIMULATED MICROGRAVITY ARE LINC COMPLEX-DEPENDENT

Srujana Neelam[1,2,3], Brian Richardson[4], Richard Barker[2], Ceasar Udave[5], Simon Gilroy[2], Mark J. Cameron[4] , Howard G. Levine[1] and Ye Zhang[1]

[1]Utilization and Life Sciences, NASA John F. Kennedy Space Center, Kennedy Space Center, Merritt Island, FL 32899, USA

[2]Department of Botany, University of Wisconsin-Madison, Madison, WI 53706, USA

[3]Universities Space Research Association, Columbia, MD 21046, USA

[4]Department of Population and Quantitative Health Sciences, Case Western Reserve University, Cleveland, OH 44106, USA

[5]Department of Biomedical Engineering, Arizona State University, Tempe, AZ 85281, USA

Citation: Neelam S, Richardson B, Barker R, Udave C, Gilroy S, Cameron MJ, Levine HG, Zhang Y. Changes in Nuclear Shape and Gene Expression in Response to Simulated Microgravity Are LINC Complex-Dependent. Int J Mol Sci. 2020 Sep 15;21(18):E6762. doi: 10.3390/ijms21186762. PMID: 32942630.

ABSTRACT

Microgravity is known to affect the organization of the cytoskeleton, cell and nuclear morphology and to elicit differential expression of genes associated with the cytoskeleton, focal adhesions and the extracellular matrix. Although the nucleus is mechanically connected to the cytoskeleton through the Linker of Nucleoskeleton and Cytoskeleton (LINC) complex, the role of this group of proteins in these responses to microgravity has yet to be defined. In our study, we used a simulated microgravity device, a 3-D clinostat (Gravite), to investigate whether the LINC complex mediates cellular responses to the simulated microgravity environment. We show that nuclear shape and differential gene expression are both responsive to simulated microgravity in a LINC-dependent manner and that this response changes with the duration of exposure to simulated microgravity. These LINC-dependent genes likely represent elements normally regulated by the mechanical forces imposed by gravity on Earth.

Keywords: LINC complex; simulated microgravity; nuclear morphology

INTRODUCTION

The nucleus is the largest organelle of the cell and homeostatic control of its shape and position are known to be crucial for cellular functions such as cell migration, cell division and the regulation of gene expression [1,2,3,4,5]. Forces generated by the cytoskeleton and external mechanical stimuli are directly transmitted to the nuclear surface through the Linker of Nucleoskeleton and Cytoskeleton (LINC) complex [3,6,7,8,9,10,11]—a series of proteins embedded in the nuclear envelope (NE). Thus, LINC complex Nesprins on the outer nuclear membrane are connected to different cytoplasmic cytoskeletal elements and to several other force-generating motor proteins. These Nesprin family proteins also have a KASH (Klarsicht, ANC-1, Syne Homology) domain that is located in the perinuclear space and connects to the SUN (Sad1 and UNC-84) proteins of the inner nuclear membrane [12]. The SUN proteins are in turn connected to chromatin through lamins. This mechanical connection from nuclear surface to chromatin is crucial in enabling the forces from the cytoskeleton and extracellular matrix (ECM) to be transmitted to both the surface and interior of the nucleus. These forces can, therefore, alter not just nuclear morphology but also internal elements such as chromatin architecture and so lead to the modulation of mechanosensitive gene expression.

Exposure of cells to microgravity is known to lead to the reorganization of the cytoskeleton and the ECM. In addition, microgravity leads to morphological changes in both the cell and its nucleus, and to altered gene expression [13,14,15,16,17,18,19,20]—all features related to LINC complex function. The LINC complex is also known to be a critical factor in the normal formation and development of tissues, such as the breast cell acinus [21]. Simulated microgravity (SMG) achieved by clinorotation disrupts features such as cell proliferation, cell adhesion, cell cycle progression and cytokine production [13,22]. These are all LINC-regulated processes, and failure in their normal functioning has been linked to aberrant developmental regulation, such as cancer. Indeed, in addition to altering all these LINC-dependent features, SMG also suppresses tumor cell growth and development in e.g., breast epithelial cell cultures [23]. However, despite such circumstantial evidence, the role of the LINC complex in cellular responses to altered gravitational environments has yet to be studied. Therefore, we have asked how the LINC complex affects changes in nuclear shape and differential gene expression in SMG. We report that the nucleus requires an intact LINC complex for its shape to be modulated by altered gravity conditions. We also identify LINC-dependent genes associated with this response, indicating that elements of the genome regulation normally seen in altered gravity requires an intact LINC complex.

RESULTS

Simulated Microgravity Elicits Changes in Nuclear Shape that Depend on the LINC Complex

As noted above, breast epithelial cells have been previously shown to exhibit responses to SMG, suggesting that they might be a useful model to study the involvement of the LINC complex in SMG-related effects. Furthermore, we have a deep understanding of nuclear shaping under normal gravity conditions in MCF 10A—a non-tumerogenic human mammary epithelial cell line [24]. This cell line was engineered to express a construct that interferes with LINC complex function (see below), making it well-suited to investigating the role of the LINC complex in SMG-related effects. Therefore, we chose MCF 10A cells, wildtype and engineered, to study how SMG may impact nuclear events, using a 3-D clinostat to simulate microgravity.

We first established whether the nuclear shape of MCF 10A cells is responsive to clinostat-induced SMG. After SMG treatment, we fixed the cells, and the nuclei were stained with Hoechst 33342, allowing us to measure any change in nuclear height, major axis and minor axis using confocal microscopy. Actin staining was used to define the shape of the cells to ensure only those with the morphology characteristic of adherence to the substrate were analyzed. We observed that the nuclear height significantly ($p < 0.05$, *t*-test) decreased after 20 h of exposure to SMG, and the nuclear spreading area increased.

To understand if this change in nuclear shape in response to SMG depends on the LINC complex, we used a doxycycline inducible MCF 10A cell line where the LINC complex function is disrupted. This inducible cell line stably expresses SS-HA-SUN1L-KDEL (signal sequence-hemaglutinin epitope tagged to SUN1 protein, with a lumenal domain-ER retrieval amino acid sequence; SUN1L-KDEL) [25]. SUN1L-KDEL acts in a dominant negative fashion disrupting the LINC complex by interfering with the connection between the endogenous KASH domain of nesprin-2 and nesprin-4 proteins and the SUN1 proteins in the nuclear envelope [3,21,25]. Cells expressing a similar doxycycline inducible construct SS-GFP-KDEL (GFP-KDEL) were used as the corresponding paired control. We confirmed that the induction of SUN1L-KDEL but not GFP-KDEL did indeed disrupt the LINC complex by immunostaining with a Nesprin-2 antibody. As noted above, Nesprin-2 localizes to the nuclear membrane in a LINC-complex-dependent fashion, and the immunofluorescence staining showed a loss of Nesprin-2 in the NE in the doxycycline induced SUN1L-KDEL cells but not in the corresponding control doxycycline-treated GFP-KDEL cells. These observations are consistent with those of Zhang et.al. who previously showed that this system is effective in disrupting the LINC function [21].

The doxycycline-induced cells were, therefore, exposed to 3-D clinorotation for 20 h. The 1G (ground) controls (i.e., identically doxycycline treated cells but not on the clinostat) were placed next to the clinostat in the same incubator. After treatment, the cells were fixed and the 3-dimensional nuclear morphology was monitored using confocal microscopy as described earlier.

The GFP-KDEL cells showed a decrease in nuclear height after 20 h (similar to the MCF 10A wild type cells) in SMG (Figure 1A and Table 1). In contrast, the SUN1L-KDEL cells showed a significant nuclear height increase when compared to 1G cells over this same time period in SMG. These results show that the normal nuclear shape change to SMG requires an intact LINC complex.

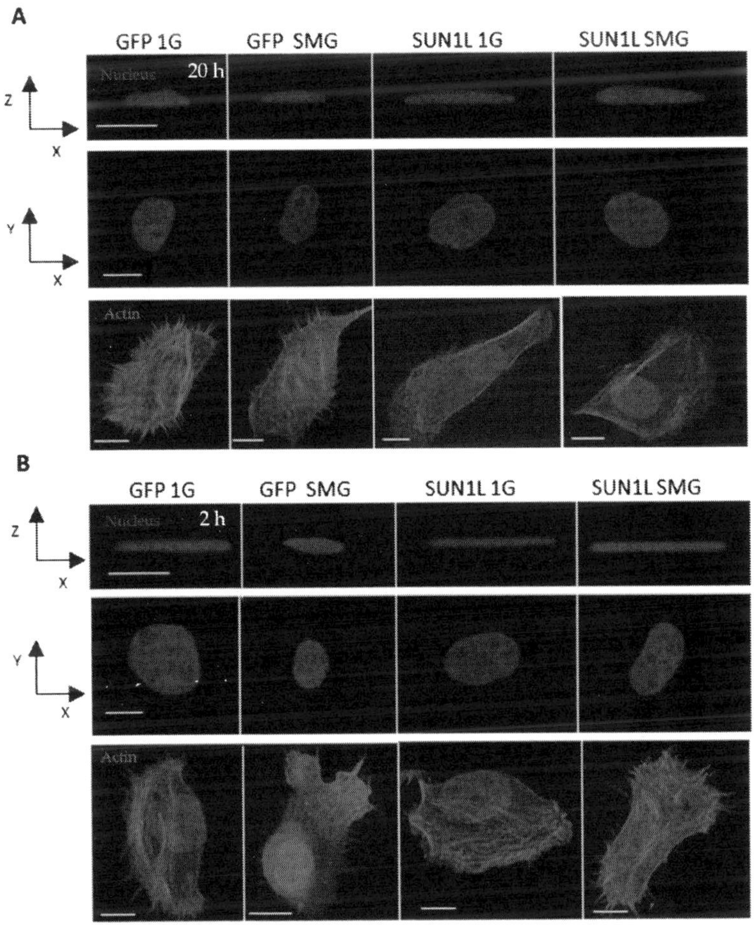

Figure 1: Nuclear shape response to simulated microgravity (SMG) depends on the LINC complex. (A) Confocal images show the nuclei (blue) and actin (red) staining (to reveal the cell morphology) of fixed GFP-KDEL (GFP) and SUN1L-KDEL (SUN1L) cells after 20 h in 1G and SMG. Top row shows the x-z view and the bottom rows show the x-y view of the corresponding nuclei and cells. Similarly, (B) shows confocal images of the nuclei and actin of fixed GFP-KDEL and SUN1L-KDEL cells in 1G and SMG condition for 2 h. Scale bars are 10 μm. Images are representative of the mean values.

Table 1: Nuclear morphology: After 20 h exposure to SMG, the nuclear height decreased in GFP-KDEL cells, while the height increased in SUN1L-KDEL cells again indicating a role of the LINC complex in the response of the nuclear shape to SMG. * $p < 0.05$, comparing same cell line to the 1G condition within the same time-point, statistically significantly different; the comparisons are with the corresponding 1G control. All values are shown as the mean ± SD.2.2. Short Duration Simulated Microgravity-Induced, LINC-Dependent Changes in Nuclear Shape.

	Cell Type	GFP-KDEL			SUN1L-KDEL		
time		n	Nuclear height (μm)	X-Y nuclear area (μm^2)	n	Nuclear height (μm)	X-Y nuclear area (μm^2)
20 h	Ground	74	2.7 ± 1.1	207 ± 89	73	2.7 ± 1.2	216 ± 111
	SMG	81	2.2 ± 1.1 *	267 ± 148 *	99	3.4 ± 1.1 *	164 ± 41 *
2 h	Ground	82	2.0 ± 0.8	234 ± 104	45	2.2 ± 0.6	212 ± 60
	SMG	68	3.0 ± 1.5 *	176 ± 107 *	43	2.6 ± 1.1	193 ± 60

We next asked whether significant LINC-dependency of nuclear responses to SMG could also be seen with a shorter SMG treatment. Therefore, to observe the impact of short duration exposure of LINC disrupted cells to SMG, we assayed responses after 2 h on the clinostat. The nuclear height increased significantly ($p < 0.05$, *t*-test), and the nuclear area decreased in GFP-KDEL cells exposed to SMG for 2h. This impact, however, was opposite to the response at the 20 h time point, i.e., decrease in nuclear height in SMG (Figure 1B, Table 1), indicating that the cells may start adapting to SMG differently at longer timepoints. The SUN1L-KDEL cells on the other hand did not show any significant difference in nuclear morphology between the 1G and SMG cells over 2 h of SMG (Table 1). These observations again suggest that the effects of SMG on nuclear shape required an intact LINC complex.

LINC-Dependent Differential Gene Expression in Simulated Microgravity

Since the changes in nuclear shape triggered by both 2 and 20 h SMG appear to require the LINC complex, we hypothesized that disruption of the LINC complex with SUN1L KDEL expression might also affect patterns of gene expression. We, therefore, used RNAseq analysis to characterize differential gene expression in GFP-KDEL and SUN1L-KDEL in SMG vs. 1G controls. Figure 2A shows a plot of the fold-change in expression level (SMG vs. 1G control) against significance of this change (*p*-value) in our initial analysis of transcriptional responses after 2 h of exposure to SMG. A larger number of genes were up- or down-regulated in SUN1L-KDEL cells in SMG (558 up, 434 down, nominal $p < 0.05$) compared to those in the GFP-KDEL cells (203 up, 179 down, nominal $p < 0.05$; Figure 2A,B).

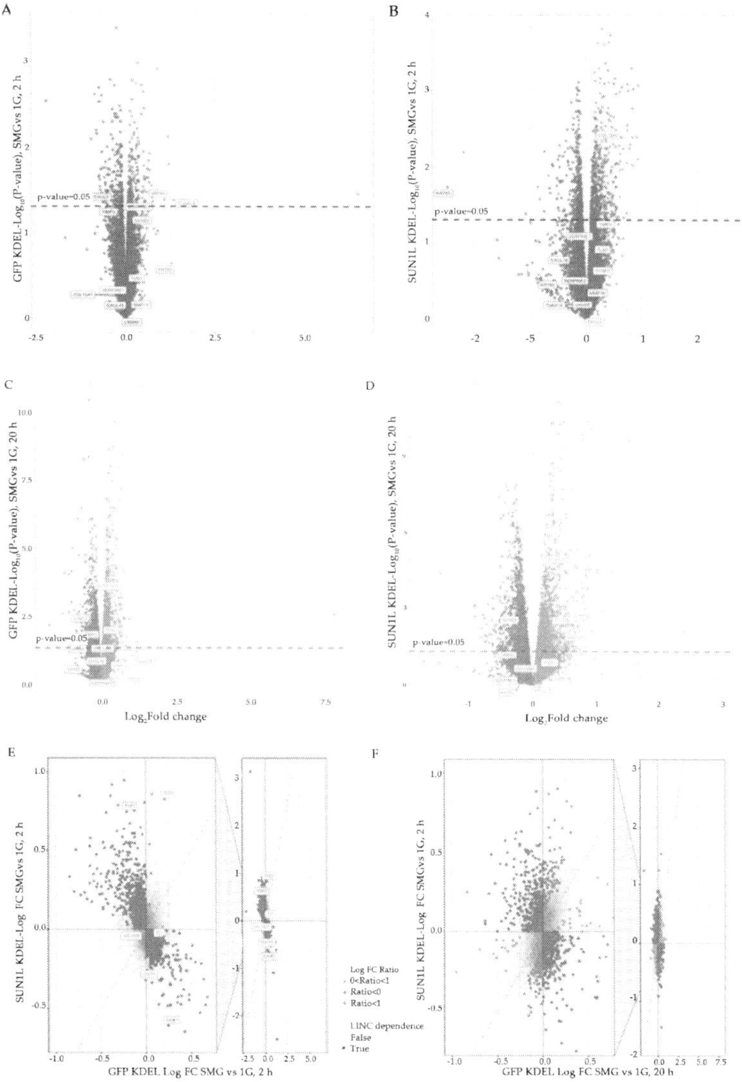

Figure 2. Transcriptome responds to SMG and varies upon LINC disruption at different exposure times. Volcano plots of differentially expressing genes between SMG and 1G with statistical significance in the *y*-axis and fold change on the *x*-axis in (A) GFP-KDEL cells after 2 h, (B) SUN1L-KDEL cells af-

ter 2 h, (C) GFP-KDEL cells after 20 h, (D) SUN1L-KDEL cells after 20 h. Upregulated genes are shown in orange and downregulated genes are shown in blue. Significantly differentially expressing genes with $p \leq 0.05$ are above the red dashed line. (E,F) summarize all the LINC-dependent genes identified through the interaction model at 2 and 20 h, respectively. The whole distribution is shown to the right with the indicated area magnified on the left. Genes with statistical significance ($p < 0.05$) are shown to be differentially regulated under the interaction of LINC disruption and SMG conditions and are highlighted with LINC dependence equal to True in the figure. Genes that are not highlighted (False) did not attain significance and are not dependent on LINC. For each gene, the log fold-change (logFC) for response to SMG vs. its 1G control in the SUNL-KDEL cells is divided by the same characterization but in the GFP KDEL controls and is color-coded on the plot. Based on this logFC ratio, these genes could be characterized as being regulated in the opposite direction for each cell line (green, Ratio < 0), regulated more strongly under the SUN1L KDEL cell line (blue, Ratio > 1) or regulated more strongly under the GFP cell line (pink, $0 <$ Ratio < 1).

We identified the top 50 most significantly differentially expressed genes (selected as showing the smallest p-value when comparing 2 h SMG to 1G control) in each condition. Nassef et al. used true microgravity (10^{-2} G) in parabolic flight to show that the expression of genes related to the cytoskeleton, extracellular matrix and chemokines is affected in breast epithelial cells exposed to true microgravity [14]. Thus, this study provides some specific genes as well as broad classes of potential genes of interest. We, therefore, closely compared the response of the individual genes in the Nassef et al. study with their responses in our study. Even though we did not see similar response in any of these genes after 2 h exposure to SMG, we found that TIMP1, SERPINE1 and RDX genes were differentially regulated after 20 h SMG exposure, similar to those reported in the Nassef study in cells after experiencing 31 parabolas. The differences in gene expression may be due to the alternating hypergravity and microgravity phases experienced during each parabola. However, in our 2 h SMG data, we did see changes in genes reflecting the broad classes of gene families reported by Nassef et al. Thus, in GFP-KDEL expressing cells, 2 h SMG led to the modulation of genes associated with the cytoskeleton (e.g., KRT85, TUBA1A), the extracellular matrix (TIMP3, SERPINF2), and chemokine response (CXCL3). These genes are highlighted in Figure 2A,B. To infer whether the LINC complex might play a role in regulating the expression of any of these genes in SMG, we compared their levels in the GFP-KDEL controls to those where the LINC complex was disrupted by SUN1L-KDEL expression. We predicted that genes for which the LINC function played a role in regulating their expression under SMG conditions would show one of several patterns of response. Thus, SMG-responsive genes for which the change in transcript level is LINC-dependent might show alteration in expression in SMG in the GFP KDEL control cells but lose this responsiveness to SMG in the SUN1L KDEL lines. Some genes might show responsiveness to SMG in the control cells but show a significantly larger or smaller change in the LINC-disrupted case, implying a role of the LINC complex in the degree of repression or activation of expression. Additionally, some genes may show no differential expression in SMG when comparing GFP KDEL cells in SMG to their 1G controls, but these genes would show differential expression in the LINC-disrupted SUN1L KDEL expressing cells. In this latter case, normal gene expression homeostasis would require LINC action, and in the LINC-disrupted cells, this regulation would be lost. Loss of this control would lead to either up- or down-regulation of their transcription, specifically in SMG, revealing a further layer of LINC-dependency to their normal transcriptional regulation under SMG conditions. Indeed, our analysis shows that the differential expression of some of these genes in SMG appears to depend on an intact LINC complex. For example, the cytoskeletal element keratin (KRT85) was upregulated by 2.13-fold in SMG in GFP-KDEL cells, but there was no significant difference in its expression levels in LINC-disrupted SUN1L-KDEL cells exposed to equivalent SMG treatment. Similarly, the chemokine response marker CXCL3 (C-X-C Motif Chemokine Ligand 3) was downregulated 1.4-fold in GFP-KDEL, but this gene lost its sensitivity to SMG in SUN1L-KDEL cells. In contrast, the LINC disruption using SUN1L-KDEL resulted in the decreased expression (5.65-fold) of KRT83 (another cytoskeletal gene belonging to the keratin family), whereas this gene did not alter due to SMG in GFP-KDEL cells. Furthermore, the ECM gene (COL12A1) showed no significant difference in GFP-KDEL cells but was upregulated by 1.22-fold in the SUN1L-KDEL expressing cells. Thus, the LINC complex appears to play roles in normally inducing some genes, while repressing other genes in response to SMG and in maintaining homeostatic control of other genes under SMG conditions.

To elucidate the role of the LINC in differential gene expression, we used an interaction model between transcriptional changes seen in SUN1L-KDEL and GFP-KDEL expressing cells with and without SMG exposure. The interaction effect is described in Figure 2E where the relative fold changes between SMG and 1G in GFP-KDEL and SUN1L-KDEL samples at 2 h are compared statistically. The majority of the LINC-dependent genes reversed their direction of change upon LINC disruption (Ratio < 0). For example, the FSIP2 gene was significantly down-regulated in GFP-KDEL cells in SMG. However, the change in expression level of this gene reversed in SUN1L-KDEL cells, with a significant upregulation. Similarly, CAGE1, a gene widely associated with cancer [26] was significantly upregulated in the GFP-KDEL cells, while it was significantly downregulated in SUN1L-KDEL cells illustrating an opposite pattern of response of the gene FSIP2 to LINC disruption. In addition to these genes, we identified genes whose sensitivity to SMG was lost (0 < Ratio < 1) and genes which gained such sensitivity (Ratio > 1) upon LINC disruption. For example, ARL14EP and ID1 are the genes which were significantly differentially expressed in GFP-KDEL cells exposed to 2h of SMG but were not altered in SUN1L-KDEL cells exposed to SMG, implying that these genes lost their sensitivity to SMG upon LINC disruption. On the other hand, PER3 and OSCP1 genes were not affected by SMG in GFP-KDEL cells but were significantly differentially expressed in SUN1L-KDEL cells exposed to 2 h of SMG. The results both imply that the LINC complex is required to maintain transcriptional homeostasis of some genes in the presence of SMG and that an intact mechanical nucleo-cytoskeletal linkage may mediate response to SMG and to regulate subsequent differential gene expression of other genes.

KEGG (Kyoto Encyclopedia of Genes and Genomes) signaling pathway analysis of the differentially expressed genes revealed that in GFP-KDEL cells exposed to 2 h of SMG, six pathways were enriched, while a larger number of pathways were enriched only in the SUN1L-KDEL cells under the same conditions (Figure 3A). We next used the Molecular Signatures Database (MSigDB) to perform gene ontology analysis. The MSigDB is a curated set of annotated gene sets which can be analyzed by Gene Set Variation Analysis (GSVA) to determine if any genes in a particular curated set (pathway, biological state, cellular locale etc.) are enriched in a genome-wide expression analysis using a ranked set of genes [27,28]. Pathway analysis of the GFP-KDEL cells in SMG for the MSigDB GSVA GO biological process dataset revealed enrichment in genes related to microtubule-based movement, the reproductive process and negative regulation of catalytic activity. A similar analysis for cellular components revealed that terms relating to nuclear specks and the kinetochore were significantly enriched. A greater number of significantly enriched GO pathways were identified in the SMG SUN1L-KDEL expressing cells, including many biological processes related to mechanoresponse systems such as: cell matrix adhesion, regulation of cell migration and negative regulation of cell migration. GO pathway analysis of cellular components revealed enrichment related to the mitochondrion, actin cytoskeleton and cell adhesion. Some of the LINC-dependent GO cellular component pathways for mechanosensing identified by the interaction effect were mitochondrion, cell junction, intercellular junction, tight junction, kinetochore and nuclear specks.

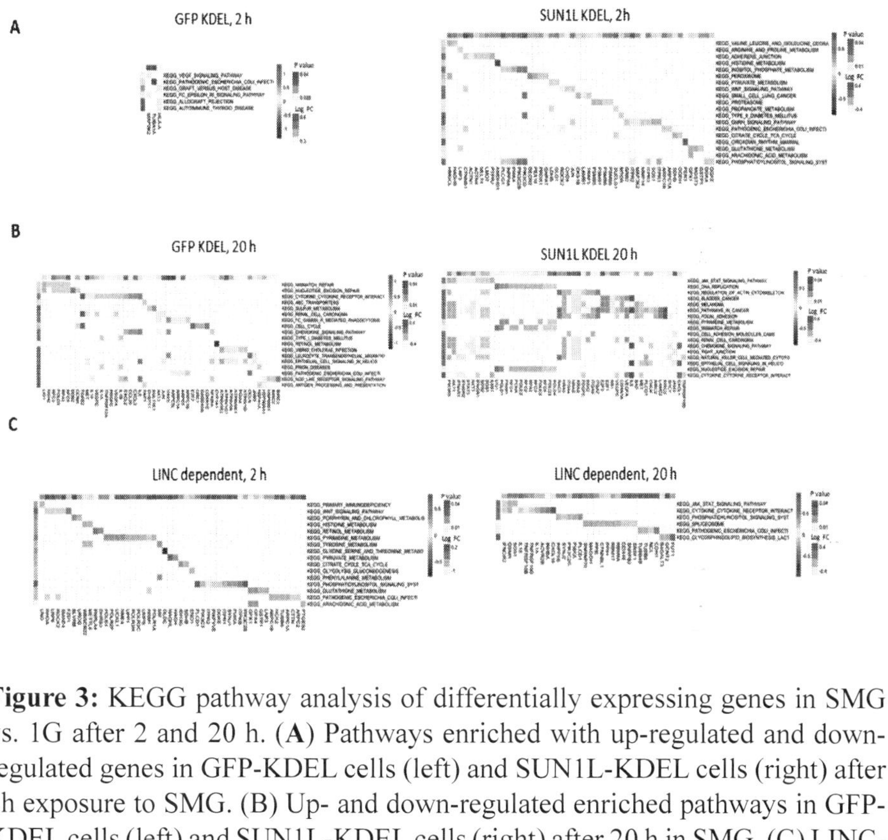

Figure 3: KEGG pathway analysis of differentially expressing genes in SMG vs. 1G after 2 and 20 h. (**A**) Pathways enriched with up-regulated and down-regulated genes in GFP-KDEL cells (left) and SUN1L-KDEL cells (right) after 2h exposure to SMG. (B) Up- and down-regulated enriched pathways in GFP-KDEL cells (left) and SUN1L-KDEL cells (right) after 20 h in SMG. (C) LINC-dependent up- and down-regulated enriched pathways in SMG after 2 (left) and 20 h (right).

LINC Dependency of SMG-Induced Changes in Gene Expression Is Also Determined by the Duration of Exposure to SMG

After 20 h in SMG, the GFP-KDEL cells had a larger number of genes that were significantly differentially expressed compared to the 2 h condition. Additionally, there was an even larger number of differentially expressed genes in SUN1L-KDEL expressing cells (1279 upregulated,

1542 downregulated, $p \leq 0.05$) than the GFP-KDEL cells (813 upregulated, 846 downregulated, ≤ 0.05) after 20 h (Figure 2C,D). In GFP-KDEL cells, the levels of expression of several genes associated with cytokines and chemokines were significantly altered, including the VEGFA gene. While the focal adhesion associated genes TLN2 and SERPINE1 (involved in the regulation of cell adhesion and spreading) were upregulated, the ECM genes TIMP3 and LAMA5 were down-regulated. The expression of the NE gene encoding for nesprin-3 (SYNE3) showed a significant decrease in expression (Figure 2C). We identified the genes whose expression depended on the duration of exposure to SMG by assessing the interaction effect between GFP-KDEL transcriptional changes at 2 and 20 h.

We also identified the unique genes and pathways that depend on the LINC complex to potentially mechanosense the gravitational forces in SMG after 20 h by evaluating the interaction effect between SUN1L-KDEL and GFP-KDEL transcriptional changes at 20 h (Figure 2F). All the above-mentioned genes were differentially expressed in SMG-treated SUN1-KDEL cells, in a similar fashion to GFP-KDEL, implying that they were not dependent on the LINC complex, with the notable exception of TIMP3. In contrast, MMP19 (encodes a collagenase related to ECM dynamics), which was not responsive to SMG in GFP-KDEL cells, gained responsiveness upon LINC complex disruption. Similarly, the chemokine gene CXCL16 was also modulated differently in SMG-treated SUN1-KDEL cells compared to the expression in GFP KDEL cells (Figure 2C,D). We also identified the genes in SUN1L-KDEL cells that were dependent on the duration of exposure to SMG by assessing the interaction effect between SUN1L-KDEL transcriptional changes at 2 and 20 h.

More KEGG pathways were enriched in GFP-KDEL cells after 20 h in SMG when compared to the effects of 2 h exposure (Figure 3A,B). For example, the cell cycle pathway was enriched in the differentially expressed transcripts in GFP-KDEL cells exposed to SMG for 20 h, whereas it was not affected in cells exposed for 2 h. In the SUN1-KDEL cells, regulation of the actin cytoskeleton was highlighted in the KEGG analysis, as were pathways associated with adhesion, such as focal adhesion, cell adhesion molecules and tight junction, implying alteration in the systems mediating cell-to-matrix and cell-to-cell contacts in SMG. The DNA replication mismatch repair pathway was also enriched, which may be linked to alterations in various cancer associated pathways/processes highlighted as being differentially regulated in the data (bladder cancer, melanoma, pathways in cancer, renal cell carcinoma). Altered cytokine receptor interaction and chemokine signaling pathways were also seen in this analysis. Thus, the LINC complex disruption in simulated microgravity may be associated with pathways related to cancer development. We also identified the KEGG pathways that are LINC-dependent in microgravity with the interaction model. It is important to note that there were a greater number of LINC-dependent pathways that were enriched at 2 h when compared to 20 h (Figure 3C).

Cluster analysis was next performed on the LINC-dependent differentially expressing genes ($p \leq 0.05$) at both 2 and 20 h SMG exposure. Using the Metascape tool (www.metascape.org), pathway and process enrichment analysis was performed using KEGG Functional Sets, KEGG Pathway, in this approach, network and clusters were developed based on the enrichment with $p \leq 0.01$ and with a minimum count of 3. In Figure 4A, the top 20 clusters are shown with each node color coded based on the cluster it belongs to. From the pie chart representation of each node (Figure 4B), we can see that clusters 1, 2 and 5 had all the nodes enriched for both the LINC-dependent 2 and 20 h SMG exposure conditions. Interestingly, Cluster 1 and Cluster 2 comprised enriched terms all related to the mitochondrian as listed in Figure 4C. Previous studies have found similar dysregulation and dysfunction of the mitochondrian in human cells exposed to simulated microgravity provided by 3-D clinostat and RPM exposures [29,30]. Our analysis clearly shows that the LINC complex may be required in regulating mitochondrion-dependent pathways and processes due to simulated microgravity at 2 and 20 h. We hypothesize that LINC may be playing a crucial role in modulating the mitochondrial response to SMG. However, more experimentation will be required to robustly test how LINC may be impacting the response of mitochondria in simulated microgravity.

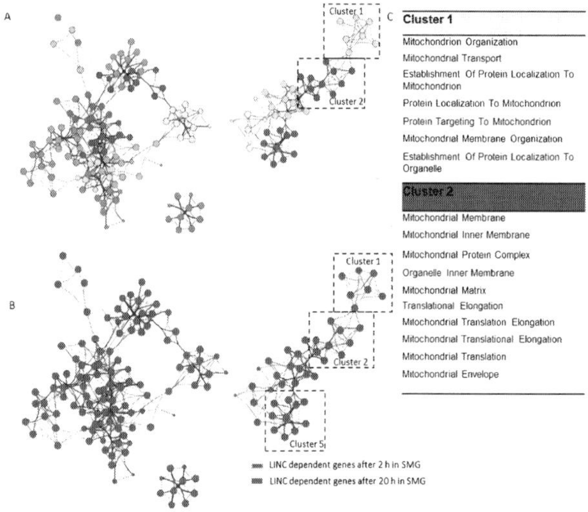

Figure 4. Network of top 20 clusters of the enriched terms. (**A**) Network representing each cluster by a different color; (B) network enrichment shown with pie chart. The size of the pie in each node depends on the gene count of the two conditions. Red indicates enrichment of pathways due to LINC-dependent genes at 20 h of SMG, and blue indicates enrichment of pathways in LINC-dependent genes at 2 h of SMG. (C) Enriched terms in Clusters 1 and 2 are all linked to mitochondria.

The GO pathway analysis of these data revealed that processes such as regulation of cell growth, cell activation and protein folding were enriched, alongside cellular components including the actin cytoskeleton, myosin complex, microtubule organization center and endoplasmic reticulum in GFP-KDEL cells. The GO analysis for of SUN1L-KDEL cells further revealed enrichment in biological processes such as response to external stimulus, response to extracellular stimulus, response to hypoxia, response to wounding and inflammatory response. The GO analysis for cellular components revealed a mix of upregulated and down-regulated processes after 20 h of SMG: for example, upregulation of actin cytoskeleton, leading edge, cell junction, tight junction and downregulation of microtubule organizing center and chromosomes. Some patterns of gene expression were consistent between responses at 2 and 20 h, including enrichment in cell adhesion and the actin cytoskeleton. The GO cellular component pathway analysis with the interaction model revealed the LINC-dependent components after 20 h in SMG mainly highlighted the tight junction, plasma membrane, intercellular junction components and mitochondrial components. The LINC-dependent GO biological pathways were also identified.

DISCUSSION

A highly reproducible suite of effects were observed in cells exposed to altered gravity relating to the reorganization of their cytoskeleton and cell adhesion proteins [31], coupled with changes in the morphologies of the cells, including their nuclei. There is a close link between nuclear shape and the cytoskeleton, with various cytoskeletal elements playing roles in regulating nuclear shape and position. Thus, actin polymerization flattens the nucleus and repositions it by pulling on it. Additionally, forces generated by microtubule motor proteins such as dynein and kinesin, and acto-myosin contractile and tensile forces are also transmitted to the nuclear surface. Intermediate filaments also form a cage-like structure around the nucleus giving it structural stability. Each of these elements are mechanically connected to the nucleus through LINC complex proteins. The mechanical integration afforded by these connections through the LINC complex plays a critical role in transferring mechanical signals from the ECM and the forces generated within the cytoplasm to the nucleus [32]. These signals regulate features such as nuclear shape, homeostatic position and rotation—all important elements for nuclear functioning. Therefore, not surprisingly, the LINC complex is important for an array of normal cellular functions such as migration, division, proliferation, differentiation and wound healing [33]. Consistent with these ideas of a central role of the LINC complex in mechanical signaling, there are also major LINC-associated diseases such as nuclear envelopathies, laminopathy, skeletal muscular dystrophy and cardiomyopathy. Abnormal regulation of LINC complex genes SUN1, SUN2 and nesprin 2 is also seen in breast cancer.

Several LINC-associated proteins such as SUN1, SUN2, Emerins and Nesprin-3 were seen to be differentially regulated in cancer cells exposed to simulated microgravity [13]. Indeed, SMG promotes breast cancer cell proliferation, invasion and adhesion, nuclear positioning, as well as cell cycle progression, apoptosis [22] and abnormal expression of cancer inducing cytokines [14]. The LINC complex is known to play a very important role in the normal formation and development of breast cell acinus and in regulating the shape of the cell and the nucleus in both the breast cell culture of single cells and of monolayers [24]. Thus, understanding how SMG affects these cellular processes holds potential for both understanding how gravitational forces might be playing roles in normal tissue development on Earth and in revealing potential practical issues and risks related to long duration spaceflight.

In the present study, we looked at breast epithelial cells (MCF 10A) in SMG with and without LINC disruption. We discovered that the changes in nuclear shape seen after 2 h and 20 h exposures to SMG depend on the LINC complex. Previous studies have shown that both the shape of cells and their nuclei are impacted by altered gravity environments [34], and our studies now provide evidence that this effect depends on a functional LINC complex. Thus, after 2 h in SMG, LINC disruption leads to a lack of nuclear shape change normally observed in cells with intact LINC function. After 20 h in SMG, the response reverses in direction, with the nucleus rounding up compared to flatter nuclei in cells with an intact LINC complex. As discussed earlier, the cytoskeleton reorganizes in microgravity, with disruption of the microtubule network, and a so-called 'loosening' of the cytokeratin network around the nucleus, and this may be the reason for the rounding up of the nucleus (increased nuclear height and decrease in x-y area) in GFP-KDEL control cells exposed to 2 h of SMG. We hypothesize that, as in SUN1L-KDEL cells, the cytoskeleton is not mechanically connected to the nuclear surface, and so such SMG-induced changes in the cytoskeleton are not able to affect the nuclear shape. After 20 h in SMG, however, the nuclear height decreased, and the x-y nuclear area increased in GFP-KDEL cells compared to the cells in 1G. This observation can be explained by the increase in cell area reported in both simulated microgravity and true microgravity experiments conducted for long durations, and that the cell shape is inextricably linked to the nuclear shape, i.e., cell spreading is driving nuclear "flattening" [24,35,36,37]. On the other hand, the SUN1L-KDEL cells showed an increase in nuclear height under the same conditions, consistent with the idea of a loss of cytoskeletal interactions that normally pull on the periphery of the nuclear envelope. It has been reported previously that the response of cell morphology to calibrated physical external stimulus in LINC disrupted cells is different when compared to the cells with intact LINC complex function [1,2]. Therefore, we may expect the nuclear morphology in the LINC-disrupted cells to also respond to the longer duration SMG very differently to the control GFP-KDEL cells. Despite these time-dependent effects, our analysis shows that both at 2 and 20 h SMG exposure times, the LINC complex plays a role in the nuclear shaping mechanism. These results support the idea that the mechanosensitivity of the nuclear shape to external stimuli strongly depends on the SUN1-Nesprin linkage in the nuclear envelope provided by the LINC complex [3].

We also looked at the role of the LINC complex in differential gene expression in response to SMG, revealing that there are LINC-dependent genes that are up- and down-regulated in response to SMG. We hypothesize that the nuclear shape regulation and gene expression dependency on the LINC complex are inter-related. Thus, gene activity depends on the location of the chromosome and its packing inside the nucleus [38], e.g., chromosome territories are positioned differently and have different interactions in breast epithelial cells (MCF 10A) compared to corresponding malignant cells (MCF 10CA1a), supporting different patterns of gene expression [39]. Nuclear shape and rigidity depends on the packing of the heterochromatin [40] and changes in nuclear shape may in turn feedback to change the level of chromatin condensation [41]. Therefore, it is highly likely that patterns of gene expression depend to some degree on nuclear shape [4,42]. The LINC complex connects the cytoskeleton to the nuclear surface; the SUN proteins in the inner nuclear envelope are connected to lamins, which are connected to chromatin. Thus, the LINC complex provides a strong candidate for linking cellular level mechanical forces to changes in nuclear gene expression. For example, interfering with the LINC complex in plant cells led to decondensation of chromocenters and silencing of transcription [43]. Hence, the LINC complex may impact gene expression in two ways: (1) by directly affecting chromatin compaction; (2) by regulating the shape of the nucleus, and both mechanisms may be in play once cells move from the 1G to a microgravity environment.

It is important to note that there are changes in cell physiology just with time growing in culture. For example, Seaman et al. have found that the nuclear morphology changes significantly over time in cultured human primary fibroblast cells [44]. Indeed, in our data both the GFP-KDEL and SUN1L-KDEL cells at 1G (Table 1) show expected decrease in nuclear area and increase in nuclear height when comparing data from 2 h and 20 h in culture. In most cell types, the shape of the nucleus depends mainly on factors such as the shape of the cell, cytoskeleton organization, and external adhesions such as cell-matrix and cell-cell adhesion [3,24,45]. Studies with human breast epithelial cells found that the shape of the nucleus is controlled by the shape of the cell [24]. Also, fibroblast cell nuclei are flatter in cells with larger spreading area than in lesser spread cells [37]. The dynamic actin cap flattens the nucleus while actomyosin contractile forces pull on the nucleus regulating its volume and the gross morphology [2,46,47]. Intermediate filaments also wrap the nucleus like a cage and act as a passive network

growth factor (Peprotech, Cranbury, NJ, USA), hydrocortisone (Sigma), insulin (Sigma) and penicillin streptomycin (Thermofisher, Grand Island, NY, USA). The cells were maintained at 37 °C in a humidified, 5% (*v/v*) CO_2 incubating chamber. The inducible cells were trypsinized and seeded in fibronectin coated chamber flasks with 30% confluency and cultured for 24 h. To induce SUN1L-KDEL for LINC disruption and GFP-KDEL for control comparison, the media were supplemented with 1 µg/mL doxycycline [21]. These cells were cultured in doxycycline for 24 h before placing them on the 3-D clinostat.

Simulated Microgravity Using 3-D Clinostat

The chamber flasks were filled completely with media containing 1 µg/mL doxycycline (Fisher) and care taken to remove all the bubbles that may cause shear stress to the cell monolayer. The chamber flask was sealed with a rubber stopper, and a syringe was used to release any excess pressure. The 3-D clinostat (Gravite, Space Bio Laboratories Ltd., Hiroshima, Japan) consisting of two perpendicular rotating axes was run in Mode C preprogrammed with an inbuilt computer and gravity acceleration sensor to measure simulate microgravity (to the order of 10^{-3} G) by rotating the inner and outer frame simultaneously at a constant angular velocity (~2 rotations per minute) [58,59,60]. Four cell culture flasks on slides (Thermofisher, Grand Island, NY, USA, 170920) were mounted at the center and were placed in a controlled environment maintaining the temperature at 37 °C for 2 and 20 h. The 1G control samples were placed in the same incubator. All the experiments were conducted twice, and each experiment had two sets of samples in both 1G and SMG conditions. A Student's *t*-test was applied for statistical comparison for nuclear morphology.

Immunostaining and Microscopy

After the cells were exposed to simulated microgravity for fixed periods of time, they were fixed with 4% paraformaldehyde for 10 min and then washed with PBS. To stain for F-actin, 1:50 Alex Fluor 594 phalloidin was used and for the nucleus, 1:250 Hoechst 33,342 (Life Technologies, Carlsbad, CA, USA) was used. Cells were then incubated for 20 min at room temperature before washing with PBS. The cells were then imaged with the 60×/1.40 NA oil immersion objective of a Nikon A1R laser scanning inverted confocal microscope (Nikon Instruments, Melville, NY, USA). To immunostain for the Nesprin-2 protein, we followed the protocol of Zhang et al. [21] and used the primary antibody anti-Nesprin-2 (Abcam, Cambridge, MA, USA, ab204308) at a 1:100 dilution and the secondary antibody goat anti-rabbit IgG (Thermofisher, Grand Island, NY, USA, A-11037) at a 1:200 dilution. Z-stack images were acquired for the cells with 0.3 μM height intervals. The images were reconstructed using the Nikon NIS elements software, and maximum intensity profiles for the nucleus were created for the x-z view. The full width half maximum (FWHM) technique in MATLAB (MathWorks, Natick, MA, USA) was used to acquire the nuclear height measurement. The maximum intensity projection of the x-y view of the nucleus was used to measure the nuclear area using the ImageJ software.

RNA seq Library Preparation, Sequencing and Data Analysis

Approximately 2×10^6 cells were frozen in 700 μL of QIAzol buffer (Qiagen, Hilden, Germany). The samples were immediately vortexed and placed on ice before storing at −80 °C. Once all samples were collected, they were randomized, and total RNA was extracted and purified using the Qiagen RNeasy Micro Kit, following recommended procedures, with on-column DNase treatment. Libraries were prepared from purified RNA (0.3 ng) using the TruSeq Stranded Total RNA Library Prep Globin (Illumina, San Diego, CA, USA) according to the manufacturer's protocol. Three runs of 75 bp paired-end sequencing were performed on pooled libraries using the Illumina NextSeq 550 System. Quality control of purified RNA and RNA libraries was carried out using the Fragment Analyzer System (Agilent Technologies, Santa Clara, CA, USA).

Raw DE multiplexed fastq paired end read files were trimmed of adapters and filtered using the program skewer [61] to discard any reads with an average phred quality score of less than 30, or a length of less than 36 nucleotides. Trimmed reads were then aligned using the HISAT2 aligner to the *Homo sapiens* NCBI reference genome assembly version GRCh38 and sorted using SAM tools [62,63]. Aligned reads were counted and assigned to gene metafeatures using the program feature Counts as part of the Subread package [64]. These count files were imported into the R programming language and were assessed for quality control, normalized and analyzed using an in-house pipeline utilizing the edge R Bioconductor library, for differential gene expression testing, and the GSVA library, for GO and KEGG pathway analysis, utilizing their respective collections from the Molecular Signatures Database (MSigDB) [28,65,66]. The GO pathway analysis is based on GSVA using the Molecular Signatures Database (MSigDB) GO collections, as opposed to the GO term enrichment, which does not take into account the ranking of the genes [27,28].

ACKNOWLEDGMENTS

This work was supported by the NASA Postdoctoral Program (NPP) administered by Universities Space Research Association (USRA) and the NASA Space Biology Program. We also extend our sincere appreciation to Tanmay Lele, University of Florida, for the kind gift of the MCF 10A engineered cells. We would like to sincerely thank Stephanie Richards and Jeffrey Richards for providing support with laboratory logistics in the Microgravity Simulation Support Facility at KSC.

REFERENCES

1. Chancellor, T.J.; Lee, J.; Thodeti, C.K.; Lele, T. Actomyosin tension exerted on the nucleus through nesprin-1 connections influences endothelial cell adhesion, migration, and cyclic strain-induced reorientation. Biophys. J. 2010, 99, 115–123.
2. Wu, J.; Kent, I.A.; Shekhar, N.; Chancellor, T.J.; Mendonca, A.; Dickinson, R.B.; Lele, T.P. Actomyosin pulls to advance the nucleus in a migrating tissue cell. Biophys. J. 2014, 106, 7–15.
3. Neelam, S.; Chancellor, T.J.; Li, Y.; Nickerson, J.A.; Roux, K.J.; Dickinson, R.B.; Lele, T.P. Direct force probe reveals the mechanics of nuclear homeostasis in the mammalian cell. Proc. Natl. Acad. Sci. USA 2015, 112, 5720–5725.
4. Thomas, C.H.; Collier, J.H.; Sfeir, C.S.; Healy, K.E. Engineering gene expression and protein synthesis by modulation of nuclear shape. Proc. Natl. Acad. Sci. USA 2002, 99, 1972–1977.
5. Wang, Y.; Nagarajan, M.; Uhler, C.; Shivashankar, G.V. Orientation and repositioning of chromosomes correlate with cell geometry-dependent gene expression. Mol. Biol. Cell 2017, 28, 1997–2009.
6. Arsenovic, P.T.; Ramachandran, I.; Bathula, K.; Zhu, R.; Narang, J.D.; Noll, N.A.; Lemmon, C.A.; Gundersen, G.G.; Conway, D.E. Nesprin-2G, a Component of the Nuclear LINC Complex, Is Subject to Myosin-Dependent Tension. Biophys. J. 2016, 110, 34–43.
7. Hieda, M. Signal Transduction across the Nuclear Envelope: Role of the LINC Complex in Bidirectional Signaling. Cells 2019, 8, 124.
8. Alam, S.G.; Zhang, Q.; Prasad, N.; Li, Y.; Chamala, S.; Kuchibhotla, R.; Kc, B.; Aggarwal, V.; Shrestha, S.; Jones, A.L.; et al. The mammalian LINC complex regulates genome transcriptional responses to substrate rigidity. Sci. Rep. 2016, 6, 38063.
9. Alam, S.G.; Lovett, D.; Kim, D.I.; Roux, K.J.; Dickinson, R.B.; Lele, T.P. The nucleus is an intracellular propagator of tensile forces in NIH 3T3 fibroblasts. J. Cell Sci. 2015, 128, 1901–1911.
10. Wang, N.; Butler, J.P.; Ingber, D.E. Mechanotransduction across the cell surface and through the cytoskeleton. Science 1993, 260, 1124–1127.
11. Starr, D.A.; Fridolfsson, H.N. Interactions between nuclei and the cytoskeleton are mediated by SUN-KASH nuclear-envelope bridges. Annu. Rev. Cell Dev. Biol. 2010, 26, 421–444.

12. Roux, K.J.; Crisp, M.L.; Liu, Q.; Kim, D.; Kozlov, S.; Stewart, C.L.; Burke, B. Nesprin 4 is an outer nuclear membrane protein that can induce kinesin-mediated cell polarization. Proc. Natl. Acad. Sci. USA 2009, 106, 2194–2199.
13. Zhao, T.; Li, R.; Tan, X.; Zhang, J.; Fan, C.; Zhao, Q.; Deng, Y.; Xu, A.; Lukong, K.E.; Genth, H.; et al. Simulated Microgravity Reduces Focal Adhesions and Alters Cytoskeleton and Nuclear Positioning Leading to Enhanced Apoptosis via Suppressing FAK/RhoA-Mediated mTORC1/NF-κB and ERK1/2 Pathways. Int. J. Mol. Sci. 2018, 19, 1994.
14. Nassef, M.Z.; Kopp, S.; Wehland, M.; Melnik, D.; Sahana, J.; Krüger, M.; Corydon, T.J.; Oltmann, H.; Schmitz, B.; Schütte, A.; et al. Real microgravity influences the cytoskeleton and focal adhesions in human breast cancer cells. Int. J. Mol. Sci. 2019, 20, 3156.
15. Crawford-Young, S.J. Effects of microgravity on cell cytoskeleton and embryogenesis. Int. J. Dev. Biol. 2006, 50, 183–191.
16. Thiel, C.S.; Paulsen, K.; Bradacs, G.; Lust, K.; Tauber, S.; Dumrese, C.; Hilliger, A.; Schoppmann, K.; Biskup, J.; Gölz, N.; et al. Rapid alterations of cell cycle control proteins in human T lymphocytes in microgravity. Cell Commun. Signal. 2012, 10, 1.
17. Thiel, C.S.; Hauschild, S.; Huge, A.; Tauber, S.; Lauber, B.A.; Polzer, J.; Paulsen, K.; Lier, H.; Engelmann, F.; Schmitz, B.; et al. Dynamic gene expression response to altered gravity in human T cells. Sci. Rep. 2017, 7, 5204.
18. Ma, X.; Pietsch, J.; Wehland, M.; Schulz, H.; Saar, K.; Hübner, N.; Bauer, J.; Braun, M.; Schwarzwälder, A.; Segerer, J.; et al. Differential gene expression profile and altered cytokine secretion of thyroid cancer cells in space. FASEB J. 2014, 28, 813–835.
19. Ulbrich, C.; Pietsch, J.; Grosse, J.; Wehland, M.; Schulz, H.; Saar, K.; Hübner, N.; Hauslage, J.; Hemmersbach, R.; Braun, M.; et al. Differential gene regulation under altered gravity conditions in follicular thyroid cancer cells: Relationship between the extracellular matrix and the cytoskeleton. Cell. Physiol. Biochem. 2011, 28, 185–198.
20. Hughes-Fulford, M.; Rodenacker, K.; Jütting, U. Reduction of anabolic signals and alteration of osteoblast nuclear morphology in microgravity. J. Cell. Biochem. 2006, 99, 435–449.
21. Qian, A.; Zhang, W.; Xie, L.; Weng, Y.; Yang, P.; Wang, Z.; Hu, L.; Xu, H.; Tian, Z.; Shang, P. Simulated weightlessness alters biological

characteristics of human breast cancer cell line MCF-7. Acta Astronaut. 2008, 63, 947–958.

22. Chen, Z.-Y.; Guo, S.; Li, B.-B.; Jiang, N.; Li, A.; Yan, H.-F.; Yang, H.-M.; Zhou, J.-L.; Li, C.-L.; Cui, Y. Effect of weightlessness on the 3D structure formation and physiologic function of human cancer cells. BioMed Res. Int. 2019, 2019, 4894083.
23. Neelam, S.; Hayes, P.R.; Zhang, Q.; Dickinson, R.B.; Lele, T.P. Vertical uniformity of cells and nuclei in epithelial monolayers. Sci. Rep. 2016, 6, 19689.
24. Crisp, M.; Liu, Q.; Roux, K.; Rattner, J.B.; Shanahan, C.; Burke, B.; Stahl, P.D.; Hodzic, D. Coupling of the nucleus and cytoplasm: Role of the LINC complex. J. Cell Biol. 2006, 172, 41–53.
25. Zhang, Q.; Narayanan, V.; Mui, K.L.; O'Bryan, C.S.; Anderson, R.H.; Kc, B.; Cabe, J.I.; Denis, K.B.; Antoku, S.; Roux, K.J.; et al. Mechanical stabilization of the glandular acinus by linker of nucleoskeleton and cytoskeleton complex. Curr. Biol. 2019, 29, 2826–2839.
26. Seaman, L.; Meixner, W.; Snyder, J.; Rajapakse, I. Periodicity of nuclear morphology in human fibroblasts. Nucleus 2015, 6, 408–416.
27. Denais, C.; Lammerding, J. Nuclear mechanics in cancer. Adv. Exp. Med. Biol. 2014, 773, 435–470. [Google Scholar]
28. Li, Y.; Lovett, D.; Zhang, Q.; Neelam, S.; Kuchibhotla, R.A.; Zhu, R.; Gundersen, G.G.; Lele, T.P.; Dickinson, R.B. Moving Cell Boundaries Drive Nuclear Shaping during Cell Spreading. Biophys. J. 2015, 109, 670–686.
29. Kim, J.-K.; Louhghalam, A.; Lee, G.; Schafer, B.W.; Wirtz, D.; Kim, D.-H. Nuclear lamin A/C harnesses the perinuclear apical actin cables to protect nuclear morphology. Nat. Commun. 2017, 8, 2123.
30. Keeling, M.C.; Flores, L.R.; Dodhy, A.H.; Murray, E.R.; Gavara, N. Actomyosin and vimentin cytoskeletal networks regulate nuclear shape, mechanics and chromatin organization. Sci. Rep. 2017, 7, 5219.
31. Wolf, K.; Te Lindert, M.; Krause, M.; Alexander, S.; Te Riet, J.; Willis, A.L.; Hoffman, R.M.; Figdor, C.G.; Weiss, S.J.; Friedl, P. Physical limits of cell migration: Control by ECM space and nuclear deformation and tuning by proteolysis and traction force. J. Cell Biol. 2013, 201, 1069–1084.
32. Friedl, P.; Wolf, K.; Lammerding, J. Nuclear mechanics during cell migration. Curr. Opin. Cell Biol. 2011, 23, 55–64.

33. Gomes, E.R.; Jani, S.; Gundersen, G.G. Nuclear movement regulated by Cdc42, MRCK, myosin, and actin flow establishes MTOC polarization in migrating cells. Cell 2005, 121, 451–463.
34. Gönczy, P. Mechanisms of asymmetric cell division: Flies and worms pave the way. Nat. Rev. Mol. Cell Biol. 2008, 9, 355–366.
35. Minc, N.; Burgess, D.; Chang, F. Influence of cell geometry on division-plane positioning. Cell 2011, 144, 414–426.
36. Zhang, X.; Xu, R.; Zhu, B.; Yang, X.; Ding, X.; Duan, S.; Xu, T.; Zhuang, Y.; Han, M. Syne-1 and Syne-2 play crucial roles in myonuclear anchorage and motor neuron innervation. Development 2007, 134, 901–908.
37. Kunze, E.; Wendt, M.; Schlott, T. Promoter hypermethylation of the 14-3-3 σ, SYK and CAGE-1 genes is related to the various phenotypes of urinary bladder carcinomas and associated with progression of transitional cell carcinomas. Int. J. Mol. Med. 2006, 18, 547–557.
38. Subramanian, A.; Tamayo, P.; Mootha, V.K.; Mukherjee, S.; Ebert, B.L.; Gillette, M.A.; Paulovich, A.; Pomeroy, S.L.; Golub, T.R.; Lander, E.S.; et al. Gene set enrichment analysis: A knowledge-based approach for interpreting genome-wide expression profiles. Proc. Natl. Acad. Sci. USA 2005, 102, 15545–15550.
39. Hänzelmann, S.; Castelo, R.; Guinney, J. GSVA: Gene set variation analysis for microarray and RNA-seq data. BMC Bioinform. 2013, 14, 7.
40. Michaletti, A.; Gioia, M.; Tarantino, U.; Zolla, L. Effects of microgravity on osteoblast mitochondria: A proteomic and metabolomics profile. Sci. Rep. 2017, 7, 15376.
41. Jeong, A.J.; Kim, Y.J.; Lim, M.H.; Lee, H.; Noh, K.; Kim, B.-H.; Chung, J.W.; Cho, C.-H.; Kim, S.; Ye, S.-K. Microgravity induces autophagy via mitochondrial dysfunction in human Hodgkin's lymphoma cells. Sci. Rep. 2018, 8, 14646.
42. Sahana, J.; Nassef, M.Z.; Wehland, M.; Kopp, S.; Krüger, M.; Corydon, T.J.; Infanger, M.; Bauer, J.; Grimm, D. Decreased E-Cadherin in MCF7 Human Breast Cancer Cells Forming Multicellular Spheroids Exposed to Simulated Microgravity. Proteomics 2018, 18, e1800015.
43. Lombardi, M.L.; Jaalouk, D.E.; Shanahan, C.M.; Burke, B.; Roux, K.J.; Lammerding, J. The interaction between nesprins and sun proteins at the nuclear envelope is critical for force transmission between the nucleus and cytoskeleton. J. Biol. Chem. 2011, 286, 26743–26753.

44. Booth, A.J.; Yue, Z.; Eykelenboom, J.K.; Stiff, T.; Luxton, G.G.; Hochegger, H.; Tanaka, T.U. Contractile acto-myosin network on nuclear envelope remnants positions human chromosomes for mitosis. eLife 2019, 8, e46902.
45. Touchstone, H.; Bryd, R.; Loisate, S.; Thompson, M.; Kim, S.; Puranam, K.; Senthilnathan, A.N.; Pu, X.; Beard, R.; Rubin, J.; et al. Recovery of stem cell proliferation by low intensity vibration under simulated microgravity requires LINC complex. NPJ Microgravity 2019, 5, 11.
46. Janmaleki, M.; Pachenari, M.; Seyedpour, S.M.; Shahghadami, R.; Sanati-Nezhad, A. Impact of simulated microgravity on cytoskeleton and viscoelastic properties of endothelial cell. Sci. Rep. 2016, 6, 32418.
47. Tauber, S.; Lauber, B.A.; Paulsen, K.; Layer, L.E.; Lehmann, M.; Hauschild, S.; Shepherd, N.R.; Polzer, J.; Segerer, J.; Thiel, C.S.; et al. Cytoskeletal stability and metabolic alterations in primary human macrophages in long-term microgravity. PLoS ONE 2017, 12, e0175599.
48. Fritz, A.J.; Sehgal, N.; Pliss, A.; Xu, J.; Berezney, R. Chromosome territories and the global regulation of the genome. Genes Chromosomes Cancer 2019, 58, 407–426.
49. Fritz, A.J.; Stojkovic, B.; Ding, H.; Xu, J.; Bhattacharya, S.; Gaile, D.; Berezney, R. Wide-scale alterations in interchromosomal organization in breast cancer cells: Defining a network of interacting chromosomes. Hum. Mol. Genet. 2014, 23, 5133–5146.
50. Stephens, A.D.; Liu, P.Z.; Kandula, V.; Chen, H.; Almassalha, L.M.; Herman, C.; Backman, V.; O'Halloran, T.; Adam, S.A.; Goldman, R.D.; et al. Physicochemical mechanotransduction alters nuclear shape and mechanics via heterochromatin formation. Mol. Biol. Cell 2019, mbcE19050286T.
51. Donnaloja, F.; Jacchetti, E.; Soncini, M.; Raimondi, M.T. Mechanosensing at the nuclear envelope by nuclear pore complex stretch activation and its effect in physiology and pathology. Front. Physiol. 2019, 10, 896.
52. Smith, E.R.; Meng, Y.; Moore, R.; Tse, J.D.; Xu, A.G.; Xu, X.-X. Nuclear envelope structural proteins facilitate nuclear shape changes accompanying embryonic differentiation and fidelity of gene expression. BMC Cell Biol. 2017, 18, 8.

53. Poulet, A.; Duc, C.; Voisin, M.; Desset, S.; Tutois, S.; Vanrobays, E.; Benoit, M.; Evans, D.E.; Probst, A.V.; Tatout, C. The LINC complex contributes to heterochromatin organisation and transcriptional gene silencing in plants. J. Cell Sci. 2017, 130, 590–601.
54. Tan, X.; Xu, A.; Zhao, T.; Zhao, Q.; Zhang, J.; Fan, C.; Deng, Y.; Freywald, A.; Genth, H.; Xiang, J. Simulated microgravity inhibits cell focal adhesions leading to reduced melanoma cell proliferation and metastasis via FAK/RhoA-regulated mTORC1 and AMPK pathways. Sci. Rep. 2018, 8, 3769.
55. Ranieri, D.; Proietti, S.; Dinicola, S.; Masiello, M.G.; Rosato, B.; Ricci, G.; Cucina, A.; Catizone, A.; Bizzarri, M.; Torrisi, M.R. Simulated microgravity triggers epithelial mesenchymal transition in human keratinocytes. Sci. Rep. 2017, 7, 538.
56. Yang, Y.; Quan, L.; Ling, Y. RBMS3 inhibits the proliferation and metastasis of breast cancer cells. Oncol. Res. 2018, 26, 9–15.
57. Leygue, E. Steroid receptor RNA activator (SRA1): Unusual bifaceted gene products with suspected relevance to breast cancer. Nucl. Recept. Signal. 2007, 5, e006.
58. Ichigu, J.; Asashima, M. Dome formation and tubule morphogenesis by Xenopus kidney A6 cell cultures exposed to microgravity simulated with a 3D-clinostat and to hypergravity. In Vitro Cell. Dev. Biol. Anim. 2001, 37, 31–44.
59. Yuge, L.; Hide, I.; Kumagai, T.; Kumei, Y.; Takeda, S.; Kanno, M.; Sugiyama, M.; Kataoka, K. Cell differentiation and p38 MARK cascade are inhibited in human osteoblasts cultured in a three-dimensional clinostat. In Vitro Cell. Dev. Biol. Anim. 2003, 39, 89–97.
60. Kawahara, Y.; Manabe, T.; Matsumoto, M.; Kajiume, T.; Matsumoto, M.; Yuge, L. LIF-free embryonic stem cell culture in simulated microgravity. PLoS ONE 2009, 4, e6343.
61. Jiang, H.; Lei, R.; Ding, S.-W.; Zhu, S. Skewer: A fast and accurate adapter trimmer for next-generation sequencing paired-end reads. BMC Bioinform. 2014, 15, 182.
62. Kim, D.; Langmead, B.; Salzberg, S.L. HISAT: A fast spliced aligner with low memory requirements. Nat. Methods 2015, 12, 357–360.
63. Li, H.; Handsaker, B.; Wysoker, A.; Fennell, T.; Ruan, J.; Homer, N.; Marth, G.; Abecasis, G.; Durbin, R. 1000 Genome Project Data

Processing Subgroup. The Sequence Alignment/Map format and SAMtools. Bioinformatics 2009, 25, 2078–2079.

64. Liao, Y.; Smyth, G.K.; Shi, W. featureCounts: An efficient general purpose program for assigning sequence reads to genomic features. Bioinformatics 2014, 30, 923–930.

65. Robinson, M.D.; McCarthy, D.J.; Smyth, G.K. edgeR: A Bioconductor package for differential expression analysis of digital gene expression data. Bioinformatics 2010, 26, 139–140.

66. Liberzon, A.; Brger, C.; Thorvaldsdóttir, H.; Ghandi, M.; Mesirov, J.P.; Tamayo, P. The Molecular Signatures Database (MSigDB) hallmark gene set collection. Cell Syst. 2015, 1, 417–425.

CHAPTER 6

STRUCTURE OF THE HUMAN CHROMOSOME INTERACTION NETWORK

Sergio Sarnataro[1], Andrea M. Chiariello[2], Andrea Esposito[2,4], Antonella Prisco[3], Mario Nicodemi[2]

[1]Institut de Ge´ne´tique et de Biologie Mole´culaire et Cellulaire, 67404 Illkirch, France

[2]Dipartimento di Fisica, Universita´ di Napoli Federico II, and INFN Napoli, CNR-SPIN, Complesso Universitario di Monte Sant'Angelo, 80126 Naples, Italy

[3]CNR-IGB, via Pietro Castellino 111, Naples, Italy

[4]Max Delbru¨ck Center for Molecular Medicine in the Helmholtz Association (MDC), Berlin, Germany

Citation: Sarnataro S, Chiariello AM, Esposito A, Prisco A, Nicodemi M (2017) Structure of the human chromosome interaction network. PLoS ONE 12(11): e0188201. https://doi.org/10.1371/ journal.pone.0188201

ABSTRACT

New Hi-C technologies have revealed that chromosomes have a complex network of spatial contacts in the cell nucleus of higher organisms, whose organisation is only partially understood. Here, we investigate the structure of such a network in human GM12878 cells, to derive a large scale picture of nuclear architecture. We find that the intensity of intra-chromosomal interactions is power-law distributed. Inter-chromosomal interactions are two orders of magnitude weaker and exponentially distributed, yet they are not randomly arranged along the genomic sequence. Intra-chromosomal contacts broadly occur between epigenomically homologous regions, whereas inter-chromosomal contacts are especially associated with regions rich in highly expressed genes. Overall, genomic contacts in the nucleus appear to be structured as a network of networks where a set of strongly individual chromosomal units, as envisaged in the 'chromosomal territory' scenario derived from microscopy, interact with each other via on average weaker, yet far from random and functionally important interactions.

INTRODUCTION

New technologies, such as Hi-C, can measure the frequency of physical contacts between DNA segment pairs genome-wide and are revealing the complex spatial organization of the mammalian genome [1–4]. A complex network of interactions exists, having functional roles as, for instance, regulatory regions, such as enhancers, often control gene expression at long genomic distances through the formation of loops with their target genes based on physical interactions. Network analyses of Hi-C contacts between DNA segments have provided several insights on chromatin organization, including the identification of their enrichment for genes and chromatin marks, the presence of gene hubs and the possibility to reconstruct 3D conformations of genomic loci (see, e.g., [5–7] and ref.s therein). It has been discovered, in particular, that each chromosome is composed by a sequence of domains (also named Topological Associated Domains, TADs [8, 9]) that are enriched for internal interactions with respect to background [8, 9] and have a hierarchy of long range contacts [10]. Identified by different computational methods (see, e.g., ref.s [8–11]), they are partially conserved across cell types. To derive a large-scale picture of the structure of nuclear organization of human chromosomes, here we investigate specifically the interaction network of the *contact domains* defined by Rao and others [11] from in-situ Hi-C data in human GM12878 cells (see Methods—Contact

Domains): contact domains are identified by the *Arrowhead algorithm* that scores the DNA regions of sites having similarly enhanced levels of Hi-C mutual interactions and, correspondingly, a drop in contacts at their borders. We focus on contact domains as they represent single architectural units and are marked by coherent epigenetic signatures. Additionally, they have a median length of 185kb (which sets the genomic resolution of our study) that is one order of magnitude larger than the size of genes, providing stability to our analysis against the noise inherent to long-range Hi-C data, without compromising too strongly on resolution.

We find that their intra-chromosomal contact networks are marked by a 'universal' power law distribution of interaction frequencies. Inter-chromosomal contacts are found to be approximateley three orders of magnitude weaker [1, 11], yet, we show that they are also far from randomly positioned along the genome. We find that their network is structurally different from the net of intra-chromosomal interactions, as the intensity of contacts is exponentially distributed. Interestingly, intra-chromosomal contacts are typically enriched between epigenomically similar domains, whereas inter-chromosomal contacts are found especially between domains rich in highly expressed genes. The picture of the nucleus emerging from our analysis (schematically shown in Fig 1) returns a scenario where chromosomes contacts form a net of networks. Single chromosomes are mainly folded in territorial units, as seen by microscopy [12], having a complex internal organization. Different territories, though, interact with each other in a non-random, functional way, as envisaged in the 'chromosome intermingling' scenario, where chromosome pairs intermingle as revealed by in situ-hybridization technique (cryoFISH) [13].

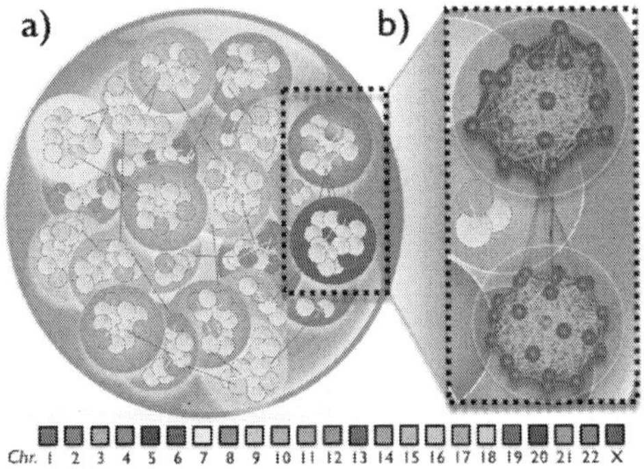

Figure 1: Our analysis of Hi-C genomic interactions shows that the nucleus of human GM12878 cells is structured as a net of networks.

Each individual chromosome form a strong intra-chromosomal network of contacts, consistent with the 'chromosomal territories' seen by microscopy. Yet, chromosomes intermingle via weaker, yet non-random and functionally important contacts, whereby distinct chromosomal networks form a global nuclear network. Panel a: A pictorial representation of the 3D organization of chromosomes (colored spheres) within the nucleus, as emerging from our network analysis. Panel b: A real network reconstruction (using the visualization tool described in [14]) of chromosomal contacts of chromosomes 19 and 20. For clarity of presentation, each sphere inside a chromosome represents ~ $1Mb$ (5 contact domains) and a high threshold is set for link visualization.

RESULTS

The System

In our analysis, the nodes of the chromosome interaction network are the contact domains, while the edges between nodes are defined by the intensity of their interactions. Human GM12878 cell chromosomes are divided in 9274 contact domains, having an exponentially distributed genomic length with a median equal to 185kb, covering roughly 70% of the entire genome [11]. They have been grouped in six categories, named 'subcompartments', A1, A2, B1, B2, B3, B4, having a median length of 300kb, which correlate with specific epigenetic patterns [11]. Subcompartments A1 and A2 belong to 'compartment A' [1] and are rich in highly expressed genes and activating chromatin marks. A1 and A2 exhibit early replication times, with A2 having lower GC content, longer genes and stronger association with the H3K9me3 histone mark than A1. B1, B2, and B3 are associated with 'compartment B': B1 correlates with facultative heterochromatin, B2 with pericentromeric heterochromatin and B3 with nuclear lamina. B4 is present only in chromosome 19 and correlates with both activating chromatin and heterochromatin. Genomic regions within a single contact domain also have correlated histone marks, while 66,5% of contact domains overlap with a single subcompartment.

The interaction between two domains is defined as the sum of their shared entries in the normalized in-situ Hi-C contact frequency matrix divided by the product of their genomic lengths (see Methods—Contact Domains). Some contact domains are contained within another contact domain. Albeit they represent only a fraction of the total (~35%), in our analysis we disregard them to avoid double counting of interactions. However, we checked that our results remain unchanged if the entire set in considered. The median length of the remaining domains is 250kb. Based on in-situ Hi-C data, the ratio of the average interaction strengths of domain pairs respectively on the same and on different chromosomes is $I_{in-cis}/I_{in-trans} = 850$, showing that the mean interaction within a chromosome is almost three orders of magnitude higher than across chromosomes [11]. Hence, we initially focus on intra-chromosomal interaction networks.

Intra-chromosomal Interaction Networks

To characterize the structure of inter-chromosomal contacts, we first measured the node average normalised degree of connectivity, n. This quantity is defined as the number of domains linked to a given domain on the same chromosome, normalized to the total number of domains on that chromosome. To establish whether two nodes are linked, we check if their interaction level is higher than a chosen minimum threshold, in order to avoid spurious effects related to noise in the data. A typical way to identify significant interactions [10] is to retain only those that fall above a threshold, so to filter the noise due to random interactions; here, we consider as threshold the value corresponding to the lowest 25-percentile of the distribution of inter-chromosomal domain interactions. The rationale behind such a choice is that measured contacts between pairs of domains on different chromosomes are more likely to be random, also considering the vastly higher number of possible pairs for the experimentally set sequencing depth. For such a set threshold we find $n = 0.998$. To check that different threshold values do not alter our scenario, we measured that when the 75-percentile is considered we get $n = 0.979$. The order of magnitude of n is unchanged even when the threshold is set based on percentiles of the distribution of intra-chromosomal interactions. Hence, at the considered genomic resolution, we find that the domains of a single chromosomes form a fully connected network, where approximately all nodes interact with all. That points out the highly dynamical nature of chromatin contacts.

We focused next on the distribution of the intensity of intra-chromosomal interactions, $P(I)$. We found that it has a power law decay over roughly two orders of magnitude, $P(I) \simeq I^{-\gamma}$ (Fig 2). The exponent is close to $\gamma \simeq 2$ and very similar in different chromosomes. This is an interesting observation because an exponent $\gamma = 2$ marks scale free networks [15], which have a special relevance as they are characterized by the presence of *hubs*, i.e., nodes with very high levels of contacts. To test the statistical significance of our analysis, we used the python library *powerlaw*, a package for the analysis of heavy-tailed distributions [16]. For every chromosome, we found that the tail of the distribution fits the best with a power law, with a negligible $p - value$. This points out that contact frequencies can extend to very large values for some genomic regions. Interestingly, the distribution, $P(I)$, and its exponent $\gamma \simeq 2$ are 'universal' across chromosomes, hinting that the structure of their domain networks is similar, as shown in Fig 2.

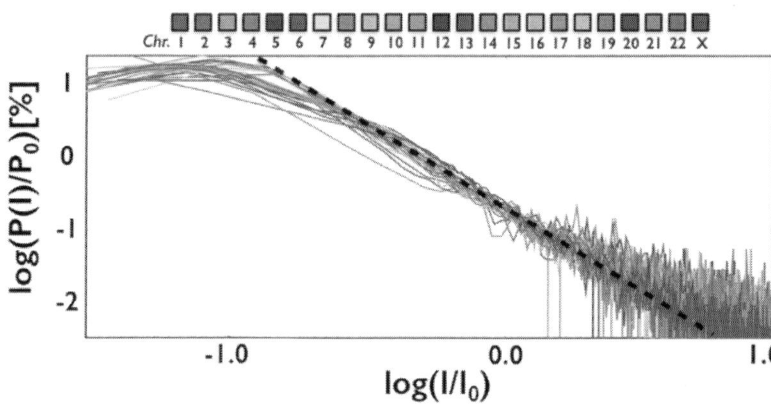

Figure 2: The distribution of intra-chromosomal interactions between contact domains (here in a $log_{10} - log_{10}$ **plot) decays as a power law,** $P(I) \simeq I^{-\gamma}$, with an exponent close to $\gamma \simeq 2$ (dashed line), which appear to be 'universal' across different chromosomes.

So, the structures of their contact networks exhibit similar quantitative features. I_0 and P_0 are factors to rescale on top of each other the data from different chromosomes.

To help dissecting the molecular nature of the determinants of contacts, we then investigated the enrichment of intra-chromosomal domain interactions across the different subcompartments, which are known to be associated each with a different set of epigenetic features [11]. In our analysis we consider only domains overlapping with a single subcompartment (77% of total), but our conclusions stay unchanged if all domains are considered and each one is assigned to a single subcompartment, e.g., by a majority rule. We repeated our analysis also for the top interacting domains, i.e., those in the top 5% of $P(I)$ distribution (see Methods—Chromosomal domain interactions across subcompartments). While single domains are prevalently overlapping with A1, followed by A2 and B1 [1], we found that the top interacting ones are more evenly spread across the different subcompartments (see Methods—Contact domain distribution across subcompartments). The heatmap in Fig 3, Panel a, reports the average interaction between domains in different subcompartments, normalized by the overall average value. Interestingly, enrichments of contacts are mainly seen along the diagonal, showing that intra-chromosomal contacts tend to be homotypical, i.e., usually occur between epigenomically homologous regions. In particular, the strongest homotypic interactions are found within the heterochromatin B1 and B2 subcompartments, and to a minor extent within A1 domains that are rich in active genes. The lamina linked B3 subcompartment has, instead, comparatively lower levels of homotypic interactions. Similar results are found when only the top interacting domains are considered (see Methods—Chromosomal domain interactions across subcompartments).

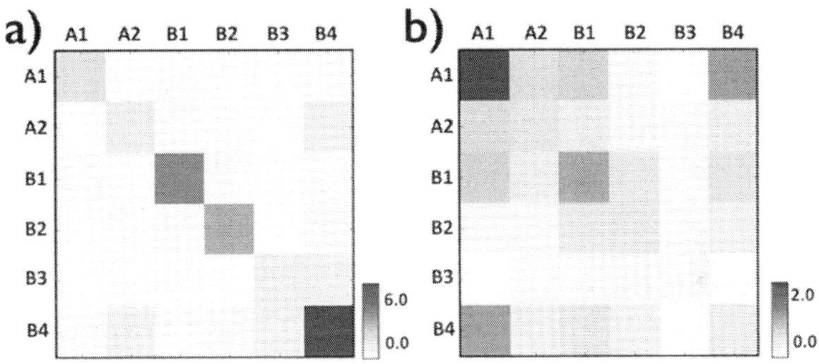

Figure 3: Heatmaps representing, across the different subcompartments A1-B4, the domain average excess interactions over background within single chromosomes (*in-cis*, Panel a) and between chromosomes (*in-trans*, Panel b).

Intra-chromosomal contacts (a) especially occur between homologous subcompartments (diagonal), while inter-chromosomal contacts (b) are particularly frequent with subcompartment A1, rich in highly expressed genes, hinting to a functional role.

Inter-chromosomal Interaction Networks

Next, we turned to the inter-chromosomal interaction network. Interestingly, it results to be also fully connected, albeit less dense than intra-chromosomal nets as its normalised average degree of connectivity is $n = 0.749$ in the case where the threshold is set, as above, to the 25th-percentile of the distribution, moving down to $n = 0.249$ for a 75th-percentile threshold.

To understand whether there are pairs of chromosomes more strongly interacting than others, we investigated their overall pairwise interactions (normalized by their genomic lengths). The heatmap in Fig 4 shows indeed that, as expected, different pairs have different degrees of contact intensities. In particular, gene richer, shorter chromosomes tend to have higher relative reciprocal interactions, consistent with the role of A1 subcompartment in establishing homotypical contacts (see below). To identify whether there are particularly isolated clusters of interacting chromosomes, we followed a standard procedure using the RSS (Residual Sum of Square) objective function [17]. More precisely, we plot the RSS quantity as a function of number of clusters k and seek for the value where the curve has a sharp slope variation (elbow method). From this analysis, we found that there are no evident preferentially isolated clusters at the length scale of contact domains ($\simeq 200kb$). Hence, while different chromosomal pairs can have different degrees of interactions, a global, single main inter-chromosomal network exists.

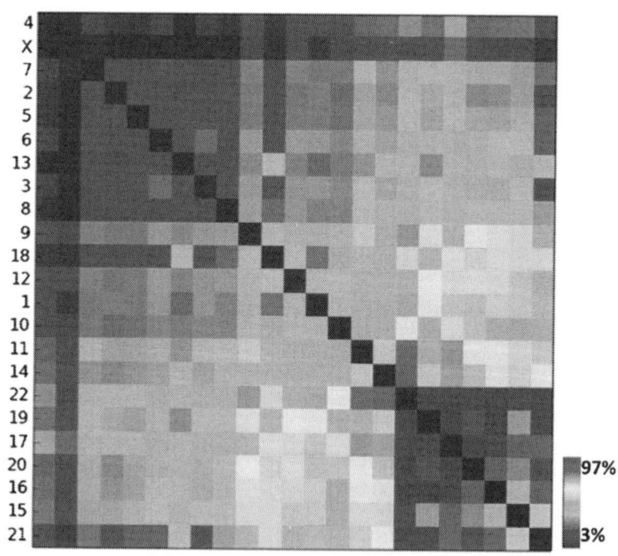

Figure 4: The heatmap of the overall contacts between chromosome pairs. While different chromosomal pairs have different degrees of interactions, the RSS analysis points out that there are no significant isolated subgroups and the system forms a single nuclear network.

Interestingly, a cross-analysis of chromosomal interactions with the sub-compartment membership (Fig 3, Panel b) reveals that domains belonging to sub-compartment A1 tend to interact mostly. Since sub-compartment A1 is rich in highly expressed genes, that suggests a functional role of highly interacting nodes in the inter chromosomal network, in agreement with previous findings ([18]). Furthermore, as discussed below, a non trivial pattern of interaction exists.

The distribution of the intensity of inter-chromosomal interactions, $P_{inter}(I)$, is similar across different chromosome pairs and has roughly an *exponential* decay, $P_{inter}(I) \simeq \exp(-I/I_0)$ (Fig 5), as would be expected for a random network [15]. To statistically prove the robusteness of this result, we fitted the tail of the distributions with an exponential function and with a power law function. We performed the analysis using the *optimize* python package. For every possible pair of chromosomes we obtained a coefficient of determination R^2 higher in the exponential case than in the power law case, and extremely close to 1 (average $R^2 = 0.989$). Moreover, in the 26% of the cases, the power law fit was not convergent. As additional check, to test if intra- and inter-chromosomal distributions were statistically different, we performed the same analysis on the intra-chromosomal datasets, finding a coefficient of determination R^2 sistematically higher for the power law fit than for the exponential one. Thus, we can conclude that the two distributions are not compatible, suggesting that the inter- and intra-chromosomal contact networks are structurally different.

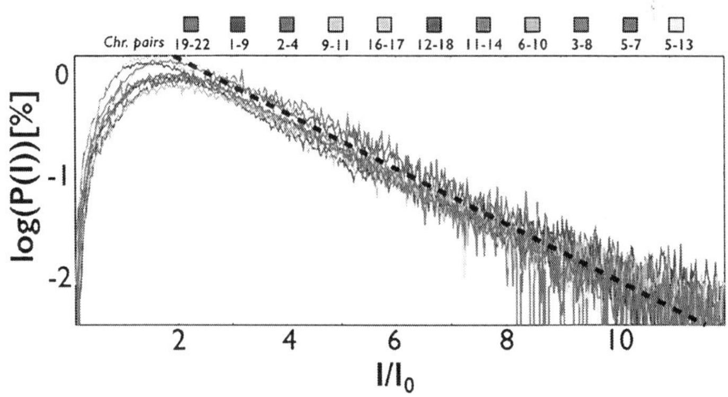

Figure 5: The distribution of interactions between contact domains on different chromosomes (here in a $lin - log_{10}$ **plot) has an exponential decay (dashed line) across chromosome pairs.** That shows that the inter-chromosomal network of contacts is structurally different from the power-law distributed intra-chromosomal networks (Fig 2).

However, we find that inter-chromosomal interactions are not randomly distributed over the genomic sequence. This is visible in the example of Fig 6 showing a heatmap with the frequency of contact between the domains of chromosome 22 v.s. 19: genomic bands with markedly strong interaction are seen, pointing out that regions enriched for contacts tend to cluster along the chromosomal sequence. For comparison, we also show a random contact heatmap generated by mixing the positions of the values in the experimental matrix so that the distribution of the intensities is unaffected.

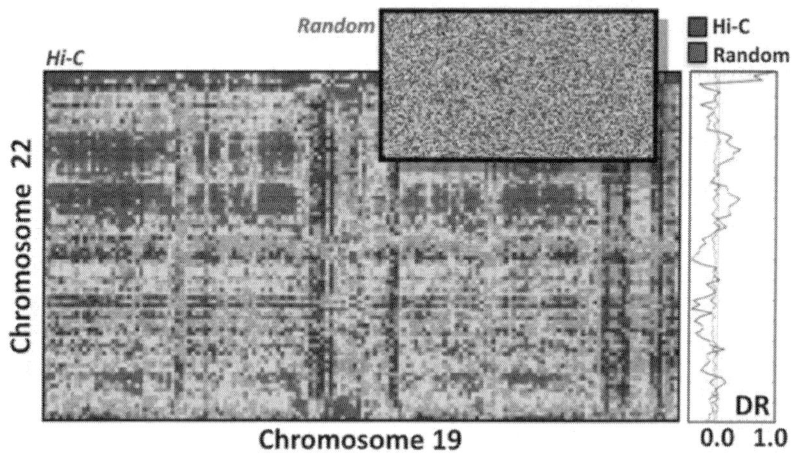

Figure 6: Heatmap of the normalized in-situ Hi-C interactions between the contact domains of chromosomes 19 and 22. Bands of regions with above background signal are visible, in contrast to the corresponding random matrix (top-right, see details in the text). The differential interaction score, *DR*, of the domains of chromosome 22 with 19 (rightmost panel) quantifies the dissimilarity with the random control and highlights the clustering of genomic locations enriched in contacts with domains on the other chromosome.

To quantify such observations, we computed the differential interaction score, *DR*, across the domains of each single chromosome. For a given domain *k*, DR_k is defined as follows: $DR_k = (S_k - \langle S \rangle)/\langle S \rangle$, where S_k is the sum of its contacts with all the other domains of the other considered chromosome (i.e. the sum of rows in the matrix of Fig 6), and $\langle S \rangle$ is the average over the matrix (see Methods—The differential interaction score, DR). The *DR* signal of chromosome 22 v.s. 19 is shown in Fig 6-right panel. As expected, *DR* varies around zero, yet its standard deviation $\sigma = 0.3$ is one order of magnitude larger than in the random bootstrapped case ($\sigma_{rand} = 0.05$, red line), reflecting the much stronger variations present in the real contacts.

We also find that the average size of genomic regions where *DR* is coherent in sign spans approximately 5.4 consecutive domains, against 2.5 in the random case, highlighting the clustering of genomic positions enriched in contacts with domains on a different chromosome. Similar results are found across other chromosome pairs.

Finally, as mentioned before, Fig 3, Panel b reports the average inter-chromosomal interaction strength of domain pairs across the different subcompartments, showing that A1 domains are enriched for interactions, especially with homologous domains. As above, similar observations apply when only the top interacting domains are considered (see Methods—Chromosomal domain interactions across subcompartments). The finding that domains in the A1 subcompartment have preferentially homotypic interactions also supports the view that inter-chromosomal contacts are far from random, as random contacts would be equally enriched across subcompartments.

DISCUSSION

Taken together, our results show that, at the 200kb resolution, chromosomes are characterised by an extended network of contacts. Intra-chromosomal interactions have a similar structure across chromosomes, marked by broad, power-law distributed interaction frequencies with an exponent $\gamma \simeq 2$ typical of scale-free nets. In facts, scale-free networks are well known in biological systems and are thought to reflect fundamental organizational principles [15].

Chromosomes are also widely interacting with each other, without isolated subgroups, yet their global interaction network is two orders of magnitude weaker in intensity than intra-chromosomal nets, and exponentially distributed. Hence, the net of inter-chromosomal contacts is structurally different from intra-chromosomal ones. A limitation of our study is that the data do not allow to distinguish the different alleles. Our analysis suggests, though, a non-random genomic positioning of inter-chromosomal interactions.

Our analysis of correlations between epigenetic signals and domain contacts, albeit limited to a few general marks, shows that intra-chromosomal interactions tend to occur between homotypic epigenomic regions and inter-chromosomal interactions are especially found between regions rich in highly expressed genes.

The relative position of chromosomes in the nucleus, based on their domain in-situ Hi-C interactions, is shown in Fig 1, Panel a, as produced by a standard tool for the 3D visualization of network graphs [14]. The zoom in Panel b shows, in particular, the reciprocal interactions of chromosome 19 and 20, along with their intra-chromosomal networks (for clarity of presentation, a high threshold is set for link visualization). Multilayer networks, or networks of networks, are thought to be important to process and spread efficiently information across different nets to produce emergent, collective behaviours in complex natural and biological systems [19, 20]. Our analysis suggests a networked organization of chromosomal interactions where the distinct nets of intra-chromosomal contacts are assembled in a weaker, yet non-random global nuclear network of nets. This important feature is known as modularity. Many network systems in nature have strong modular aspects [21]. An important reason explaining such a peculiar behaviour lies in the high stability that those networks exhibit as response to external perturbations, as recently highlighted by important studies [22]. Hence, our finding could hint towards the modular and structurally stable structure of the 3D architecture of the human genome within the cell nucleus. Our early investigations of the network features of chromosomal interactions may raise the attention of the scientific community on such an interesting topic and pave the way to more extensive bioinformatics and computational analyses to better understand the underlying features and mechanisms leading to the observed pattern in the considered framework, as known for other networked systems ([23, 24])

The emerging large-scale picture of chromosome contact networks in the cell nucleus is complemented by current higher-resolution studies employing models from polymer physics, which try to explain the molecular mechanisms shaping the 3D conformation of specific genomic loci [25–31]. Brief reviews of such models can be found in ref.s [32–34]. Our results are derived from human GM12878 cell data, however, Hi-C interaction patterns are broadly conserved across different cell types in higher mammals [2–4, 8–10], hinting towards a broader validity of our conclusions. In particular, the picture emerging from our quantitative investigation of Hi-C data reconciles the so-called "chromosomal territory scenario" [12] with the "chromosomal intermingling sceanio" [13], as in the nuclear network of contacts chromosomes tend indeed to form strong territorial units, but have also relevant inter-chromosomal interactions.

METHODS

Hi-C dataset

We employed the primary in-situ Hi-C dataset produced by Rao and others [11], relative to human cells, cell line GM12878, MAPQGE30 at 10Kb of resolution. The row data were normalized by the Knight-Ruiz method [35].

Contact Domains

We used the list of contact domains provided by Rao and others [11], obtained by their *Arrowhead Algorithm*. As discussed in the previous sections of the paper, we excluded from our analysis *internal* contact domains, i.e. those completely contained in other contact domains, to avoid double counting of contacts.

In Fig 7 we report the distribution of the lengths of the contact domains, considering all of them (left panel) and removing the internal ones (right panel). We have 9274 contact domains in total, having an average length of 258Kb and a median length of 185Kb. After removing the internal domains, we are left with 5370 contact domains, having an average length of 320Kb and a median length of 250Kb. Different mechanisms of polymer physics [36–40] could contribute to the assembly of contact domains [11, 25–29, 41–43], while additional effects, such as confinement, crowding, entanglement could have a role as found in other complex fluids (see, e.g., [44–53] and ref.s therein).

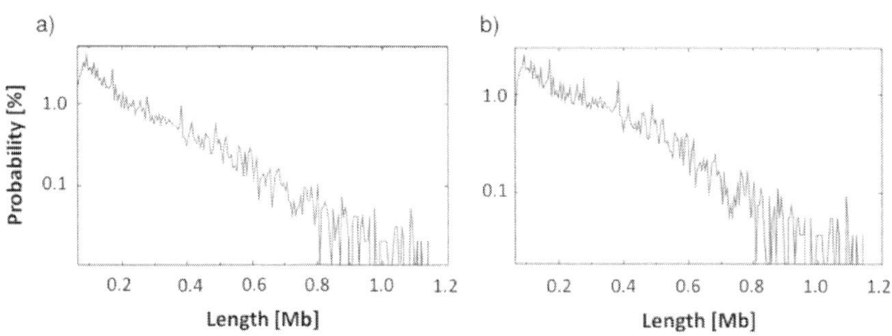

Figure 7: Panel a: Distribution of the lengths of all the contact domains, i.e. the probability to find a contact domains with a given genomic length, represented in a log-lin scale. Panel b: The same analysis for the case where 'internal domains' are removed from the list.

To analyze the structure of the domain network, we considered the average frequency of contacts between all pairs of domains. The average frequency of contacts, $I_{a,b}$, between domain a and b is defined as the sum of their shared entries in the normalized in-situ Hi-C matrix divided by the product of their genomic lengths.

Chromosomal Domain Interactions Across Subcompartments

To investigate the enrichment of intra- and inter-chromosomal domain interactions across the different subcompartments, we computed the average interaction between domains in different subcompartments, $H_{i,j}$, normalized by the overall average value. More precisely, the values $H_{i,j}$ in each bin of the heatmaps relative to Fig 3, are obtained by the following expression: $H_{i,j} = (< H(i,j) > - H_0)/H_0$, where $< H(i,j) >$ is the average frequency of contact between the contact domains belonging to the subcompartments i and j. Here $H_0 = (\sum < H(i,j) >)/N_c$ is the average value across subcompartments, and N_c is the number of the couples of subcompartments.

We did the same analysis considering only the top interacting contact domains *in-cis* and *in-trans*, i.e. we considered only the 5% of all contacts domains having the highest average frequency of interaction with the other contact domains, respectively on a given chromosome and on an other chromosome (for all the possibile couples of chromosomes). The results are reported in Fig 8

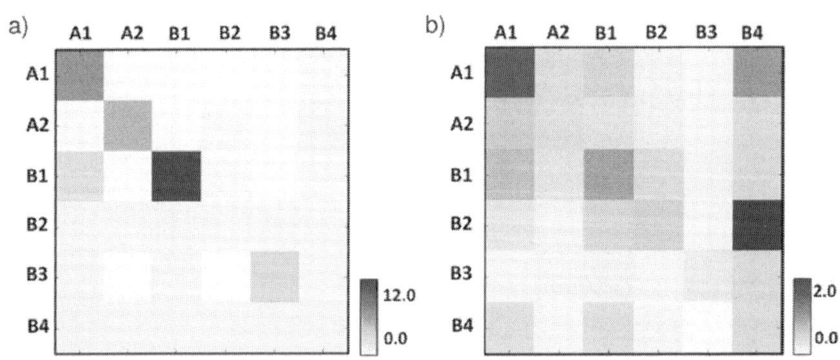

Figure 8: Heatmap representing the average frequency of interactions between 'top interacting' contact domains (see text). Panel a: *in-cis* and Panel b: *in-trans*, belonging to the different subcompartments.

Contact Domain Distribution Across Subcompartments

We also analyzed how domains are distributed across the compartments. In this analysis, we considered the domains overlapping with a single subcompartment (77% of total), but our conclusions stay unchanged if all domains are considered and each one is assigned to a single subcompartment by a majority rule. In Fig 9 we represented two cases: all domains and only the *top interacting domains*, i.e. the 5% of the domain with the higest average frequency of interaction, for *in-cis* and *in-trans* cases (left and right panel respectively). While considering all the contact domains we have a prevalence of domains in compartment A1, with A2 and B1 following, if we consider only the top interacting domains the most abundant group is in compartment A2. Top interacting domains are present in compartment B for about 35%, while there is only a small percentage if we consider all the domains.

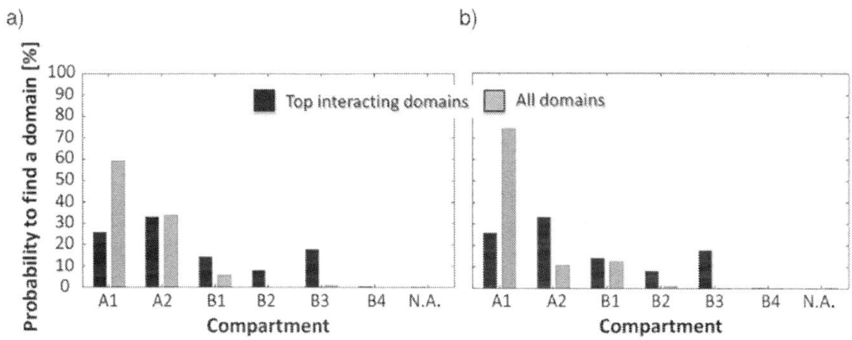

Figure 9: Panel a: Fraction of domains in each subcompartment *in-cis* (N.A. means *Not Assigned*) in two cases: for all domains (yellow) and for the 'top interacting' domains in-cis (blue). Panel b: The same analysis *in-trans*.

The Differential Interaction Score, DR

The differential interaction score, *DR*, is defined in the in section about inter-chromosomal interactions network. It was computed for all chromosome pairs, i.e., for all the domains of a given chromosome with respect to its interactions with the domains of all other single chromosomes. For example, Fig 6 refers to the *DR*'s of the domains of chromosome 22 in their interaction with those on chromosome 19.

Finally, for each chromosome, we also measured the average size of genomic regions where *DR* is coherent in sign, i.e., we calculated the number of contact domains between two consecutive crossings of the zero by *DR*. The distribution of sizes is reported in Fig 10 for the real and the bootstrapped matrices in the case of chromosome 19 v.s 22 discussed in the previous sections of the paper. In both cases an exponential fit is overlaid on the data. Similar findings are obtained for other chromosome pairs.

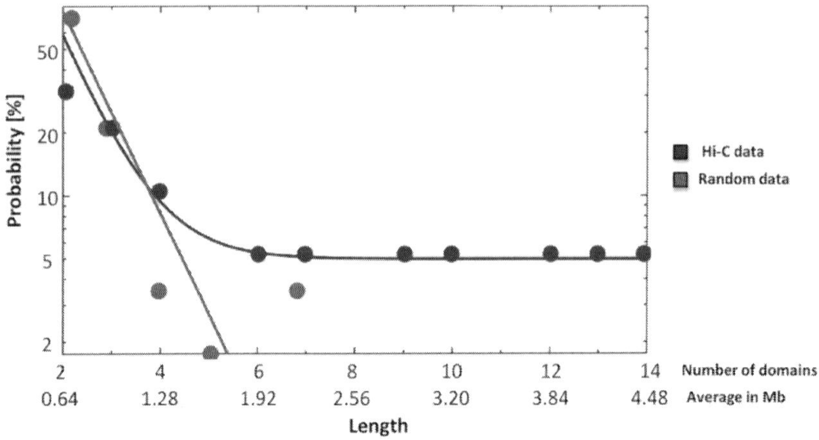

Figure 10: Distribution of the sizes of genomic regions where *DR* **is coherent in sign in the random bootstrapped (red) and Hi-C real data (blue), as a function of the number of domains or, correspondingly, in Mb (the average length of each contact domain is 320Kb).** The superimposed curves are exponential fits.

ACKNOWLEDGMENTS

We thank C. Annunziatella and A. I. Rerra for useful discussions. Work supported by computer resources at INFN, CINECA, and Scope at the University of Naples.

REFERENCES

1. Lieberman-Aiden E, Van Berkum NL, Williams L, Imakaev M, Ragoczy T, Telling A, et al. Comprehensive mapping of long-range interactions reveals folding principles of the human genome. Science. 2009;326(5950):289–293. doi: 10.1126/science.1181369
2. Tanay A, Cavalli G. Chromosomal domains: epigenetic contexts and functional implications of genomic compartmentalization. Current opinion in genetics & development. 2013;23(2):197–203. doi: 10.1016/j.gde.2012.12.009
3. Bickmore WA, van Steensel B. Genome architecture: domain organization of interphase chromosomes. Cell. 2013;152(6):1270–1284. doi: 10.1016/j.cell.2013.02.001
4. Dekker J, Mirny L. The 3D genome as moderator of chromosomal communication. Cell. 2016;164(6):1110–1121. doi: 10.1016/j.cell.2016.02.007
5. Kruse K, Sewitz S, Babu MM. A complex network framework for unbiased statistical analyses of DNA—DNA contact maps. Nucleic acids research. 2012;41(2):701–710. doi: 10.1093/nar/gks1096
6. Kaufmann S, Fuchs C, Gonik M, Khrameeva EE, Mironov AA, Frishman D. Inter-chromosomal contact networks provide insights into Mammalian chromatin organization. PloS one. 2015;10(5):e0126125 doi: 10.1371/journal.pone.0126125
7. Morlot JB, Mozziconacci J, Lesne A. Network concepts for analyzing 3D genome structure from chromosomal contact maps. EPJ Nonlinear Biomedical Physics. 2016;4(1):2 doi: 10.1140/epjnbp/s40366-016-0029-5
8. Dixon JR, Selvaraj S, Yue F, Kim A, Li Y, Shen Y, et al. Topological domains in mammalian genomes identified by analysis of chromatin interactions. Nature. 2012;485(7398):376–380. doi: 10.1038/nature11082
9. Nora EP, Lajoie BR, Schulz EG, Giorgetti L, Okamoto I, Servant N, et al. Spatial partitioning of the regulatory landscape of the X-inactivation centre. Nature. 2012;485(7398):381–385. doi: 10.1038/nature11049
10. Fraser J, Ferrai C, Chiariello AM, Schueler M, Rito T, Laudanno G, et al. Hierarchical folding and reorganization of chromosomes are linked to transcriptional changes in cellular differentiation. Molecular systems biology. 2015;11(12):852 doi: 10.15252/msb.20156492

11. Rao SS, Huntley MH, Durand NC, Stamenova EK, Bochkov ID, Robinson JT, et al. A 3D map of the human genome at kilobase resolution reveals principles of chromatin looping. Cell. 2014;159(7):1665–1680. doi: 10.1016/j.cell.2014.11.021
12. Lanctôt C, Cheutin T, Cremer M, Cavalli G, Cremer T. Dynamic genome architecture in the nuclear space: regulation of gene expression in three dimensions. Nature Reviews Genetics. 2007;8(2):104–115. doi: 10.1038/nrg2041
13. Branco MR, Pombo A. Intermingling of chromosome territories in interphase suggests role in translocations and transcription-dependent associations. PLoS biology. 2006;4(5):e138 doi: 10.1371/journal.pbio.0040138
14. Theocharidis A, Van Dongen S, Enright AJ, Freeman TC. Network visualization and analysis of gene expression data using BioLayout Express3D. Nature protocols. 2009;4(10):1535–1550. doi: 10.1038/nprot.2009.177
15. Barabasi AL, Oltvai ZN. Network biology: understanding the cell's functional organization. Nature reviews genetics. 2004;5(2):101–113. doi: 10.1038/nrg1272
16. Alstott J, Bullmore E, Plenz D. powerlaw: a Python package for analysis of heavy-tailed distributions. PloS one. 2014;9(1):e85777 doi: 10.1371/journal.pone.0085777
17. Manning C, Raghavan P, Schütze H. Introduction to information retrieval/Christopher D. Cambridge University Press, Cambridge, England; 2009.
18. Spilianakis CG, Lalioti MD, Town T, Lee GR, Flavell RA. Interchromosomal associations between alternatively expressed loci. Nature. 2005;435(7042):637–645. doi: 10.1038/nature03574
19. Gao J, Buldyrev SV, Stanley HE, Havlin S. Networks formed from interdependent networks. Nature physics. 2011;8(1):nphys2180 doi: 10.1038/nphys2180
20. D'Agostino G, Scala A. Networks of networks: the last frontier of complexity. Springer; 2014.
21. Guimera R, Sales-Pardo M, Amaral LA. Classes of complex networks defined by role-to-role connectivity profiles. Nature physics. 2007;3(1):63–69. doi: 10.1038/nphys489

22. Gilarranz LJ, Rayfield B, Liñán-Cembrano G, Bascompte J, Gonzalez A. Effects of network modularity on the spread of perturbation impact in experimental metapopulations. Science. 2017;357(6347):199–201. doi: 10.1126/science.aal4122
23. Valverde S, Cancho RF, Sole RV. Scale-free networks from optimal design. EPL (Europhysics Letters). 2002;60(4):512 doi: 10.1209/epl/i2002-00248-2
24. Dorogovtsev SN, Mendes JF. Evolution of networks: From biological nets to the Internet and WWW. OUP; Oxford; 2013.
25. Jost D, Carrivain P, Cavalli G, Vaillant C. Modeling epigenome folding: formation and dynamics of topologically associated chromatin domains. Nucleic acids research. 2014;42(15):9553–9561. doi: 10.1093/nar/gku698
26. Sanborn AL, Rao SS, Huang SC, Durand NC, Huntley MH, Jewett AI, et al. Chromatin extrusion explains key features of loop and domain formation in wild-type and engineered genomes. Proceedings of the National Academy of Sciences. 2015;112(47):E6456–E6465. doi: 10.1073/pnas.1518552112
27. Chiariello AM, Annunziatella C, Bianco S, Esposito A, Nicodemi M. Polymer physics of chromosome large-scale 3D organisation. Scientific reports. 2016;6:29775 doi: 10.1038/srep29775
28. Goloborodko A, Marko JF, Mirny LA. Chromosome compaction by active loop extrusion. Biophysical journal. 2016;110(10):2162–2168. doi: 10.1016/j.bpj.2016.02.041
29. Annunziatella C, Chiariello AM, Bianco S, Nicodemi M. Polymer models of the hierarchical folding of the Hox-B chromosomal locus. Physical Review E. 2016;94(4):042402 doi: 10.1103/PhysRevE.94.042402
30. Barbieri M, Xie SQ, Triglia ET, Chiariello AM, Bianco S, de Santiago I, et al. Active and poised promoter states drive folding of the extended HoxB locus in mouse embryonic stem cells. Nature Structural & Molecular Biology. 2017;. doi: 10.1038/nsmb.3402
31. Scialdone A, Cataudella I, Barbieri M, Prisco A, Nicodemi M. Conformation regulation of the X chromosome inactivation center: a model. PLoS computational biology. 2011;7(10):e1002229 doi: 10.1371/journal.pcbi.1002229

32. Bianco S, Chiariello AM, Annunziatella C, Esposito A, Nicodemi M. Predicting chromatin architecture from models of polymer physics. Chromosome Research. 2017; p. 1–10.
33. Barbieri M, Scialdone A, Piccolo A, Chiariello AM, di Lanno C, Prisco A, et al. Polymer models of chromatin organization. Frontiers in genetics. 2013;4 doi: 10.3389/fgene.2013.00113
34. Chiariello AM, Bianco S, Piccolo A, Annunziatella C, Barbieri M, Pombo A, et al. Polymer models of the organization of chromosomes in the nucleus of cells. Modern Physics Letters B. 2015;29(09):1530003 doi: 10.1142/S0217984915300033
35. Knight PA, Ruiz D. A fast algorithm for matrix balancing. IMA Journal of Numerical Analysis. 2013;33(3):1029–1047. doi: 10.1093/imanum/drs019
36. Nicodemi M, Pombo A. Models of chromosome structure. Current opinion in cell biology. 2014;28:90–95. doi: 10.1016/j.ceb.2014.04.004
37. Barbieri M, Chotalia M, Fraser J, Lavitas LM, Dostie J, Pombo A, et al. Complexity of chromatin folding is captured by the strings and binders switch model. Proceedings of the National Academy of Sciences. 2012;109(40):16173–16178. doi: 10.1073/pnas.1204799109
38. Barbieri M, Scialdone A, Gamba A, Pombo A, Nicodemi M. Polymer physics, scaling and heterogeneity in the spatial organisation of chromosomes in the cell nucleus. Soft Matter. 2013;9(36):8631–8635. doi: 10.1039/c3sm51436f
39. Brackley CA, Taylor S, Papantonis A, Cook PR, Marenduzzo D. Nonspecific bridging-induced attraction drives clustering of DNA-binding proteins and genome organization. Proceedings of the National Academy of Sciences. 2013;110(38):E3605–E3611. doi: 10.1073/pnas.1302950110
40. Nicodemi M, Prisco A. Thermodynamic pathways to genome spatial organization in the cell nucleus. Biophysical journal. 2009;96(6):2168–2177. doi: 10.1016/j.bpj.2008.12.3919
41. Brackley CA, Brown JM, Waithe D, Babbs C, Davies J, Hughes JR, et al. Predicting the three-dimensional folding of cis-regulatory regions in mammalian genomes using bioinformatic data and polymer models. Genome biology. 2016;17(1):59 doi: 10.1186/s13059-016-0909-0
42. Nicodemi M, Jensen HJ. Creep of superconducting vortices in the limit of vanishing temperature: A fingerprint of off-equilibrium

dynamics. Physical review letters. 2001;86(19):4378 doi: 10.1103/PhysRevLett.86.4378

43. Nicodemi M, Prisco A. Symmetry-breaking model for X-chromosome inactivation. Physical review letters. 2007;98(10):108104 doi: 10.1103/PhysRevLett.98.108104
44. Grebenkov DS, Ciamarra MP, Nicodemi M, Coniglio A. Flow, ordering, and jamming of sheared granular suspensions. Physical review letters. 2008;100(7):078001 doi: 10.1103/PhysRevLett.100.078001
45. Coniglio A, Nicodemi M. The jamming transition of granular media. Journal of Physics: Condensed Matter. 2000;12(29):6601.
46. Caglioti E, Coniglio A, Herrmann H, Loreto V, Nicodemi M. Segregation of granular mixtures in the presence of compaction. EPL (Europhysics Letters). 1998;43(5):591 doi: 10.1209/epl/i1998-00402-x
47. Tarzia M, de Candia A, Fierro A, Nicodemi M, Coniglio A. Glass transition in granular media. EPL (Europhysics Letters). 2004;66(4):531 doi: 10.1209/epl/i2004-10015-y
48. Nicodemi M, Fierro A, Coniglio A. Segregation in hard-sphere mixtures under gravity. An extension of Edwards approach with two thermodynamical parameters. EPL (Europhysics Letters). 2002;60(5):684 doi: 10.1209/epl/i2002-00363-0
49. Nicodemi M, Coniglio A. Macroscopic glassy relaxations and microscopic motions in a frustrated lattice gas. Physical Review E. 1998;57(1):R39 doi: 10.1103/PhysRevE.57.R39
50. Nicodemi M. Force correlations and arch formation in granular assemblies. Physical review letters. 1998;80(6):1340 doi: 10.1103/PhysRevLett.80.1340
51. Ciamarra MP, De Vizia MD, Fierro A, Tarzia M, Coniglio A, Nicodemi M. Granular species segregation under vertical tapping: Effects of size, density, friction, and shaking amplitude. Physical review letters. 2006;96(5):058001 doi: 10.1103/PhysRevLett.96.058001
52. Cataudella V, Franzese G, Nicodemi M, Scala A, Coniglio A. Critical clusters and efficient dynamics for frustrated spin models. Physical review letters. 1994;72(10):1541 doi: 10.1103/PhysRevLett.72.1541
53. Nicodemi M, Panning B, Prisco A. A thermodynamic switch for chromosome colocalization. Genetics. 2008;179(1):717–721. doi: 10.1534/genetics.107.083154

CHAPTER 7

NUCLEAR SHIELD: A MULTI-ENZYME TASK-FORCE FOR NUCLEUS PROTECTION

Raffaele Fabrini[1], Alessio Bocedi[2], Valentina Pallottini[3], Lorena Canuti[4], Michele De Canio[2], Andrea Urbani[2,5], Valeria Marzano[2,5], Tommaso Cornetta[3], Pasquale Stano[3], Anna Giovanetti[6], Lorenzo Stella[1], Antonella Canini[4], Giorgio Federici[2], Giorgio Ricci[1]

[1]Department of Chemical Sciences and Technologies, University of Rome Tor Vergata, Rome, Italy
[2]Department of Internal Medicine, University of Rome Tor Vergata, Rome, Italy
[3]Department of Biology, University of Roma Tre, Rome, Italy
[4]Department of Biology, University of Rome Tor Vergata, Rome, Italy
[5]S. Lucia Research Institute IRCCS, Rome, Italy
[6]Institute of Radiation Protection, ENEA-CR Casaccia, Rome, Italy

Citation: Fabrini R, Bocedi A, Pallottini V, Canuti L, De Canio M, et al. (2010) Nuclear Shield: A Multi-Enzyme Task-Force for Nucleus Protection. PLoS ONE 5(12): e14125. doi:10.1371/journal.pone.0014125

ABSTRACT

Background

In eukaryotic cells the nuclear envelope isolates and protects DNA from molecules that could damage its structure or interfere with its processing. Moreover, selected protection enzymes and vitamins act as efficient guardians against toxic compounds both in the nucleoplasm and in the cytosol. The observation that a cytosolic detoxifying and antioxidant enzyme *i.e.* glutathione transferase is accumulated in the perinuclear region of the rat hepatocytes suggests that other unrecognized modalities of nuclear protection may exist. Here we show evidence for the existence of a safeguard enzyme machinery formed by an hyper-crowding of cationic enzymes and proteins encompassing the nuclear membrane and promoted by electrostatic interactions.

Methodology/Principal Findings

Electron spectroscopic imaging, zeta potential measurements, isoelectrofocusing, comet assay and mass spectrometry have been used to characterize this surprising structure that is present in the cells of all rat tissues examined (liver, kidney, heart, lung and brain), and that behaves as a "nuclear shield". In hepatocytes, this hyper-crowding structure is about 300 nm thick, it is mainly formed by cationic enzymes and the local concentration of key protection enzymes, such as glutathione transferase, catalase and glutathione peroxidase is up to seven times higher than in the cytosol. The catalytic activity of these enzymes, when packed in the shield, is not modified and their relative concentrations vary remarkably in different tissues. Removal of this protective shield renders chromosomes more sensitive to damage by oxidative stress. Specific nuclear proteins anchored to the outer nuclear envelope are likely involved in the shield formation and stabilization.

Conclusions/Significance

The characterization of this previously unrecognized nuclear shield in different tissues opens a new interesting scenario for physiological and protection processes in eukaryotic cells. Selection and accumulation of protection enzymes near sensitive targets represents a new safeguard modality which deeply differs from the adaptive response which is based on expression of specific enzymes.

INTRODUCTION

In eukaryotic cells different types of biologic machineries contribute to protect DNA from molecules that could damage its structure or interfere with its processing. The nuclear envelope is a first important mechanical barrier that opposes the interaction of toxic compounds with the genetic material [1]. A second one is represented by specific protection enzymes and molecules (*i.e.* glutathione, vitamin A, C and E) able to eliminate many dangerous compounds. A third protection mechanism is formed by specific transcription factors mediated pathways [2]. Among the many toxic and dangerous compounds for the nucleus, a prominent killer role is due to compounds that produce oxidative (ROS), nitrosative (RNS) and alkylative stress. Catalase (CAT), glutathione peroxidase (GPX) (scavengers of H_2O_2) and superoxide dismutase (SOD) (which eliminates $HO_2^{\bullet}$ radicals) are the most important antioxidant enzymes that counteract in many cells the killer activity of ROS. Recently an active antioxidant role has been described for heme oxygenase-2 in specific cell lines [3], [4] and for DNA polimerase iota, an enzyme which has intranuclear localization [5]. Glutathione transferases (GSTs), a superfamily of enzymes grouped in at least eight gene-independent classes in mammals, are also involved in the cell protection against alkylating compounds and organic peroxides. These enzymes catalyze the conjugation of glutathione (GSH) to the electrophilic centre of toxic alkylating compounds [6] and the Alpha class isoenzymes display a selenium-independent glutathione peroxidase activity [6]. We have also demonstrated that GSTs is involved in the cell defence against excess nitric oxide (NO) sequestering this free radical in a harmless iron complex bound to the active site [7]. Thus GST represents a multifunctional enzyme involved in the protection against ROS, RNS as well as against electrophilic agents. While this enzyme can be up-regulated in case of electrophilic or ROS stress [8], [9], the intracellular concentrations of CAT and GPX in various tissues cannot be increased in case of oxidative stress conditions [10], [11]. Despite the absence of a general adaptive response, a permanent optimization of the defence power could be reached by increasing the local concentration of protection enzymes near sensible intracellular targets like the nucleus. Possible existence of this novel defence strategy is suggested by a few observations: a curious presence of GSTs near the nucleus has been reported many years ago in immunohistochemical and non-aqueous cell fractionation studies [12], [13], [14]. More recently, we have observed a relevant accumulation of Alpha class GSTs near the nuclear membrane of the rat hepatocytes [15], a phenomenon revealed by the expedient of avoiding exogenous salts or buffers during the purification of the nuclear fraction [15]. The presence of salts or buffers, usually employed for nuclear preparations, easily detached these proteins from the membrane, a finding suggesting a predominant contribution of electrostatic interactions in this binding [15]. Importantly, the use of a specific fluorescent probe for GST signalled its perinuclear accumulation even in intact cells [15]. However, all these previous studies did not verify if other key protection enzymes beside GSTs are present. The possibility that an unrecognized more complex enzyme organization near the nucleus may exist is a stimulating proposal worthy to be investigated.

The present study explores the following possibilities: a) that beside GSTs, additional protection enzymes may be associated to the nuclear envelope, b) that this phenomenon may not be restricted to hepatocytes, c) that it has a specific protection function. We demonstrate here for the first time that a surprising selected multi-enzyme machinery is present near the nuclear envelope of cells from many tissues forming a sort of enzyme task-force which contributes to DNA protection.

RESULTS

First Evidence and Physical Characterization of the Nuclear Shield

In a first experimental approach, nuclei were purified in the absence of exogenous salts, and the proteins bound electrostatically to the nuclear membrane were detached by increasing the ionic strength (see Materials and Methods). Specifically, in rat liver 2.7 mg of nuclear bound proteins per gram of fresh tissue were extracted (Table 1), including, as expected, the Alpha class GSTs (see below). This amount represents about 3% of all cytosolic proteins. Nitrogen electron spectroscopic imaging (ESI) of nuclei purified in the absence of exogenous salts confirmed the presence of a thick protein structure surrounding the nuclear membrane, which appears like a "nuclear shield" and almost vanishes upon mild salt treatment (Figure 1, A and B). Statistical examination of ESI images at varying cut sections is consistent with an estimated thickness of the shield of about 300±70 nm. Given that the nuclear membrane available is 0.022 m^2/g of liver [15], [16] (see Materials and Methods), the shield region has a protein density of about 0.4 g/cm^3, a value twice as high as the one in the cytosolic milieu (~0.2 g/cm^3), but only half as much as the one in a protein crystal (~1 g/cm^3) [17]. In other words, this structure, rather than a compact multilayer [15], may be represented as a region of perinuclear hyper-crowding which is undetectable by means of standard morphological microscopy (Figure 1C).

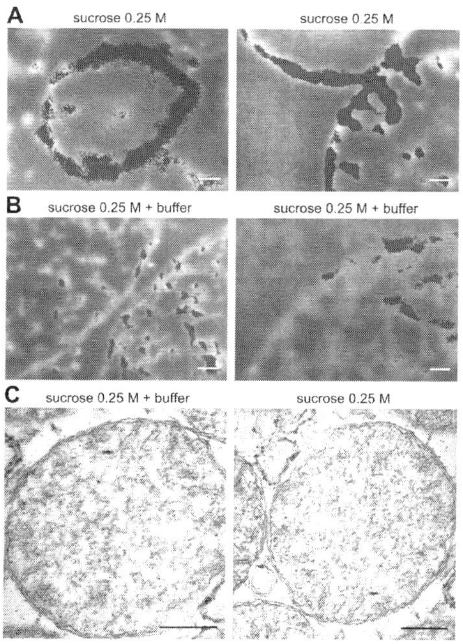

Figure 1: ESI experiments on shielded and de-shielded rat liver nuclei. (A) ESI of shielded nuclei. Images were obtained as described in Materials and Methods. The average of thickness of the red area showing the protein shield is 300±70 nm. Scale bar, 0.3 μm. (B) ESI of de-shielded nuclei. Scale bar, 0.3 μm. (C) TEM micrographs of shielded and de-shielded nuclei. Samples were treated and stained as reported in Materials and Methods. Scale bar, 0.3 μm.

Table 1: Nuclear shield proteins.

Organ	nuclear shield proteins (mg/g of tissue)	cytosolic proteins (mg/g of tissue)	membrane area/g of tissue (m^2/g)	nuclear shield proteins/ membrane area (mg/m^2)
LIVER	2.7±0.4	85±5	0.022*	120±20
KIDNEY	2.6±0.3	70±6	0.027±0.001	100±10
HEART	3.8±0.2	31±5	0.032±0.001	119±7
LUNG	5.0±0.2	60±4	0.019±0.001	260±20
BRAIN	2.7±0.3	24±7	0.051±0.002	54±6

Zeta Potential Measurements and Nuclear Shield Re-formation

Zeta potential, a sensitive function of the interface nature of suspended particles [18], [19], added further details on the protein organization of the shield. At increasing ionic strength, the change in zeta potential of isolated nuclear fractions parallels the GST detachment and proceeds without apparent discontinuity (Figure 2A), The zeta potential perturbation observed at very low ionic strength is likely due to a nuclear disaggregation process as suggested by the light scattering analysis (Figure 2B). We observed that a spontaneous and partial change of the zeta potential, accompanied by a parallel detachment of about 50% of the shield proteins, occurs even without addition of salts by simply incubating a dilute nuclear fraction in 0.25 M sucrose (Figure 2C). On the other hand, de-shielded nuclei extensively washed with 0.25 M sucrose are still able to re-constitute about 50% of the original nuclear shield when incubated with a cytosolic fraction. The process is fast but not instantaneous showing a $t_{1/2}$ of 2.5 minutes (Figure 2D). The observed time dependent and spontaneous detachment of the nuclear shield (up to 50%) in 0.25 M sucrose suggests that the shield, as it appears immediately after nuclei preparation, could not be an artefact due to the use of sucrose. Obviously, the possibility that positively charged proteins present in the cytosol may be linked to a negative counterpart (*i.e.* proteins or phospholipid layers) depends on the relative competition between the cationic proteins and other positively charged ions (mainly K^+). Thus, the relative concentration of these cationic objects drives protein attachment or detachment. The modality employed to prepare the nuclear fraction in 0.25 M sucrose does not perturb the reciprocal concentration of all cationic competitors and thus it appears a correct procedure to visualize specific electrostatic interactions like those which promote the nuclear shield. In addition, the existence of an artefact should be signalled by an increase in the shield when the relative amount of sucrose (compared to the cytosolic volume) is increased. Conversely, protein content of the nuclear shield remained almost unchanged when the nuclei were prepared from liver homogenate under different dilution conditions (*i.e.* 1:3, 1:6, 1:10 and 1:20 gram of tissue/ml of 0.25 M sucrose) (data not shown). Other previous evidences confirm the real existence of the shield; the perinuclear accumulation of GST has been demonstrated in intact cells (thus without sucrose) by using a specific fluorescent probe for this enzyme [15] and a similar evidence has been obtained using immunostaining procedures [12], [13], [14].

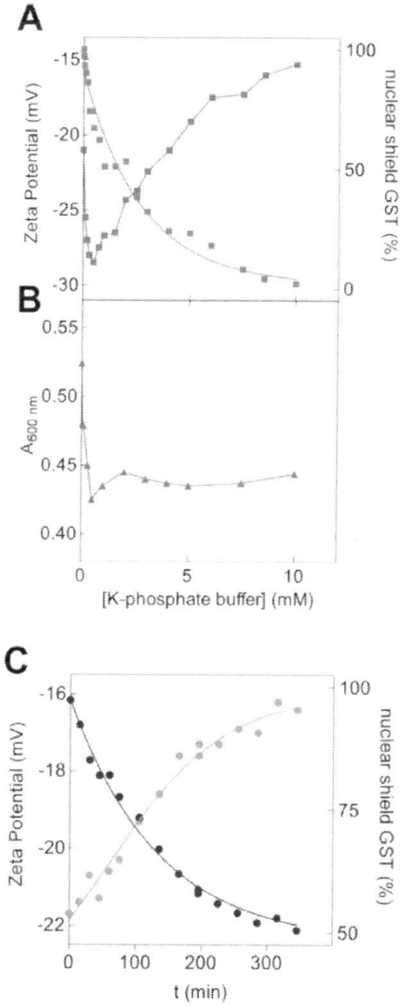

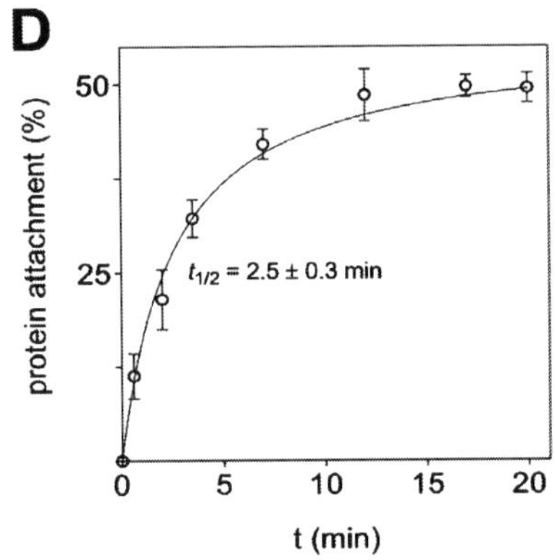

Figure 2: Zeta potential and shield re-formation. Shielded nuclei were diluted with 20 volumes of 0.25 M sucrose. (A) Zeta potential changes (*pink squares*) and GST detachment (*green squares*) due to the addition of potassium phosphate buffer, pH 7.4. Zeta potential perturbation at very low buffer concentration (below 1 mM) is due to a nuclear disaggregation process as suggested in (B) by a similar perturbation of light scattering at 600 nm (*pink triangles*). (C) Time dependent zeta potential changes and nuclear shield GST detachment in 0.25 M sucrose. Shielded nuclei suspended in 0.25 M sucrose were incubated at 25°C. At variable times the amount of GST released in solution (*blue circles*) and zeta potential (*light blue circles*) were measured. (D) Kinetic of shield re-formation. De-shielded nuclei in 0.25 M sucrose were divided into aliquots; each aliquot was resuspended in rat liver cytosolic fraction and incubated at 25°C. At fixed times, the mixture was centrifuged. The re-formation of the shield was measured either on the basis of protein content that can be detached with 50 mM NaCl. All experiments were performed in triplicate. Error bars, s.d.

Proteins of the Nuclear Shield Display Peculiar Acid-base Properties

We next examined whether the protein composition of the nuclear shield is similar to the one of the cytosolic fraction or rather only selected proteins are enriched in this perinuclear region. Isoelectric focusing experiments first suggested that this structure is formed prevalently by positively charged proteins (80%) while, in agreement with previous observations [20], the cytosol is populated predominantly by acidic proteins (60%) (Figure 3, A–C). Furthermore, a more stringent analysis of the basic region of the chromatograms indicated that the cationic proteins of the shield, grouped in discrete *pI* ranges, display a different distribution when compared to the cytosolic pool with an evident selection of the proteins with higher *pI* values (*pI*>8.0) (Figure 3D). The average *pI* of all shield proteins is ~8.0 while the average *pI* of the cytosol is ~6.0, a value close to that reported in previous studies (*pI*~5.5) [21].

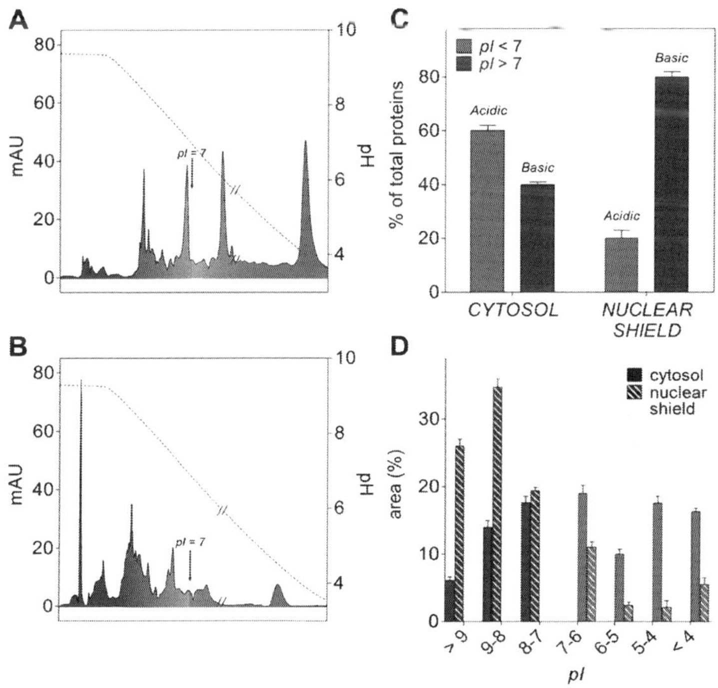

Figure 3: Isoelectrofocusing experiments.(A) Combined chromatogram of the cytosolic proteins obtained from two runs at different pH ranges (*i.e.* pH 9.0–6.0 and pH 7.0–4.0). The double slash shows the connection zone of the two chromatograms. Blue area indicates the basic proteins and red area the acidic proteins. Dotted line represents the pH gradient. (B) Combined chromatogram of the nuclear shield proteins. (C) Acid-base properties of nuclear shield and cytosolic proteins. (D) Percent of acidic and basic proteins grouped in discrete *pI* ranges as derived by the experiments shown in A and B.

Key Antioxidant Enzymes are Present in the Nuclear Shield

The identification of specific enzyme activities in the nuclear shield could be diagnostic to clarify the physiological role of this structure. Beside the Alpha class GSTs – a well known efficient barrier against alkylating compounds, organic peroxides and nitric oxide [15], [22] – we detected significant amounts of additional anti-oxidant enzymes such as CAT and GPX (Table 2 and Figure 4A). In the nuclear shield all these enzymes have specific activities (U/mg of total proteins) similar and even higher than those found in the cytosol (five times higher for GST and three times for CAT) (Figure 4B) and increased local concentrations (seven times higher for GST and four times for CAT) (Figure 4C). The normalization of the activities of these enzymes to the corresponding nuclear membrane area is an additional parameter here termed "defence potentiality" of the shield (Table 2 and Figure 4D). For comparison, we examined the presence, in the nuclear shield, of enzymes that do not have a specific "protective role", such as L-lactate dehydrogenase (LDH), L-alanine amino transferase (ALT), and creatine kinase (CK); these proteins are scarcely present in the shield and display very much lower specific activities than in the cytosol (Figure 4B). Curiously, the CAT identified in the shield is a peculiar cationic form (*pI*=7.7) that is not found in the cytosol where a few anionic forms (*pI*=5.8÷6.2), also present in large amounts in peroxisomes, are the predominant isoenzymes [23]. CAT is encoded by a single gene and its theoretical *pI* value, calculated on the basis of the amino acid composition, is 7.5 (see Materials and Methods) close to the one of the isoenzyme found in the shield, while the acidic cytoplasmic form is the result of post-translational modifications [24]. The shield contains only traces of an additional anti-oxidant enzyme *i.e.* SOD but its presence in this region could be pleonastic due to its cytosolic and nucleoplasmic localization [25]. Notably, the catalytic activities of GST, CAT and GPX are very similar when these enzymes are packed into the nuclear shield or free in solution (Figure 4E), demonstrating that active site accessibility and functional flexibility are not impaired by the increased protein density. Mass spectrometry analysis of the nuclear shield extract and of the cytosol disclosed additional details. Beside a clear confirmation of the prominent selection of cationic proteins in the shield, this approach indicated that, in addition to GST, CAT and GPX, many other enzymes contribute to the nuclear shield. Keeping in mind that mass spectrometry analysis allows to identify only proteins that display proper fragmentation patterns and volatility properties (*e.g.* GPX, whose activity was unambiguously identified, could not be detected either in the shield or in the cytosol), 78 distinct proteins have been identified in the shield, 39 of which are almost exclusive of this structure. Of note, 72% of the shield proteins have, in their native oligomeric structures, molecular masses higher than the cut-off of the nuclear pores (kDa=40÷45) [26], [27], indicating that they are unlikely to represent a contamination due to nuclear proteins. Beside the postulated antioxidant role inferred from the observed enrichment of GST, CAT, and GPX, the nuclear shield may also have additional functions as hinted from the variegated enzyme mesh identified by mass spectrometry approach. In this respect it is worth pointing out that typical "metabolic enzymes" such as malate dehydrogenase and glyceraldehyde-3-phosphate dehydrogenase, identified in the nuclear shield, have been shown to be also part of transcriptional complexes and implicated in transcription regulation [28], [29].

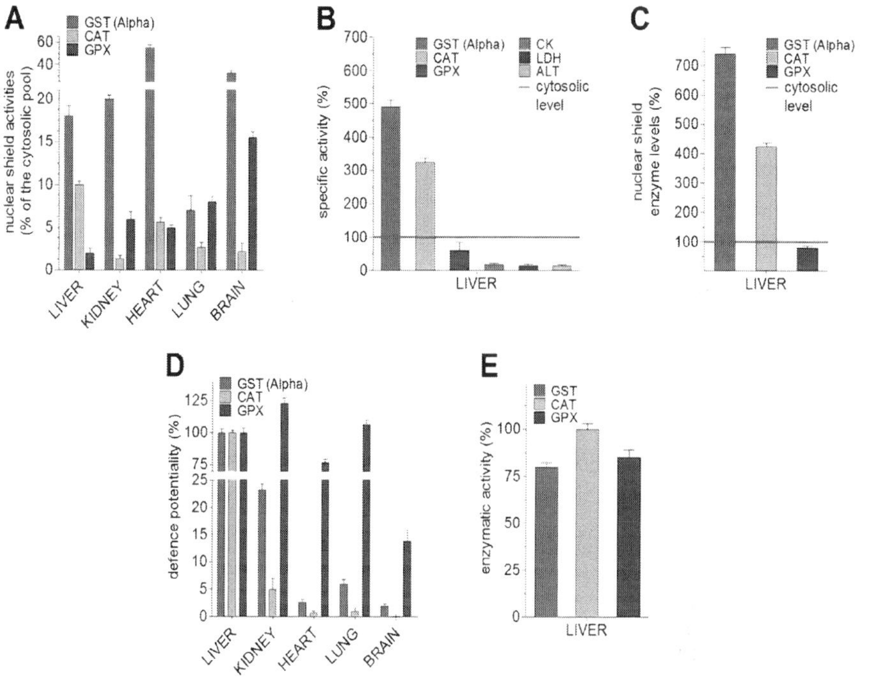

Figure 4: Quantification of antioxidant enzymes in the nuclear shield. Activity values reported in (A–E) are the means of 5 independent tissue preparations. Calculations in (A and C) were made assuming as reference the activity of the cytosolic Alpha GST found in distinct rat tissues (see Materials and Methods). (A) Nuclear shield activities of GST, CAT, GPX reported as percentages of the corresponding cytosolic activities. Error bars, s.d. (B) Specific activities of the antioxidant enzymes (GST, CAT and GPX) and non-antioxidant enzymes (CK, LDH and ALT) in the shield of rat liver. Specific activities of the corresponding cytosolic enzymes are taken as 100%; error bars, s.d. (C) Nuclear shield GST, CAT, and GPX concentrations compared with their cytosolic levels taken as 100%. A cytosol volume of 2630 μm^3 for each hepatocyte and a shield volume of 67 μm^3 (0.3 μm×222 μm^2) were assumed for calculations [53]. Errors, s.d. (D) "Defense potentiality" of the nuclear shield. GST, CAT, and GPX activities found in the nuclear shield and normalized for nuclear membrane area (U/m^2 of

membrane area) of liver, kidney, heart, lung, and brain (from Table 2) were compared to those of liver taken as 100%. Error bars, s.d. (E) Catalytic activity of GST, CAT and GPX measured when they are free in solution (100%) or packed in the nuclear shield (see Materials and Methods); error bars, s.d.

Table 2: Antioxidant enzymes of the nuclear shield.

organ	cytosolic activity (U/g of tissue)				nuclear shield activity (U/g of tissue)			shield defense potentiality* (U/m^2 of membrane area)		
	GST	GST (Alpha)**	CAT	GPX	GST (Alpha)†	CAT	GPX	GST(Alpha)†	CAT	GPX
LIVER	90±10	39±4	29000±5000	14±2	7±2	3000±600	0.28±0.04	300±80	140000±30000	13±2
KIDNEY	12±4	9±3	15000±2000	8±2	1.8±0.6	200±50	0.4±0.1	70±20	7000±2000	15±4
HEART	2.9±0.8	0.4±0.1	600±100	6±1	0.25±0.09	34±8	0.3±0.1	8±3	1100±200	9±3
LUNG	10±3	5±1	800±100	4±1	0.35±0.06	26±9	0.26±0.07	18±3	1400±500	14±4
BRAIN	6±2	0.9±0.3	370±50	0.6±0.1	0.3±0.1	10±3	0.09±0.03	6±2	200±60	1.8±0.6

Activities of GST, CAT and GPX were measured as reported in Materials and Methods. Values reported are the means of 5 independent tissue preparations. The standard deviation of the five different replicas are reported for each measurement. The units of the cytosolic GSTs (first column) are the sum of the activity contributions of all enzyme isoforms.

*"Defense potentiality" is defined as the amount of enzyme unit normalized per membrane area.

**Units of cytosolic Alpha class GST were calculated from previous studies (see Materials and Methods).

GST units in the shield of various tissues were tentatively ascribed to Alpha class GST as occurs for the liver [15].

DNA Protection

Whatever possible additional roles played by such a variegated enzymatic composition of the shield, the clear enrichment of DNA-protective enzymes (Figure 4C) suggests for this newly identified structure a role as a guardian of genomic integrity. Comet assay, allowing the assessment of DNA breaks, performed on shielded and de-shielded nuclei, showed a direct evidence that the nuclear shield represents an efficient protection barrier for the genetic material, both in the presence and in the absence of oxidizing compounds (Figure 5A). A quantification of the damage produced by H_2O_2 on DNA has been obtained by measuring the tail length (Figure 5B). Furthermore, a clear indication that perinuclear Alpha GSTs may act as an efficient trap even against non-oxidizing toxic compounds like nitric oxide derivatives has been described previously in intact hepatocytes treated with NO donors [7].

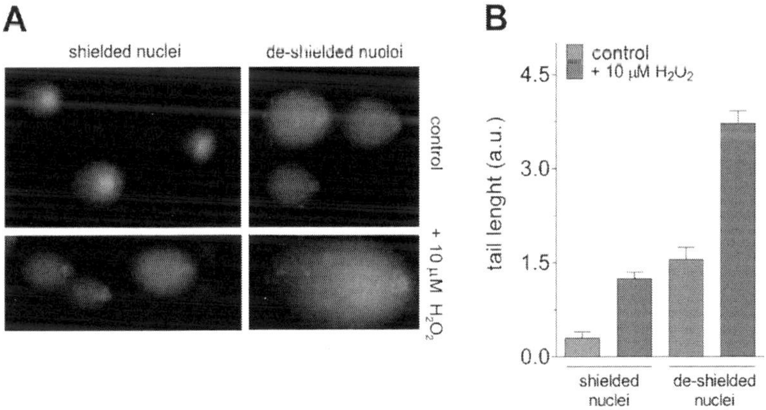

Figure 5: Protection of DNA from oxidative damage in the presence or absence of the nuclear shield.

(A) Comet assay on shielded and de-shielded nuclei exposed to 10 μM H_2O_2 for three minutes. Experimental details are reported under Materials and Methods. (B) Statistical analysis of comet assay. Error bars, s.d.

Nuclear Shield Compositions Vary in Cells of Different Tissues

This protein structure is not restricted to the liver cells but it is also present in the perinuclear regions of cells of tissues as diverse as kidney, heart, lung and brain (Table 1, Table 2 and Figure 4A). The most and the less populated nuclear shields were found in the lung and in the brain, respectively (Table 1). This difference might be related to the fact that lung cells are more exposed to exogenous toxic compounds and, conversely, the brain is a protected tissue by virtue of the hematoencephalic barrier. Interestingly, each tissue shows different and characteristic levels of antioxidant enzymes in the shield; for example, while the liver shield has a relevant amount of CAT and little GPX, the opposite it is true in the kidney, lung and brain (Figure 4A).

Nuclear Proteins are the Electrostatic Counterpart for the Cationic Shield

Only the nuclear membrane displays propensity to form the protein shield. Membranes of different intracellular organelles, such as mitochondria, microsomes and lysosomes, lack this enzymatic protection [15]. Given that all intracellular membranes have similar, albeit non identical, lipid composition, it is unlikely that subtle differences may be responsible for the unique presence of the shield on the nuclear membrane. We speculate that some negatively charged proteins, exclusively present in the nuclear membrane and firmly bound to the envelope, may provide the necessary electrostatic potential to attract and stabilize the protein shield. Proteolytic experiments support this hypothesis; incubation of de-shielded nuclei with trypsin or protease K completely inhibits the re-formation of the nuclear shield (Figure 6A).

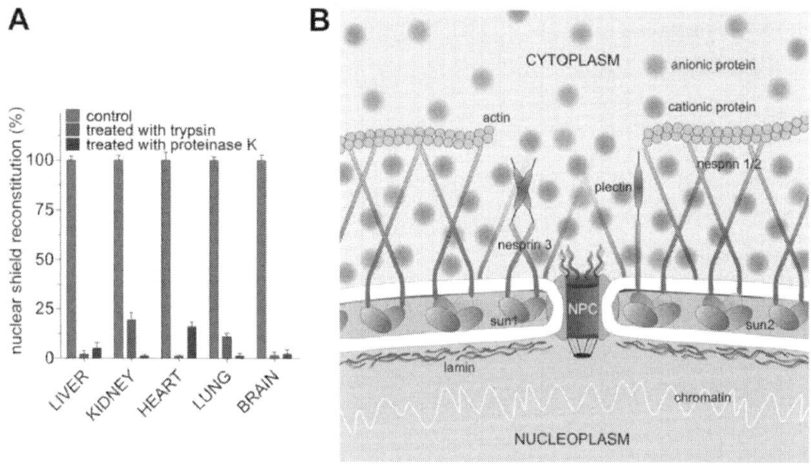

Figure 6: Proteolysis experiments and tentative pictorial representation of the nuclear shield. (A) Effect of proteolysis by trypsin and proteinase K on the nuclear shield reconstitution (see Materials and Methods). Error bars, s.d. (B) Tentative pictorial representation of nuclear shield on the outer nuclear membrane, also based on the models shown in refs. [32], [33]. Proportions between anionic and cationic proteins in the shield and in the cytosol shown in the picture are those obtained experimentally.

DISCUSSION

For the first time a particular enzyme-network is identified which is electrostatically associated to the outer nuclear membrane. In the last years many studies have been performed to detail composition and role of the inner and outer nuclear envelopes and their enzymatic equipments. For example, nuclear pores and other integral membrane protein complexes (belonging to the inner membrane) have been found to play a fundamental role in the dynamic organization of the genome, positioning in DNA repair, recombination and stability [30]. On the other hand, recent findings point to important structural roles for nesprins and plectins, giant rod-like proteins anchored exclusively to the outer nuclear membrane [31]. These proteins stretch out into the cytosol and are involved in the nucleus positioning. The nuclear shield described in the present study implements the functions correlated to the outer nuclear membrane. The existence in different rat cells of a perinuclear multi-enzyme structure mainly formed by selected cationic proteins including key protection enzymes (Figure 3 and Table 2) indicates that this particular region may act as guard against oxidative damage. This peculiar finality can be only revealed on the basis of classical biochemical studies instead of immunocytochemical assays that cannot verify the enzymatic competence of the detected enzymes. Indeed, antibody works made in the past [12], [13], [14] clearly indicated the accumulation of GST in the perinuclear region but they were unable to assess any biological activity. The presence of active forms of GST, CAT and GPX, as determined in this paper, designs this perinuclear region a sort of hyper-filter where the detoxifying power is enhanced up to seven times with respect to the cytosol (Figure 4C). A simple comet assay visualizes an increased DNA protection against ROS favoured by this structure (Figure 5). Additional evidences point to a specific finality of the shield and not to a casual enzyme assemblage. For example, this structure appears more prominent in tissues more exposed to toxic compounds (lung) while it is less populated in the highly protected brain (Table 1). Furthermore, in tissues where CAT is abundant, GPX (which has similar detoxifying activity against H_2O_2) must be redundant and, accordingly, it is scarcely represented in the shield (Table 2). Obviously, the protection role may be just one of many other possible functions inherent in this structure. Beside CAT, GST and GPX, the large variety of different enzymes that composes the shield represents an unexplored mine for future investigations in this direction. We also note that most of the shield enzymes have molecular masses higher than the one allowing the entrance into the nucleus and thus this machinery may be a clever system to approach these enzymes to the chromosomes.

The biochemical and physical characterization of this novel cellular structure outlines a new level of cell organization, mediated by weak, relatively unspecific, electrostatic interactions, that somewhat contrasts a picture of the cell delineated by well organized organelles interspersed in a relatively homogeneous milieu. The particular protein density of the shield (only two times higher than in the cytosol) as calculated through ESI measurements, allows to assimilate this structure to an hyper-crowding which cannot be detected by classical microscopy (Figure 1C). This may explain why its presence remained masked until now. Interestingly, the lack of shield re-formation after proteolytic treatment of the nucleus (Figure 2D) indicated that some negatively charged proteins (specifically present on the outer nuclear envelope) could play a crucial role in the shield formation. Only a few membrane proteins may be indicated as possible shield scaffold. Among them, nesprin-1, nesprin-2 and plectin are giant proteins (ranging from 2800 to 6900 residues) showing rod-like structures about 300–400 nm long [31] (a length similar to the shield thickness), and exclusively present as a sort of negatively charged network on the outer nuclear membrane surface [32], [33]. These proteins are involved in the nucleus positioning and display conspicuous net negative charge *i.e.* about −70 for plectin, −100 for nesprin-1 and −240 for nesprin-2 (see Materials and Methods) that may represent a good electrostatic counterpart for the positively charged proteins of the nuclear shield. A tentative pictorial model of this structure is shown in Figure 6B.

If the nuclear shield is a functionally important structure, as proposed here, genetic alterations that reduce its thickness or modify its enzyme composition should affect the protective role and possibly result in pathological phenotypes. Intriguingly, many neurodegenerative pathologies like Alzheimer, Parkinson, Huntington and amyotrophic lateral sclerosis are characterized by typical aggregations of misfolded or damaged proteins in the perinuclear region termed aggresomes [34]. The possibility that such structures alter the protective nuclear shield described in this study is just one of the hypotheses that must be verified in the immediate future.

MATERIALS AND METHODS

Reagents

All reagents used in this study were from Sigma-Aldrich Inc. (St. Louis, USA) and used without further purification. Trypsin and Protease K (Sigma-Aldrich) employed in the proteolytic experiments were from bovine pancreas (11,400 U/mg) and from *Tritirachium album* (13.1 U/mg), respectively.

Animals

Male *Wistar* rats were anaesthetised with sodium pentobarbital (50 mg/kg body weight, injected *i.p.*) before rapid killing by cervical dislocation minimizing sufferings and subsequent liver, kidney, heart, lung and brain dissections. Experiments were carried out with the approval by the Ethic Committee of the University of Roma Tre in accordance to the ethical guidelines for animal research of the Italian Ministry of Health (permit number: 246/H10) D.Lvo 116/92.

Nuclei Preparations

After perfusion with 0.25 M sucrose and heparine to remove blood, different tissues (liver, kidney, heart, lung and brain) from male rats were excised, minced and homogenized in a teflonglass Potter Elvehjem homogenizer, in 0.25 M sucrose (10 ml per gram of tissue). After a brief centrifugation at 300×*g* (3 min) to remove unbroken cells and periplasmic membranes, the homogenate was centrifuged at 1000×*g* for 10 min to isolate the nuclear fraction [15]. The resulting supernatant, centrifuged at 100,000×*g* for 30 min, represents the "cytosolic extract". The nuclear pellet was washed three times with 10 ml of 0.25 M sucrose and re-suspended in 10 ml of 0.25 M sucrose. This suspension represents the "shielded" nuclei fraction. It contains less than 2% of contaminating structures as judged by microscopy analysis. Even less contamination was observed by measuring marker enzymes of cytosolic, mitochondrial, lysosomal and microsomal origin [15]. De-shielded nuclei were obtained by incubating the shielded nuclei (coming from 1 gram of fresh tissue) with 50 mM NaCl (or 10 mM potassium phosphate buffer, pH 7.4). After centrifugation at 10,000×*g* for 5 min, 4°C, the pellet was re-suspended in 10 ml of 0.25 M sucrose. The total protein content of cytosolic fraction and nuclear shield fraction were determined by the method of Lowry [35].

Nuclear Membrane Surface

Nuclear membrane area of cell from different rat tissues were calculated on the basis of the total phospholipidic content. Phospholipids were determined according to literature protocols [36], [37], [38]. The extraction step started from 800 μl of a nuclear suspension of different tissues prepared as above described. In a glass tube 4 ml of a solution $CHCl_3$/MeOH 2:1 v/v were mixed with the nuclear sample, vortexed for one min and centrifuged at 2,000 rpm for 2 minutes. The aqueous phase was discarded. Chromatographic controls were made by using a TLC slice of silica-gel POLYGRAM SIL UV 254 (0.25 mm gel with fluorescence probe), eluted with a solution $CHCl_3$/MeOH/H_2O/NH_3 - 25% 65/30/4/2 v/v. The organic phase was mixed with 1 ml of NaCl 0.9% at 0°C, vortexed and centrifuged at 2,000 rpm for 2 minutes. The washing aqueous phase was discarded and the organic phase was stored at −20°C. To quantify the phospholipidic content, 80 µl of organic phase were evaporated (110°C, 15 min), then 70 µl of H_2SO_4/$HClO_4$ 1:1 v/v were added and the sample was heated at 240°C for 30 minutes. 1.6 ml of ascorbic acid/ammonium molibdate 0.83%/0.2% were added, vortexed and incubated at 45°C for 30 minutes. The blue color due to the complex formed was quantified spectrophotometrically at 820 nm. The calibration curve derived from a standard solution of 1 mM KH_2PO_4.

For quantitative analysis, the total membrane surface area of a single rat liver hepatocyte is assumed 110,000 μm^2 [16]. Considering that 1 gram of fresh liver contains about 10^8 cells and that the outer nuclear membrane is 0.2% of the total hepatocyte membranes, it results that the available outer nuclear surface area is about 0.022 m^2 per gram of fresh liver. The membrane area/g of tissue for kidney, lung, heart, and brain was determined by evaluating the phospholipidic content as reported above and by comparing this value with the phospholipidic content found in rat liver.

Nitrogen Electron Spectroscopic Imaging (ESI) and Morphological Microscopy

Shielded and de-shielded nuclei from rat hepatocytes suspended in 0.25 M sucrose were treated with a fixative solution containing 3% glutaraldehyde for six hours. Nuclei were post-fixed for 2 hours in 1% osmium tetroxide in the same buffer, dehydrated with ethanol, and embedded in epon 812 resin [39] (TAAB, England). Thin and ultra-thin (<40 nm) unstained sections were collected in uncoated 200 mesh copper grids and observed with a Zeiss CEM transmission electron microscope at 80 kV. Staining of the ultra-thin sections with lead citrate and uranyl acetate (usually performed to visualize better sub-cellular structures in transmission electron microscopy) was omitted to avoid the interference with the ESI and electron energy loss spectroscopy analyses. For the localization of N in the specimens, the electron spectroscopic images were taken at ΔE=410 eV, just above the ionization edge (IE) of N (Nk ΔE=401 eV), to detect the total N signal, and at ΔE=377 eV, the pre-ionization edge (PIE) of N, as a reference carrying information on the background [39], [40]. The unit test area was circular with a variable diameter ranging from 1.3 to 8 mm, which was anyway smaller than the selected microscopic field and inversely proportional to the degree of microscopic magnification. Net N images were obtained by using a digital image analyzer with an interactive built analysis system: the N map obtained by subtracting the PIE from the IE images was recorded with a highly sensitive camera.

Enzymatic Assays and Catalytic Activities

GST, CAT, GPX, and SOD activities were assayed both in cytosolic and nuclear shield fractions following usual assay conditions [15], [41], [42], [43]. Cytosolic and nuclear shield fractions were treated with dithiothreitol (DTT) to preserve GPX enzymatic activity [42]. Enzymatic assays for CK, LDH, and ALT were carried out on a Modular P800 device (Roche) [44]. The enzymatic units of cytosolic GSTs are the sum of the activities of all enzyme isoforms. Conversely, as demonstrated previously for the liver nuclei [15], the activities of GSTs found in the shield of the different rat tissues can be related mainly to the Alpha class GSTs. The cytosolic Alpha GST abundances are calculated on the basis of previous studies on liver [45], kidney [46], heart [47], lung [48], and brain [49] GSTs.

Comparison of the catalytic activities of GST, CAT and GPX bound to the nuclear shield or free in solution was made on shielded nuclei aliquots (suspended in 0.25 M sucrose), in the absence or in the presence of 50 mM NaCl. Activities were measured by usual spectrophotometric methods [15], [41], [42] but in the absence of any buffer.

Zeta Potential Measurements

Zeta potential values were obtained by using a Laser-Doppler microelectrophoresis using a Zetasizer 5000 instrument (Malvern, UK). Measurement cell was aligned before analysis by using reference latex beads (−50±5 mV) (Malvern, UK). Zeta potentials were calculated by the integrated Malvern proprietary software (v. 1.36) by using the Smoluchowsky model (F(ka)=1.5). Reported values refer to the average of five independent measurements (variation coefficient ca. 5–10%). Shielded nuclei samples from rat liver in 0.25 M sucrose, kept constantly at 4°C, were treated with variable amounts of 0.1 M potassium-phosphate buffer, pH 7.4 (or of 1 M NaCl). The volume of the added buffer did not exceed 2.5% v/v. Samples were then poured into the measurement cell and measured after temperature equilibration at 25°C (about 1 minute). Preliminary experiments have shown that nuclei do not sediment significantly in such time interval.

Nuclear Shield Re-formation

Nuclei from rat liver were de-shielded with 50 mM NaCl and washed three times with 0.25 M sucrose. The nuclear suspension was divided into aliquots, each aliquot was resuspended in rat liver cytosolic fraction and incubated at 25°C. At fixed times, the mixture was centrifuged. The re-formation of the shield was measured either on the basis of protein detached from the nuclei with 50 mM NaCl or, alternatively, by measuring the decrease of the total protein concentration of the cytosol after incubation with the de-shielded nuclei. Experiments were performed in triplicate.

FPLC Chromatofocusing

Isoelectrofocusing experiments were performed on an AKTA Purifier system (Amersham Biosciences, Inc.), equipped with a pump system (P-900), spectroscopy unit (UV-900), pH-meter/conductimeter (pH/C-900), and a fraction collector (Frac-900). The chromatography runs were performed on a Mono P 5/200 GL (HR 5/20) column (Amersham Biosciences Inc.), with 1 ml/min flow rate. The pH range explored in two different chromatographic runs were from 9.0 to 6.0 using a start buffer ethanolamine-CH_3COOH 0.025 M pH 9.4 and a Polybuffer 96-CH_3COOH pH 6.0 (Amersham Biosciences, Inc.), and from 7.0 to 4.0 by a start buffer bis/Tris-Iminodiacetic acid 0.025 M pH 7.4 and a Polybuffer 74-Iminodiacetic acid pH 4.0 (Amersham Biosciences, Inc.). Rat hepatocyte cytosol extract and nuclear shield extract were loaded at a concentration of 0.3 mg/ml. The final chromatograms were recorded and analyzed by the software Unicorn 5.2 (Amersham Biosciences, Inc.).

Theoretical *pI* Calculations

Theoretical *pI* estimation and electrostatic calculations were performed at pH 7.0 by the Protein Calculator Server v. 3.3 at www.scripps.edu. Catalase *pI* estimation (accession number: AAB42378). Proteins selected for net charge calculations: plectin (accession number: CAA42169), nesprin-1 (accession number: AAL47053) and nesprin-2 (accession number: XP_001080795).

Mass Spectrometry Analysis

Proteins derived from rat liver cytosolic fraction and nuclear shield fraction were precipitated with 50% ethanol, 25% methanol, 25% acetone and dissolved in 6 M urea and 100 mM Tris pH 7.9. After reduction with 10 mM DTT and alkylation with 20 mM iodacetamide, protein samples were digested 50:1 (w/w) with sequence grade trypsin (Promega, Madison, WI, USA) at 37°C overnight. The reaction was stopped by adding a final concentration of 0.1% trifluoroacetic acid. Samples were diluted with 0.1% formic acid, 3% acetonitrile in water at a concentration of 0.36 μg/μl, and 0.72 μg of protein digestion were loaded on column for peptide separation. Peptides were trapped on a 5 μm Symmetry C18 trapping column 180 μm×20 mm (Waters Corp., Milford, MA, USA) and separated using a 175 min reversed phase gradient at 250 nl/min (3 to 40% acetonitrile in water over 125 min) on a nanoACQUITY UPLC™ System (Waters), utilizing a 1.7 μm BEH 130 C18 NanoEase™ 75 μm×25 cm nanoscale LC column (Waters). The lock mass ([Glu1]-Fibrinopeptide B/μl, 500 fmol/μl) was delivered from the auxiliary pump of the with a constant flow rate of 200 nl/min. The separated peptides were mass analyzed by a hybrid quadrupole orthogonal acceleration time-of-flight mass spectrometer (Q-Tof Premier™, Waters Corp.) directly coupled to the chromatographic system and programmed to step between low (4 eV) and high (15–40 eV) collision energies on the gas cell, using a scan time of 1.5 per function over 50–1990 m/z (Expression mode: data independent parallel parent and fragment ion analysis [50]). Continuum LC-MS data from three replicates experiments for both rat liver fractions were processed using the software ProteinLynx Global Server v2.3 (Waters). Protein identifications were obtained with the embedded ion accounting algorithm of the software [51] and searching on the UniProtKB/Swiss-Prot protein knowledgebase (release 57.15 of 02-March-10 containing 515203 sequence entries, with taxonomical restriction: *Rattus norvegicus*, 7483 sequence entries). The search parameters were: automatic tolerance for precursor ions and for product ions, minimum 3 fragment ions matched per peptide, minimum 7 fragment ions matched per protein, minimum 2 peptides matched per protein, 1 missed cleavage, carbamydomethylation of cysteine as fixed modification and oxidation of methionine as variable modification.

Comet Assay

Rat liver nuclei suspension in 0.25 M sucrose and the corresponding de-shielded nuclei were treated with H_2O_2 (10 μM) for three minutes. The assay was performed starting with 20 μl of nuclei suspension (~10.000–20.000 nuclei) mixed with 180 μl of 1.7% low melting agarose in sucrose 0.25 M and immediately pipetted onto a frosted glass microscope slide pre-coated with a layer of 1% normal melting point agarose, prepared in PBS lacking Ca^{2+} and Mg^{2+}. Three slides were prepared for each experimental point. The agarose was allowed to set at 4°C for the necessary time. After, slides were placed on a horizontal electrophoresis unit containing fresh buffer (1 mM EDTA, 300 mM NaOH pH 13.0). Electrophoresis was then conducted in fresh electrophoresis buffer (pH 13.0) for 15 min at 25 V and 300 mA (0.8 V/cm) at 4°C. Subsequently, the slides were gently washed in neutralization solution (0.4 M Tris-HCl, pH 7.5) for 5 min and fixed in 100% fresh methanol for 3 min. Slides were stained with 50 μl ethidium bromide (20 μg/ml) and covered with a coverslip. The main protocol for comet assay was based according by usual method [52]. Stained nucleoids were scored visually using a fluorescence microscope (Leica) equipped with a camera COHU (20× magnification). Three slides were analyzed for each experimental point and comet images on each slide were acquired using the 'I.A.S.' automatic image analysis software purchased from Delta Sistemi (Rome, Italy). The comet images were digitalized and a statistical assessment of tail length was conducted.

Proteolysis Experiments

Nuclei from one gram of five different rat tissues were de-shielded with 50 mM NaCl, washed three times with 10 ml of 0.25 M sucrose and resuspended in 0.25 M sucrose (10 ml). Two samples for each tissue (1 ml) were incubated for 2 hours at 25°C with trypsin and proteinase K (1.0 mg/ml final concentration). A third sample was incubated only with 50 mM NaCl, and a fourth sample taken as simple control (sample D). After incubation the samples were washed three times in 0.25 M sucrose. Proteolyzed and control nuclei from liver, kidney, heart, lung and brain were incubated with their corresponding cytosolic fractions for 20 minutes at 25°C, and then washed three times with 0.25 M sucrose. The amount of re-constituted shield was evaluated by measuring the proteins or Alpha GST detached by 50 mM NaCl.

ACKNOWLEDGMENTS

We thank J.Z. Pedersen, G. Cesareni, M. Piacentini, and G. Melino for critical reading the manuscript and for helpful discussion. We are indebted with our lab members A. Amodio and F. Basili for experimental help.

REFERENCES

1. Franke WW, Scheer U, Krohne G, Jarasch E-D. The nuclear envelope and the architecture of the nuclear periphery. J Cell Biol. 1981;91:39s–50s.
2. Dhakshinamoorthy S, Long DJ, 2nd, Jaiswal AK. Antioxidant regulation of genes encoding enzymes that detoxify xenobiotics and carcinogens. Curr Top Cell Regul. 2000;36:201–216.
3. Ma N, Ding X, Doi M, Izumi N, Semba R. Cellular and subcellular localization of heme oxygenase-2 in monkey retina. J Neurocytol. 2004;33:407–415.
4. Basuroy S, Bhattacharya S, Tcheranova D, Qu Y, Regan RF, et al. HO-2 provides endogenous protection against oxidative stress and apoptosis caused by TNF-alpha in cerebral vascular endothelial cells. Am J Physiol Cell Physiol. 2006;291:C897–C908.
5. Petta TB, Nakajima S, Zlatanou A, Despras E, Sarasin A, et al. Human DNA polymerase iota protects cells against oxidative stress. EMBO J. 2008;27:2883–2895.
6. Sheehan D, Meade G, Foley VM, Dowd CA. Structure, function and evolution of glutathione transferases: implications for classification of non-mammalian members of an ancient enzyme superfamily. Biochem J. 2001;360:1–16.
7. Pedersen JZ, De Maria F, Turella P, Federici G, Mattei M, et al. Glutathione transferases sequester toxic dinitrosyl-iron complexes in cells. J Biol Chem. 2007;282:6364–6371.
8. Kaspar JW, Niture SK, Jaiswal AK. Nrf2:INrf2 (Keap1) signaling in oxidative stress. Free Rad Biol Med. 2009;47:1304–1309.
9. Eggler AL, Gay KA, Mesecar AD. Molecular mechanisms of natural products in chemoprevention: induction of cytoprotective enzymes by Nrf2. Mol Nutr Food Res. 2008;52:S84–S94.
10. Han ES, Muller FL, Perez VI, Qi W, Liang H, et al. The in vivo gene expression signature of oxidative stress. Physiol Genomics. 2008;34:112–126.
11. Desaint S, Luriau S, Aude J-C, Rousselet G, Toledano MB. Mammalian antioxidant defences are not inducible by H2O2. J Biol Chem. 2004;279:31157–31163.
12. Abei M, Harada S, Tanaka N, McNeil M, Osuga T. Immunohistochemical localization of human liver glutathione S-transferase (GST) isozymes with special reference to polymorphic GST1. Biochim Biophys Acta. 1989;995:279–284.

13. McCusker FM, Phillips MF, Boyce SJ, Mantle TJ. Glutathione S-Transferases and Drug Resistance. In: Hayes JD, Pickett CB, Mantle TJ, editors. London: Taylor & Francis Ltd; 1990. pp. 419–429.
14. Soboll S, Gruendel S, Harris J, Kolb-Bachofen V, Ketterer B, et al. The content of glutathione and glutathione S-transferases and the glutathione peroxidase activity in rat liver nuclei determined by a non-aqueous technique of cell fractionation. Biochem J. 1995;311:889–894.
15. Stella L, Pallottini V, Moreno S, Leoni S, De Maria F, et al. Electrostatic association of glutathione transferase to the nuclear membrane. J Biol Chem. 2007;282:6372–6379.
16. Alberts B, et al. Molecular Biology of the Cell. Garland, New York: 2002. 661
17. Cantor CR, Shimmel PR. Biophysical Chemistry, vol. II. San Francisco: Freeman, W. N. & Co; 1980. 461
18. Hunter RJ. Zeta Potential in Colloid Science. New York: Academic Press; 1981.

1. Delgado AV, González-Caballero F, Hunter RJ, Koopal K, Lyklema J. Measurement and interpretation of electrokinetic phenomena. Pure Appl Chem. 2005;77:1753–1805.
2. Dice JF, Godberg AL. Relationship between in vivo degradative rates and isoelectric points of proteins. Proc Natl Acad Sci USA. 1975;72:3893–3897.
3. Bickmore WA, Sutherland HGE. Addressing protein localization within the nucleus. EMBO J. 2002;21:1248–1254.
4. Hurst R, Bao Y, Jemth P, Mannervik B, Williamson G. Phospholipid hydroperoxide glutathione peroxidase activity of human glutathione transferases. Biochem J. 1998;332:97–100.
5. Bulitta C, Ganea C, Fahimi HD, Völkl A. Cytoplasmic and peroxisomal catalases of the guinea pig liver: evidence for two distinct proteins. Biochim Biophys Acta. 1996;1293:55–62.
6. Cao C, Leng Y, Kufe D. Catalase activity is regulated by c-Abl and Arg in the oxidative stress response. J Biol Chem. 2003;278:29667–29675.
7. Crapo JD, Oury T, Rabouille C, Slot JW, Chang LY. Copper,zinc superoxide dismutase is primarily a cytosolic protein in human cells. Proc Natl Acad Sci USA. 1992;89:10405–10409.

8. Terasaki M, Campagnola P, Rolls MM, Stein PA, Ellenberg J, et al. A new model for nuclear envelope breakdown. Mol Biol Cell. 2001;12:503–510.
9. Peters R. Nucleo-cytoplasmic flux and intracellular mobility in single hepatocytes measured by fluorescence microphotolysis. EMBO J. 1984;3:1831–1836.
10. Lee SM, Kim JH, Cho EJ, Youn HD. A nucleocytoplasmic malate dehydrogenase regulates p53 transcriptional activity in response to metabolic stress. Cell Death Diff. 2009;16:738–748.
11. Zheng L, Roeder RG, Luo YS. S phase activation of the histone H2B promoter by OCA-S, a coactivator complex that contains GAPDH as a key component. Cell. 2003;114:255–266.
12. Mekhail K, Moazed D. The nuclear envelope in genome organization, expression and stability. Nat Rev Mol Cell Biol. 2010;11:317–328.
13. Simpson JG, Roberts RG. Patterns of evolutionary conservation in the nesprin genes highlight probable functionally important protein domains and isoforms. Biochem Soc Trans. 2008;36:1359–1367.
14. Lu W, Gotzmann J, Sironi L, Jaeger VM, Schneider M, et al. Sun1 forms immobile macromolecular assemblies at the nuclear envelope. Biochim Biophys Acta. 2008;1783:2415–2426.
15. Dauer WT, Worman HJ. The nuclear envelope as a signaling node in development and disease. Dev Cell. 2009;17:626–638.
16. Olzmann JA, Li L, Chin LS. Aggresome formation and neurodegenerative diseases: therapeutic implications. Curr Med Chem. 2008;15:47–60.
17. Lowry OH, Rosebrough NJ, Farr AL, Randall RJ. Protein measurement with the Folin phenol reagent. J Biol Chem. 1951;193:265–275.
18. Folch J, Lees M, Stanley GHS. A simple method for the isolation and purification of total lipides from animal tissues. J Biol Chem. 1957;226:497–509.
19. Ames BN. Assay of inorganic phosphate, total phosphate and phosphatases. Methods Enzymol. 1966;8:115–118.
20. Gober KH, Guenther BR, Luenebach EN, Repplinger G, Wiedemann M. Phospholipids Handbook. In: Cevc G, Dekker M, editors. New York: 1993. pp. 46–47.

21. Canini A, Albertano P, Grilli Caiola M. Sub-cellular localization of calcium in Azolla–Anabaena symbiosis by chlorotetracycline, ESI and EELS. Botanica Acta. 1993;106:146–153.
22. Reimer L, Fromm I, Rennkamp R. Operations modes of electron spectroscopic imaging and electron energy-loss spectroscopy in a transmission electron microscope. Ultramicroscopy. 1988;1:1–5.
23. Beers RF, Sizer IW. A spectrophotometric method for measuring the breakdown of hydrogen peroxide by catalase. J Biol Chem. 1952;195:133–140.
24. Mannervik B. Glutathione peroxidase. Methods Enzymol. 1985;113:490–495.
25. Marklund S, Marklund G. Involvement of superoxide anion radical in the autoxidation of pyrogallol and a convenient assay for superoxide dismutase. Eur J Biochem. 1974;47:469–474.
26. Horowitz GL, Zaman Z, Blanckaert NJC, Chan DW, DuBois JA, et al. Modular analytics: a new approach to automation in the clinical laboratory. J Autom Methods Manag Chem. 2005;2005:8–25.
27. Yeh HI, Hsieh CH, Wang LY, Tsai SP, Hsu HY, et al. Mass spectrometric analysis of rat liver cytosolic glutathione S-transferases: modifications are limited to N-terminal processing. Biochem J. 1995;308:69–75.
28. Ostlund Farrants AK, Meyer DJ, Coles B, Southan C, Aitken A, et al. The separation of glutathione transferase subunits by using reverse-phase high-pressure liquid chromatography. Biochem J. 1987;245:423–428.
29. Ishikawa T, Sies H. The isozyme pattern of glutathione S-transferases in rat heart. FEBS Lett. 1984;169:156–160.
30. Robertson IG, Jensson H, Guthenberg C, Tahir MK, Jernström B, et al. Differences in the occurrence of glutathione transferase isoenzymes in rat lung and liver. Biochem Biophys Res Commun. 1985;127:80–86.
31. Johnson JA, el Barbary A, Kornguth SE, Brugge JF, Siegel FL. Glutathione S-transferase isoenzymes in rat brain neurons and glia. J Neurosci. 1993;13:2013–2023.
32. Geromanos SJ, Vissers JP, Silva JC, Dorschel CA, Li GZ, et al. The detection, correlation, and comparison of peptide precursor and product ions from data independent LC-MS with data dependant LC-MS/MS. Proteomics. 2009;9:1683–1695.

33. Li GZ, Vissers JP, Silva JC, Golick D, Gorenstein MV, et al. Database searching and accounting of multiplexed precursor and product ion spectra from the data independent analysis of simple and complex peptide mixtures. Proteomics. 2009;9:1696–1719.
34. Singh NP, McCoy MT, Tice RR, Schneider EL. A simple technique for quantitation of low levels of DNA damage in individual cells. Exp Cell Res. 1988;175:184–191.
35. Weibel ER, Stäubli W, Gnägi HR, Hess FA. Correlated morphometric and biochemical studies on the liver cell. J Cell Biol. 1969;42:68–91.

CHAPTER 8

IS THE CELL NUCLEUS A NECESSARY COMPONENT IN PRECISE TEMPORAL PATTERNING?

Jaroslav Albert[1,2], Marianne Rooman[1]

[1]BioModeling, BioInformatics & BioProcesses, Université Libre de Bruxelles, Brussels, Belgium
[2]Applied Physics Research Group, Vrije Universiteit Brussel, Brussels, Belgium

ABSTRACT

One of the functions of the cell nucleus is to help regulate gene expression by controlling molecular traffic across the nuclear envelope. Here we investigate, via stochastic simulation, what effects, if any, does segregation of a system into the nuclear and cytoplasmic compartments have on the stochastic properties of a motif with a negative feedback. One of the effects of the nuclear barrier is to delay the nuclear protein concentration, allowing it to behave in a switch-like manner. We found that this delay, defined as the

Citation: Albert J, Rooman M (2015) Is the Cell Nucleus a Necessary Component in Precise Temporal Patterning? PLoS ONE 10(7): e0134239. doi:10.1371/journal.pone.0134239

time for the nuclear protein concentration to reach a certain threshold, has an extremely narrow distribution. To show this, we considered two models. In the first one, the proteins could diffuse freely from cytoplasm to nucleus (simple model); and in the second one, the proteins required assistance from a special class of proteins called importins. For each model, we generated fifty parameter sets, chosen such that the temporal profiles they effectuated were very similar, and whose average threshold time was approximately 150 minutes. The standard deviation of the threshold times computed over one hundred realizations were found to be between 1.8 and 7.16 minutes across both models. To see whether a genetic motif in a prokaryotic cell can achieve this degree of precision, we also simulated five variations on the coherent feed-forward motif (CFFM), three of which contained a negative feedback. We found that the performance of these motifs was nowhere near as impressive as the one found in the eukaryotic cell; the best standard deviation was 6.6 minutes. We argue that the significance of these results, the fact and necessity of spatio-temporal precision in the developmental stages of eukaryotes, and the absence of such a precision in prokaryotes, all suggest that the nucleus has evolved, in part, under the selective pressure to achieve highly predictable phenotypes.

INTRODUCTION

Biological cells house an enormous number of diverse molecules that interact with the cells' environment and among each other. Such a complex network of interactions makes the evolution of any molecule stochastic, be it the change in its position, conformation, or copy number. The effects stochasticity has on the phenotype of a cell are many, ranging from significant to negligible; in some cases they are detrimental [1], in others they play an important role [2]. For example, a population of *Escherichia coli* may survive the attack from an antibiotic drug thanks to random switching between two phenotypes—one sensitive, the other resistant to the drug [3]. Sporulation in yeast is another example; in this case the concentration level of a single protein determines the fate (death or survival) of a cell [4]. In eukaryotes, too, one can observe noise-induced subpopulations at work, such as was done for the immune T-cells [5]. Even in the early development of eukaryotes, where spatio-temporal precision is essential, stochastic variations can be of service, as was exemplified by the segmentation clock [6]: the oscillating expression of the gene *her1* is sustained by virtue of copy number fluctuations.

As essential as these mechanisms may appear, observation, experimentation and modeling of gene expression suggest that important genes must be (and are) tightly regulated [7–9]. And in fact, numerous mechanisms of noise reduction and noise filtering have been discovered. Negative transcriptional [10, 11] and translational [8, 12] autoregulation, non-linear protein degradation [13], multimer formation [14–16] are just a few examples worth mentioning here. Especially in the developmental stages of eukaryotes is noise reduction desirable: the spatio-temporal execution of the genetic program must be as noise free as possible [17, 18].

Previously, studies on the regulation of noise in development have focused mainly on stabilization of newly differentiated cell states [17], noise filtering of morphogen gradients [18], how cellular boundaries are formed and maintained [19], and how the segmentation clock, e. g. in Zebrafish, can lead to synchronized oscillations due to cell signaling [20, 21]. In this paper, we investigate how the segregation of the nuclear space from that of the cytoplasm affects the timing accuracy of gene expression. The effects of such division have been investigated in previous studies, e. g. in the context of Notch signaling and synchronized oscillations [20], the Hes1 and p53-Mdm2 pathways [22], and how the intrinsic noise, coupled with the effects of the nuclear membrane, can explain heterogeneity in embryonic stem cell differentiation [23]. Motifs with autoregulation have also been studied with emphasis on the spatial degree of freedom [24]. While all these studies do incorporate the presence of the nuclear membrane, they do not focus specifically on the question of whether such separation can lead to enhanced timing accuracy in gene transcription. It is this niche that we try to fill in this paper.

What we consider are two models of protein transport from cytoplasm to nucleus. In the first model, we allow the protein to diffuse freely across the nuclear envelope. The second model involves a three-gene motif comprising of the gene of interest and two other genes that code for importin-α and importin-β—two out of a class of proteins responsible for facilitating nuclear import of other proteins. Both models include a negative feedback. The most simplistic version of this motif, one which does not involve the nuclear envelope and hence importins, has been extensively studied and is well understood: the time for a gene to reach a steady state is fast and the noise diminished compared to the situation with no negative feedback. In order to prolong the time to reach a steady state, a positive feedback has been the traditional device; however, this leads to an increase of noise. We demonstrate that the inclusion of the cell nucleus can also lead to a prolonged

Let us write down the set of equations governing the dynamics of mRNA and protein concentrations under the above mentioned assumptions. Using the same notation for the cytoplasmic and nuclear mRNA and cargo protein as in the previous section, the equations of motion read:

$$\begin{aligned}
\dot{X}_n &= r(1 - P_1 - P_2) - \lambda X_n, \\
\dot{X}_c &= \lambda X_n - kX_c, \\
\dot{Y}_n &= \kappa Y_c - qY_n, \\
\dot{Y}_c &= KX_c + (d + 2q_i)Z_3 - a_2 Z_2 Y_c - qY_c,
\end{aligned} \tag{3}$$

where Z_2 is the importin-α–importin-β heterodimer, and Z_3 is the importin-α–importin-β–cargo complex. The coefficient a_2 is the binding rate of Y_c to Z_2, and d the rate at which they dissociate. As before, κ is the transport rate into the nucleus, but this time it is Z_3 that is being transported. Lastly, q_i stands for the degradation rate of both types of importins.

$$\begin{aligned}
\dot{P}_1 &= w_1 Y_n(1 - P_1 - P_2) + \tilde{w}_{-1}P_2 - w_{-1}P_1 - w_1 Y_n P_1 \\
\dot{P}_2 &= w_1 Y_n P_1 - \tilde{w}_{-1}P_2,
\end{aligned} \tag{4}$$

The equations for the α- and β-importins in the cytoplasm and nucleus $Y_{\alpha n}$, $Y_{\alpha c}$, $Y_{\beta n}$, $Y_{\beta c}$ and the corresponding mRNAs, $X_{\alpha n}$, $X_{\alpha c}$, $X_{\beta n}$, $X_{\beta c}$ from which they are translated, read:

$$\begin{aligned}
\dot{X}_{\alpha n} &= r_i - \lambda X_{\alpha n}, \\
\dot{X}_{\alpha c} &= \lambda X_{\alpha n} - k_i X_{\alpha c}, \\
\dot{X}_{\beta n} &= r_i - \lambda X_{\beta n}, \\
\dot{X}_{\beta c} &= \lambda X_{\beta n} - k_i X_{\beta c}, \\
\dot{Y}_{\alpha n} &= \gamma_2 Y_{\alpha c} - \gamma_1 Y_{\alpha n} + \kappa Z_3 - q_i Y_{\alpha n} \\
\dot{Y}_{\alpha c} &= -\gamma_2 Y_{\alpha c} + \gamma_1 Y_{\alpha n} + K_i X_{\alpha c} + (d + q_i)Z_2 + q_i Z_2 - aY_{\alpha c}Y_{\beta c} - q_i Y_{\alpha c} \\
\dot{Y}_{\beta n} &= \gamma_2 Y_{\beta c} - \gamma_1 Y_{\beta n} + \kappa Z_3 - q_i Y_{\beta n} \\
\dot{Y}_{\beta c} &= -\gamma_2 Y_{\beta c} + \gamma_1 Y_{\beta n} + K_i X_{\beta c} + (d + q_i)Z_2 + q_i Z_2 - aY_{\alpha c}Y_{\beta c} - q_i Y_{\beta c}.
\end{aligned} \tag{5}$$

The coefficients $\gamma_1 = \gamma$ and $\gamma_2 = v\gamma$ are the free diffusion rates for importins, with $v = V_n/V_c$ being the ratio of the volume of the nucleus to the volume of the cytoplasm. For the protein complexes, we have

$$\begin{aligned}
\dot{Z}_2 &= aY_{\beta c}Y_{\alpha c} - (d + 2q_i)Z_2, \\
\dot{Z}_3 &= a_2 Z_2 Y_c - (d + 2q_i + q + \kappa)Z_3.
\end{aligned} \tag{6}$$

The importin-α- importin-β heterodimer binds to (dissociates from) Y_c at the rate of a_2 (d).

Let us now select values for all the parameters and solve numerically Eqs (3), (5) and (6). Fig 2 shows the temporal profiles of all proteins and their complexes in a) cytoplasm and b) the nucleus. All three genes arc turned on at $t = 0$; this may be the result of a signaling molecule or the unraveling of the chromatin(s) harboring these genes. The most relevant feature for our purposes is the delayed production of Y_n. For the first forty minutes or so its concentration remains low, and then begins rapidly increasing. This switch-like behavior is common in many biological systems. In what follows, we look at the stochastic behavior of this system and how the delays in the production of Y_n change from cell to cell; stated otherwise, we analyze the noise.

Stochastic Model

Our goal in this section is not only to look at noise in the system defined by the parameters in Figs 2 and 3, but also to study how much the noise changes when we change these parameters. We will only consider those sets of parameters that give rise to a phenotype similar to the profile of Y_n in Fig 3. We will call this profile the reference curve and denote it by Y_r. There is nothing particularly special about this reference cure other than being more or less a typical concentration profile.

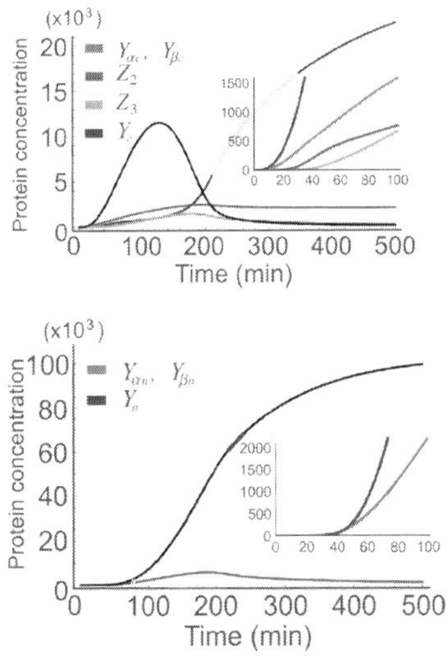

Figure 3: Evolution of average protein concentrations. A close-up view of the figure is given on the right-hand side A) Protein concentrations in the cyto-

plasm. At t = 0 all three genes are turned on. The cargo protein in the cytoplasm (black) begins to accumulate first, followed by the importin α and β (red). At about t = 20 min, importins α and β start forming heterodimers (blue), and 20 min after that, the concentration of importin-α–β-cargo complex (green) begins to rise. B) Protein concentrations in the nucleus. Both the cargo protein (black) and the importins α and β (red) start to accumulate at about t = 40 min.

Since we are interested mostly in the transitional part of Y_r, we impose that at t = 150 min the concentrations of all generated profiles be within 5% of $Y_r(t = 150$ min), and that their slopes fall within 5% of $\dot{Y}_r$ (t = 150 min). Since we consider the steady state concentrations less important, we give them a latitude of 30%. Although we are less concerned with the profiles of all the other species, we do not want them behaving in a manner that is improbable, such as reaching unusually large steady state concentrations. To that end, we ensure that the sum of steady state concentrations of all species (proteins and their complexes) corresponding to a particular parameter set is no greater than 30% of the sum of steady state concentrations arising from the reference parameter set responsible for Y_r. This last constraint is related to the amount of energy the cell needs to expend in order to maintain the steady state concentrations. We will discuss this point further in the "Discussion" section. The fifty profiles we obtain for the two models are shown in Fig 4A and 4B. The range of values for each parameter, along the references from which they were taken, are listed in Table 1.

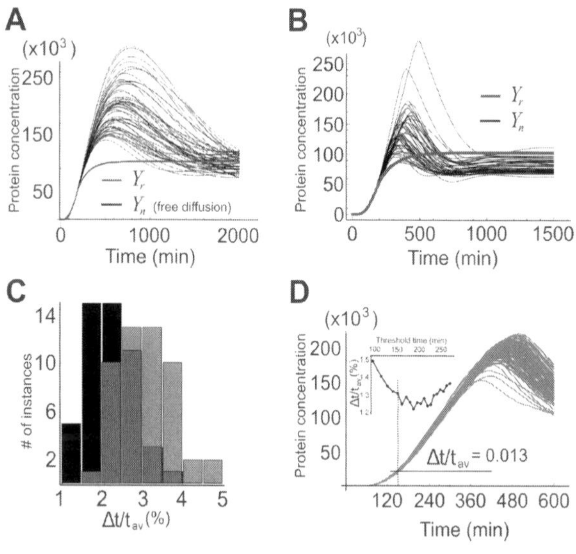

Figure 4: Results for the simple and the extended models.

Profiles for the nuclear concentration corresponding to the 50 parameter sets for A) the simple model, and B) the extended model. In red is the reference curve Y_r. C) Distribution of threshold times for the simple model (green) and the extended model (black). D) Superposition of 100 realizations for the best case offered by the extended model: $\Delta t/t_{av} = 0.013$. The smaller plot within shows values of $\Delta t/t_{av}$ for different threshold times t_{av}.

Table 1. Range of values for all parameters.

Parameter	Range	Reference
r, r_i	1–10/min	[34, 35]
K, K_i	maximum: 11/min/mRNA	[34, 35]
k	0.5–0.005/min	[34, 35]
q	0.5–0.005/min	[34, 35]
λ	maximum 0.0534/min	[27, 36]
q_i	0.5775–6.79 × 10^{-3}/min	[37]
γ	0.01–0.2/min	[30]
a	0.01975–1.225 × 10^{-4}/min	[38]
a_2	$0.306a_2$	[38]
d	0.015–0.102/min	[38]
κ	0.0248–0.426/min	[29]
V_n(Nuclear volume)	4μm^3	[29]
V_C(Cytoplasmic volume)	40μm^3	[29]
Number of nuclear pore complexes	80–170	[29]

Parameters (in symbols), their range, and the reference from which they were taken.

The next step is to simulate the reactions of Fig 1 using the Gillespie algorithm for each parameter set. In the standard notation, for the simple models, these reactions are:

$$\begin{array}{rcl}
E + Y_n & \xrightarrow{w_1} & P_1 \\
P_1 & \xrightarrow{w_{-1}} & E + Y_n \\
P_1 + Y_n & \xrightarrow{w_1} & P_2 \\
P_2 & \xrightarrow{w_{-1}} & P_1 + Y_n \\
E & \xrightarrow{r} & E + X_n \\
X_n & \xrightarrow{\lambda} & X_c \\
X_c & \xrightarrow{k} & \phi \\
Y_n & \xrightarrow{q} & \phi \\
Y_c & \xrightarrow{q} & \phi \\
X_c & \xrightarrow{K} & X_c + Y_c \\
Y_c & \xrightarrow{\kappa} & Y_n
\end{array} \qquad (7)$$

and for the extended model they read:

$$
\begin{aligned}
E + Y_n &\xrightarrow{w_1} P_1 \\
P_1 &\xrightarrow{w_{-1}} E + Y_n \\
P_1 + Y_n &\xrightarrow{w_1} P_2 \\
P_2 &\xrightarrow{\tilde{w}_{-1}} P_1 + Y_n \\
E &\xrightarrow{r} E + X_n \\
X_n &\xrightarrow{\lambda} X_c \\
X_c &\xrightarrow{k} \phi \\
Y_n &\xrightarrow{q} \phi \\
Y_c &\xrightarrow{q} \phi \\
X_c &\xrightarrow{K} X_c + Y_c
\end{aligned}
\tag{8}
$$

$$
\begin{aligned}
Y_{\alpha c} + Y_{\beta c} &\xrightarrow{a} Z_2 \\
Z_2 &\xrightarrow{d} Y_{\alpha c} + Y_{\beta c} \\
Z_2 &\xrightarrow{q_i} Y_{\alpha c} \\
Z_2 &\xrightarrow{q_i} Y_{\beta c} \\
Z_2 + Y_c &\xrightarrow{a_2} Z_3 \\
Z_3 &\xrightarrow{d} Y_c + Z_2 \\
Z_3 &\xrightarrow{\kappa} Y_{\alpha n} + Y_{\beta n} + Y_n \\
Z_3 &\xrightarrow{q} Z_2 \\
Z_3 &\xrightarrow{q_i} Y_{\beta c} + Y_c \\
Z_3 &\xrightarrow{q_i} Y_{\alpha c} + Y_c \\
Y_{\alpha n} &\xrightarrow{\gamma_1} Y_{\alpha c}
\end{aligned}
\tag{9}
$$

$$
\begin{aligned}
Y_{\alpha c} &\xrightarrow{\gamma_2} Y_{\alpha n} \\
Y_{\beta c} &\xrightarrow{\gamma_2} Y_{\beta n} \\
E_i &\xrightarrow{r_i} E_i + X_{\alpha n} \\
X_{\alpha n} &\xrightarrow{\lambda} X_{\alpha c} \\
X_{\alpha c} &\xrightarrow{k_i} \phi \\
Y_{\alpha n} &\xrightarrow{q_i} \phi \\
Y_{\alpha c} &\xrightarrow{q_i} \phi \\
X_{\alpha c} &\xrightarrow{K_i} X_{\alpha c} + Y_{\alpha c}
\end{aligned}
$$

$$
\begin{aligned}
E_i &\xrightarrow{r_i} E_i + X_{\beta n} \\
X_{\beta n} &\xrightarrow{\lambda} X_{\beta c} \\
X_{\beta c} &\xrightarrow{k_i} \phi \\
Y_{\beta n} &\xrightarrow{q_i} \phi \\
Y_{\beta c} &\xrightarrow{q_i} \phi \\
X_{\beta c} &\xrightarrow{K_i} X_{\beta c} + Y_{\beta c}
\end{aligned}
\qquad (10)
$$

where E and E_i stand for the unoccupied promoter of the genes that code for X_n, $X_{\alpha n}$ and $X_{\beta n}$. From these reactions, one can obtain Eqs (1), (2), (3), (5) and (6) by using the standard assumption that the average of any product can be approximated by the product of the individual averages, e. g. $\langle Y_{\alpha c} Y_{\beta c} \rangle \approx \langle Y_{\alpha c} \rangle \langle Y_{\beta c} \rangle$.

What is most interesting to us is the distribution of times at which the protein concentrations of Y_n cross a certain threshold. The concentration threshold of 26.5×10^3 is reached after the average time $t_{av} = 150$ min, as can be worked out from Eqs (3), (5) and (6). As with the reference curve Y_r t being 150 min is not particularly special, but such a delay does conform to what is observed in the developmental stages of eukaryotes [34, 35]. The results for all 50 sets, for both the simple and the extended models, is plotted in Fig 4C as a histogram. Appearing on the horizontal axis is the standard deviation of the threshold times Δt divided by the average time $t_{av} \approx 150$ min. The vertical axis shows the number of cases, out of 50, corresponding to a specific range of $\Delta t / t_{av}$. The superposition of histograms corresponding to the two models reveals that the extended model does better than the simple model. This may seem counterintuitive, given that the presence of importins should introduce an additional source of intrinsic noise which is not present in the simple model. Nevertheless, the degree of noise filtering achieved by both models is equally impressive. Fig 4D shows the best case scenario, belonging to the extended model, for various threshold times. There is another interesting feature to be observed here: 1) the noise during the transition, more specifically between 80 and about 300 min, is very small (see Fig 4C). 2) the noise at equilibrium is quite large, giving a Fano factor for the concentration Y_n of 1540. This may seem surprising: negative autoregulation is known to decrease noise. However, a limit on this effect has been shown to exist when delays are included in stochastic models [39, 40]. Although we did not add delays in our model explicitly, they are implicit in the fact that between transcription and promoter association there are several reactions that must take place. Looking at the histogram in Fig 4C, it is tempting to conclude that accuracy in timing is built into this type of system. We defer a more extensive discussion of this point for the "discussion" section.

Delay Generating Circuits in Prokaryotes

Coherent Feed-forward Motif

We would like to turn attention now to gene networks which do not involve a physical barrier between different reactions, such as those found in prokaryotes, and ask the question: can these types of network mimic the temporal profile of Y_r of Fig 3, and if so, can they do it while maintaining such low level of noise? This question is too hypothetical to be graced with a definite answer; the number of possible motifs is too large to explore. However, we can start by looking at some delay generating motifs common to prokaryotes and see if they could be modified, e.g. by tuning their parameters, to minimize noise while keeping a temporal profile similar to Y_r. Ideally, this (typically three-gene) motif should resemble the one of Fig 1B as closely as possible, i.e. the concentration of the protein for which gene C codes can grow only when genes A and B are on and their gene products abundant. The most common motif that fits the bill is the coherent feed-forward motif (CFFM), occurring at a frequency of 70% in *E. coli*[25]. It can come in two flavors, as depicted in Fig 5. In both cases, initiation of gene C can begin only when the transcription factors of both gene A and gene B are bound to the promoter of this gene; however, they may bind separately, one after the other (Fig 5B), or as a heterodimer (Fig 5C). In what follows, we explore both cases. For now we do not include a negative feedback; this feature will be added in the next section.

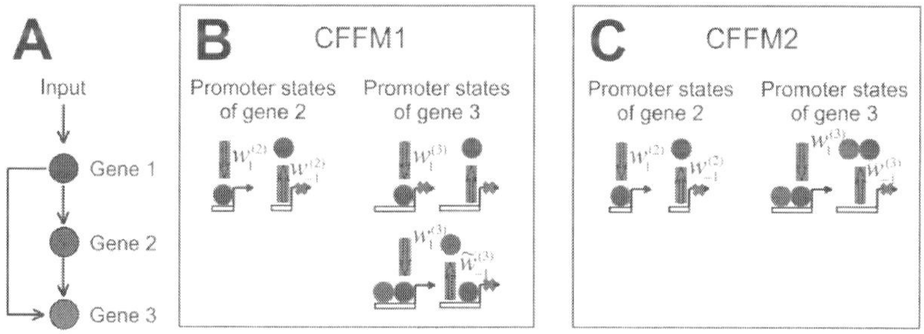

Figure 5: Two types of the coherent feed-forward motif. A) A schematic of the CFFM. The transcription factor of gene A (TFA) enables transcription of gene B, and TFA and TFB together enable transcription of gene C. B) First, TFA binds to the promoter of C; then, either it dissociates or TFB binds to the promoter next to TFA. TFB can bind to the promoter only when TFA is already there. Only when TFA and TFB are both in place can transcription of gene C be initiated. C) TFA and TFB form a heterodimer which then binds the promoter of gene 3, initiating its transcription.

We are interested in making a comparison between the system shown in Fig 1B and the two versions of the CFFM of Fig 5. In particular, the temporal profiles of Y_r and the protein belonging to gene C should be close. To make this comparison as fair as possible, we apply the same conditions and constraints as in the previous section. The reader should refer to this section and merely replace Eqs (3), (5) and (6) with the following equations:

$$\begin{aligned}
\dot{X}_1 &= r_1 - k_1 X_1 \\
\dot{X}_2 &= r_2 S - k_2 X_2 \\
\dot{X}_3 &= r_3 P_2 - k_3 X_3 \\
\dot{Y}_1 &= K_1 X_1 - q_1 Y_1 \\
\dot{Y}_2 &= K_2 X_2 - q_2 Y_2 \\
\dot{Y}_3 &= K_3 X_3 - q_3 Y_3
\end{aligned} \quad (11)$$

$$\begin{aligned}
\dot{S} &= w_1^{(2)} Y_1(1-S) - w_{-1}^{(2)} S \\
\dot{P}_1 &= w_1^{(3)} Y_1(1-P_1-P_2) + \tilde{w}_{-1}^{(3)} P_2 - w_1^{(3)} Y_2 P_1 - w_{-1}^{(3)} P_1 \\
\dot{P}_2 &= w_1^{(3)} Y_2 P_1 - \tilde{w}_{-1}^{(3)} P_2
\end{aligned}$$

for CFFM1, and

$$\begin{aligned}
\dot{X}_1 &= r - k_1 X_1 \\
\dot{X}_2 &= rS - k_1 X_2 \\
\dot{X}_3 &= r_3 P - k_3 X_3 \\
\dot{Y}_1 &= KX_1 - qY_1 \\
\dot{Y}_2 &= KX_2 - qY_2 \\
\dot{Y}_3 &= K_3 X_3 - q_3 Y_3 \qquad (12) \\
\dot{Z} &= aY_1 Y_2 + w_{-1}^{(3)} P - w_1^{(3)} Z(1-P) - (b+2q)Z \\
\dot{S} &= w_1^{(2)} Y_1(1-S) - w_{-1}^{(2)} S \\
\dot{P} &= w_1^{(3)} Z(1-P) - w_{-1}^{(3)} P
\end{aligned}$$

for CFFM2. As before, the variables X and Y denote mRNA and protein concentrations respectively, with the subscript indicating the gene, and Z refers to the heterodimer comprising of Y_1 and Y_2. For CFFM1, the association and dissociation rates between the protein (either monomer or heterodimer) and a gene's promoter are given by $w_1^{(i)}$ and $w_{-1}^{(i)}$ respectively, with the upper index indicating the gene: i = A,B,C. The rate at which a protein encoded by gene A binds with that of gene B is represented by a, while the rate of their dissociation is given by b. In deriving Eqs (11) and (12), we made the same assumptions (those relevant to the present system) listed at the beginning of section "Deterministic model". When all the steps of the previous section are carried out, the results are as presented in Fig 6.

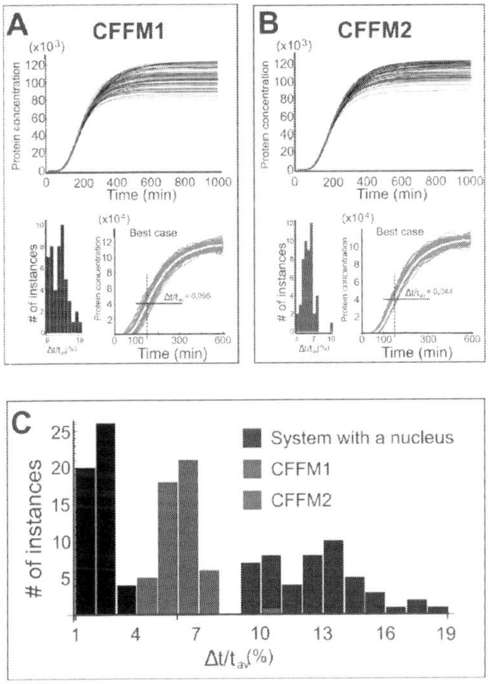

Figure 6: Results for CFFM1 and CFFM2. Shown here are the profiles of all fifty cases (top) and the resulting threshold times distributions (bottom), including the best case for A) CFFM1 and B) CFFM2. C) Threshold times distribution for the system with a nucleus (black) superimposed on CFFM1 (blue) and CFFM2 (red).

The difference between the system of Fig 1B and the two types of the CFFM is immediately apparent: the distribution of $\Delta t/t_{av}$ for CFFM1 and CFFM2 is shifted to the right; the first is also more spread out. Interestingly, the best case for CFFM2 is nearly as good as the worst case for the system with a nucleus and importins.

Variations on the coherent feed-forward motif.

Since in the nucleus bearing system the gene of interest auto-repressed itself, it seems only fair that we add such a feature to the CFFM. Instead of considering only the case where gene C represses itself, we look at two other variations on negative feedback within the CFFM: gene C represses gene A; and gene C represses gene B. The equations governing the promoter occupancy of the gene being repressed are the following:

1. Gene C represses gene A (CFFM1(-)):

$$\begin{aligned} \dot{P}_1 &= w_1^{(1)} Y_3 (1 - P_1 - P_2) + \tilde{w}_{-1}^{(1)} P_2 - w_1^{(1)} Y_3 P_1 - w_{-1}^{(1)} P_1 \\ \dot{w}_2 &= w_1^{(1)} Y_3 P_1 - \tilde{w}_{-1}^{(1)} P_2 \end{aligned} \tag{13}$$

2. Gene C represses gene B (CFFM2(-)):

$$\begin{aligned} \dot{S} &= w_1^{(2)} Y_1 (1 - S - P_1 - P_2) - w_{-1}^{(2)} p_1 \\ \dot{P}_1 &= w_1^{(2)} Y_3 (1 - S - P_1 - P_2) + \tilde{w}_{-1}^{(2)} P_2 - w_1^{(2)} Y_3 P_1 - w_1^{(2)} P_1 \\ \dot{P}_2 &= w_1^{(2)} Y_3 P_1 - \tilde{w}_{-1}^{(2)} P_2 \end{aligned} \tag{14}$$

3. Gene C represses itself (CFFM3(-)):

$$\begin{aligned} \dot{P}_1 &= w_1^{(3)} Y_1 (1 - P_1 - P_2 - P_3 - P_4) + w_{-1}^{(3)} P_2 - w_1^{(3)} Y_2 P_1 - w_{-1}^{(3)} P_1 \\ \dot{P}_2 &= w_1^{(3)} Y_2 P_1 - \tilde{w}_{-1} P_2 \\ \dot{P}_3 &= w_1^{(3)} Y_3 (1 - P_1 - P_2 - P_3 - P_4) + \tilde{w}_1^{(3)} P_4 - w_1^{(3)} Y_3 P_3 - w_{-1}^{(3)} P_3 \\ \dot{P}_4 &= w_1^{(3)} Y_3 w_1 - \tilde{w}_{-1}^{(3)} P_4 \end{aligned} \tag{15}$$

The new variables P_1 and P_2 refer to the promoter occupancy of gene C by one and two copies of Y_3 respectively. Repeating the now familiar procedures of sections "Stochastic model" and "Coherent feed-forward motif", renders the results shown in Fig 7. The profile for the best case, corresponding to the motif CFFM2(-), is shown in Fig 7D. The quantity $\Delta t/t_{av}$ in this case is 2 times greater than what it is for the worst case for the system with a nucleus, which is $\Delta t/t_{av} = 0.05$, and 6 times greater than its best case (Fig 4B). The eukaryotic motif wins out again.

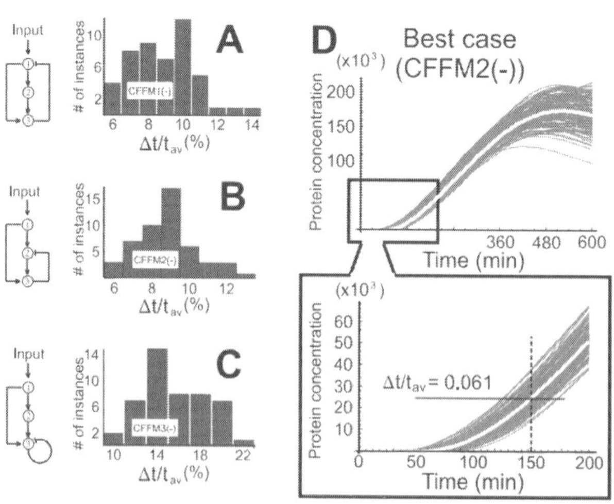

Figure 7: Three variations on the coherent feed-forward motif. The histograms in A, B and C show the distributions for the three variations on the CFFM the

schematics of which are to the left. In all three cases, the transcription factors bind to the different promoters one at a time: no (hetero) dimers. D) Best case, corresponding to CFFM2(-) with $\Delta t/t_{av} = 0.061$.

DISCUSSION

We have shown that in the eukaryotic cell, a negative feedback combined with an import of proteins from cytoplasm to nucleus, either active or diffusive, can not only lead to a delayed switch-like transition, but also to a highly precise timing of this transition. Conventionally, a negative feedback has been used to speed up the transition into a steady state, not slow it down. However, the presence of a nuclear barrier ensures that proteins inside the nucleus do not amass until the proteins in the cytoplasm gain abundance. Once they do, the nucleic proteins start growing at a large rate, giving rise to a switch-like behavior. As for the timing accuracy, this is likely a result of the fact that when a process, e.g. protein synthesis, consists of many steps, the uncertainty in "when" does the last step occur is inversely proportional to the number of steps [41]. In the present case, going from mRNA to nucleic protein, takes four steps, for the simple model, and five steps for the extended model (refer to Fig 1).

We note that the results of the extended model rely on the requirement that all three genes are turned on at the same time. In real systems this may not always be the case. In many situations the importins may have reached steady state well before the gene of interest is turned on. In fact, experimental studies reveal that neither case is strictly true; the importins themselves are regulated by other molecules, making their concentration vary over time [31–33]. For the sake of completion, however, we did consider a situation where, at time $t = 0$, all the gene products pertaining to importins are in a steady state. As an example, we took the parameters corresponding to the best case scenario, shown in Fig 4D, and modified them such that the temporal profile of Y_n under the new conditions matches the reference profile Y_r according to the constraints of section "Stochastic model". To have a good match, we only had to let $q \rightarrow q/2.4$ and $a_2 \rightarrow a_2/50$. Fig 8 shows the result. The temporal precision, while still very good, is diminished by about 30%. This result does not necessarily mean that the precision is always diminished for the latter, as the parameters for the best case scenario may be completely different. An analysis similar to the one carried out in section "Stochastic model" will be necessary in order to determine whether starting out with importins, and the mRNAs from which they are translated, already in a steady state follow the distribution similar to the ones in Fig 4C. We leave this analysis for the future. The proven capability to control importins experimentally [42] makes our results testable.

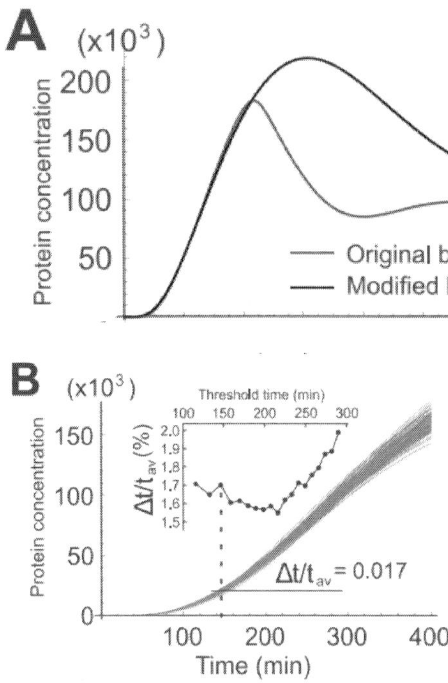

Figure 8: Importins in a steady state at t = **0.** A) Average protein concentrations for Y_n. The red curve corresponds to the best case discussed in Fig 4. The black curve is the profile obtained when the concentrations $X_{\alpha n}$, $X_{\alpha c}$, $X_{\beta n}$, $X_{\beta c}$, $Y_{\alpha n}$, $Y_{\alpha c}$, $Y_{\beta n}$, $Y_{\beta c}$ and Z_2 start out in a steady state, and $q \rightarrow q/2.4$ and $a_2 \rightarrow a_2/50$. B) 100 profiles generated by the Gillespie algorithm. At $t = 150$, $\Delta t/t_{av} = 0.017$.

With regards to the prokaryotic cell, we asked whether a genetic motif could be constructed such that its protein output would mimic the output of the nucleic protein. We considered two types of the coherent feed-forward motif, and three variations on it, and showed that while all of these motifs can be tuned to reproduce the temporal profile of the nucleic protein concentration, they cannot compete with the eukaryotic motif when it comes to temporal precision. The difference is not merely noticeable; it is remarkable. In the best case scenario, provided by the motif CFFM2, the relative deviation from the average threshold time amounted to 4.4%, a value 3.5 times greater than that found in the eukaryotic cell.

These findings suggest that one of the functions of the cell nucleus is to regulate fluctuations in protein abundance during transitions into steady states. From our limited number of cases, it appears that cells without a nucleus (and importins) do not have the means to reduce these fluctuations enough in order to effectuate reliable changes in any genetic program downstream. There may be, of course, larger motifs consisting of many genes that may cater to one specific gene in order to reduce noise associated with its corresponding protein. However, such gene regulatory networks, while conceivable, would demand a lot of energy to function. Research suggests [43] that early in the history of eukaryotes, energy was an important selection factor; networks that required less energy were more likely to persist. This begs the question: could organisms with complex spatio-temporal patterns have evolved without a nucleus? Or is the fact that in billions of years prokaryotes have not evolved into such organisms just a coincidence?

ACKNOWLEDGMENTS

J. A. would like to thank Ekaterina Ejkova for her technical support.

REFERENCES

1. Pilpel Y. Noise in biological systems: pros, cons, and mechanisms of control. Methods Mol. Biol. 2011; 759: 409–25.
2. Eldar A, Elowitz MB. Functional roles for noise in genetic circuits. Nature Reviews. 2010; 467: 09326
3. Balaban NQ, Merrin J, Chait R, Kowalik L, Leibler S. Bacterial persistence as a phenotypic switch. Science. 2004; 305: 1622–1625. pmid:15308767
4. Nachman I, Regev A, Ramanathan S. Dissecting timing variability in yeast meiosis. Cell. 2007; 131(3): 544–56. pmid:17981121
5. Fienerman O, Veiga J, Dorfman JR, Germain RN, Altan-Bonnet G. Variability and robustness in T cell activation from regulated heterogeneity in protein levels. Science. 2008; 321: 1081–1084.
6. Lewis J. Autoinhibition with transcriptional delay: a simple mechanism for the zebrafish somitogenesis oscillator. Curr. Biol. 2003; 13 1398–408. pmid:12932323
7. Fraser HB, Hirsh AE, Giaever G, Kumm J, Eisen MB. Noise minimization in eukaryotic gene expression. PLOS Biol. 2004; 2(6): 0834.
8. Joshi A, Beck Y, Michoel T. Post-transcriptional regulatory networks play a key role in noise reduction that is conserved from micro-organisms to mammals. FEBC Journal. 2012; 279: 3501–3512.
9. Little SC, Tikhonov M, Gregor T. Precise Developmental Gene Expression Arises from Globally Stochastic Transcriptional Activity. Cell. 2013; 154: 789–800. pmid:23953111
10. Dublanche Y, Michalodimitrakis K, Kümmere N, Foglierini N, Serrano L. Noise in transcription negative feedback loops: simulation and experimental analysis. Mol. Syst. Biol. 2006; 2: 41 pmid:16883354
11. Lestas I, Vinnicombe G, Paulsson J. Fundamental limits on the suppression of molecular fluctuations. Nature. 2010; 467: 09333
12. Swain PS. Efficient attenuation of stochasticity in gene expression through post-transcriptional control. J. Mol. Biol. 2010; 344: 965–976.
13. Komorowski M, Miekisz J, Stumpf MPH. Decomposing Noise in Biochemical Signaling Systems Highlights the Role of Protein Degradation. Biophys. J. 2013; 104: 1783–1793. pmid:23601325

14. Rooman M, Albert J, Duerinckx M. Stochastic noise reduction upon complexification: Positively correlated birth-death type systems. J. Theor. Biol. 2014; 354: 113–123. pmid:24632443
15. Bundchuh R, Hayot R, Jayaprakash F. The role of dimerization in noise reduction of simple genetic networks. J. Theor. Biol. 2003; 220:161–9.
16. Ghim CM, Almaas E, Jayaprakash F. Genetic noise control via protein oligomerization. BMC Syst. Biol. 2008; 2:
17. 94. pmid:1898069717.Arias AM, Hayward P. Filtering transcriptional noise during development: concepts and mechanisms. Nature Reviews. 2006; 7: 34–44. pmid:16369570
18. Houchmandzadeh B, Wieschaus E, Leibler S. Establishment of developmental precision and proportions in the early Drosophila embryo. Nature. 2002 415: 798–802. pmid:11845210
19. Dahman C, Oates AC, Brand M. Boundary formation and maintenance in tissue development Nature Reviews. 2011; 12: 43–55.
20. Terry J, Sturrock M, Dale JK, Maroto M, Chaplain MAJ. A spatio-temporal model of Notch signaling in the zebrafish segmentation clock: conditions for synchronized oscillatory dynamics PLoS ONE. 2011; 6(2):e16980 pmid:21386903
21. Mather W, Hasty J, Tsimring LS. Synchronization of Degrade-and-Fire Oscillations via a Common Activator PRL 2014; 113: 128102
22. Sturrock M, Terry JA, Xirodimas DP, Thompson AM, Chaplain MAJ. Influence of nuclear membrane, active transport and cell shape on the Hes1 and p53-Mdm2 pathways: Insights from spatio-temporal modelling Bulletin of Math. Biol. 2012; 74: 1531–1579
23. Sturrock M, Hellander A, Matzavinos A, Chaplain MAJ. Spatial stochastic modeling of the Hes1 gene regulatory network: intrinsic noise can explain heterogeneity in embryonic stem cell differentiation Journal of the Royal Society Interface. 2013; 10:20120988
24. Sturrock M, Murray PJ, Matzavinos A, Chaplain MAJ. Mean field analysis of a spatial stochastic model of a gene regulatory network J Math. Biol. pmid:25323318
25. Mangan S, Zaslaver A, Alon U. The coherent feedforward loop serves as a sign-sensitive delay element in transcription networks. J. Mol. Biol. 2003; 334: 197–204 pmid:14607112
26. Köhler A, Hurt E. Exporting RNA from the nucleus to the cytoplasm. Nature Reviews. 2007; 8: 761–773. pmid:17786152

27. Vargas DY, Ray A, Marras SAE, Kramer FR, Tyagi S. Mechanism of mRNA transport in the nucleus. PNAS. 2005; 102: 47: 17008–17013. pmid:16284251
28. Macara IG. Transport into and out of the nucleus. Microbiol. Mol. Biol. Rev. 2005; 65(4): 570–594.
29. Timney BL, Tetenbaum-Novatt J, Agate DS, Williams R, Zhang W, Chait BT, et al. Simple kinetic relationship and nonspecific competition govern nuclear import rates in vivo. J. Cell. Biol. 2006; 175: 579–593. pmid:17116750
30. Riddick G, Macara IG. A system analysis of importin-α−β mediated nuclear protein import. J. Cell. Biol. 2005; 168: 1027–1038. pmid:15795315
31. Yasuhara N, Oka M, Yoneda Y. The role of the nuclear transport system in cell differentiation. Semin Cell Dev Biol. 2009; 20(5): 590–9. pmid:19465141
32. You P, Peng Z, Wang Y, Tao T. Expression and subcellular distribution of imp13 are regulated in brain development. In Vitro Cell Dev Biol Anim., 2013; 49(5): 346–353. pmid:23605716
33. Okada N, Ishigami Y, Suzuki T, Kaneko A, Yasui K, Fukutomi R, et al. Importins and exportins in cellular differentiation. J. Cell. Mol. Med. 2008; 12, No 5B: 1863–1871. pmid:18657223
34. Bolouri H, Davidson EH. Transcriptional regulatory cascades in development: Initial rates, not steady states, determine network kinetics. PNAS. 2003; 100(16): 9371–9376. pmid:12883007
35. de-Loen SB-T, Davidson EH. Modeling the dynamics of transcriptional gene regulatory networks for animal development. Dev. Biol. 2009; 325: 317–328.
36. Audibert A, Weil D, Dautry F. In vivo kinetics of mRNA splicing and transport in mammalian cells, Mol. Cell. Biol. 2002; 22(19): 6706–6718. pmid:12215528
37. Zienkiewicz J, Armitage A, Hawiger J. Targeting nuclear import shuttles, importins/karyopherins alpha by peptide mimicking the NF κB1/p50 nuclear localization sequence. J. Am. Heart. Assoc. 2013 2 e000386. pmid:24042087
38. Catimel B, Teh T, Fontes MR, Jennings JG, Jans DA, Howlett GJ, et al. Biophysical characterization of interactions involving importin-α

during nuclear import. J. Biol. Chem. 2001; 276: 34189–34198. pmid:11448961

39. Zavala E, Marquez-Lago TT. Delays Induce Novel Stochastic Effects in Negative Feedback Gene Circuits Biophys. J. 2014; 106(2):467–478.
40. Bratsun D, Volfson D, Tsimring LS, Hasty J. Delay-induced stochastic oscillations in gene regulation PNAS. 2005; 102(41): 14593–14598. pmid:16199522
41. Bel G, Munsky B, Nemenman I. The simplicity of completion time distributions for common complex biochemical processes. Phys Biol. 2010; 7, 016003.
42. Quensel C, Friedrich B, Sommer T, Hartmann E, Kohler M. In vivo analysis of importin alpha proteins reveals cellular proliferation inhibition and substrate specificity. Mol Cell Biol. 2004; 24(23): 10246–55. pmid:15542834
43. Wagner A. Energy Constraints on the Evolution of Gene Expression. Mol. Biol. and Evol. 2005; 22(6): 1365–1374.

CHAPTER 9

STRENGTH IN DIVERSITY: NUCLEAR EXPORT OF VIRAL RNAS

Jón Pol Gales[1], Julie Kubina[1,2], Angèle Geldreich[1] and Maria Dimitrova[1]

[1]Institut de Biologie Moléculaire des Plantes, The French National Center for Scientific Research (CNRS) UPR2357, Université de Strasbourg, F-67084 Strasbourg, France
[2]SVQV UMR-A 1131, INRAE, Université de Strasbourg, F-68000 Colmar, France

ABSTRACT

The nuclear export of cellular mRNAs is a complex process that requires the orchestrated participation of many proteins that are recruited during the early steps of mRNA synthesis and processing. This strategy allows the cell

Citation: Gales, J.P.; Kubina, J.; Geldreich, A.; Dimitrova, M. Strength in Diversity: Nuclear Export of Viral RNAs. Viruses 2020, 12, 1014.

to guarantee the conformity of the messengers accessing the cytoplasm and the translation machinery. Most transcripts are exported by the exportin dimer Nuclear RNA export factor 1 (NXF1)–NTF2-related export protein 1 (NXT1) and the transcription–export complex 1 (TREX1). Some mRNAs that do not possess all the common messenger characteristics use either variants of the NXF1–NXT1 pathway or CRM1, a different exportin. Viruses whose mRNAs are synthesized in the nucleus (retroviruses, the vast majority of DNA viruses, and influenza viruses) exploit both these cellular export pathways. Viral mRNAs hijack the cellular export machinery via complex secondary structures recognized by cellular export factors and/or viral adapter proteins. This way, the viral transcripts succeed in escaping the host surveillance system and are efficiently exported for translation, allowing the infectious cycle to proceed. This review gives an overview of the cellular mRNA nuclear export mechanisms and presents detailed insights into the most important strategies that viruses use to export the different forms of their RNAs from the nucleus to the cytoplasm.

Keywords: virus; nucleus; viral RNA nuclear export; viral ribonucleoprotein complexes

INTRODUCTION

The hallmark difference between eukaryotic and prokaryotic life is the presence of the nucleus. Limited by the double lipidic bilayer nuclear envelope, the nucleus physically separates the genetic material from the cytoplasm, decoupling the transcriptional and translational steps of gene expression. The nucleus remains connected to the cytoplasm through the nuclear pore complex (NPC), the gatekeeper allowing the nucleocytoplasmic transport of selected macromolecules [1,2,3].

Viruses represent the most abundant and genetically diverse biological entities on Earth (reviewed in [4]) and infect virtually every organism. Interestingly, when comparing the known viromes of eukaryotes and prokaryotes, there is a notable difference in the abundance of the types of viral genomes [5]. In prokaryotes, the majority of viruses possess a double-stranded DNA genome, allowing them to easily exploit the host's DNA replication and transcription machineries [5]. In eukaryotes, the situation is the opposite. The majority of viruses have a positive-strand RNA genome and are capable of replicating within the cytoplasm, negating the need for access to the nucleus [5]. DNA viruses are comparatively rare, which is likely at least partially due to the appearance of the nuclear envelope, which created new evolutionary challenges for viruses requiring access to the nucleus. Among these challenges is the initial nuclear import of the viral genome, the nuclear export of viral RNAs for translation, and the trafficking of viral proteins to and from the nucleus. Once these challenges are overcome, however, the nucleus represents a new evolutionary niche that some viruses have exploited with remarkable success. The most prominent example might be retroviruses, which, to date, have been found to infect only eukaryotes [5].

The nuclear import of viral genomes and the nucleocytoplasmic transport of viral proteins have recently been reviewed in, for example, [6,7,8]. This review presents a brief overview of the literature on host nuclear mRNA export pathways before summarising the diverse mechanisms by which viruses exploit these pathways to complete their replication cycles. We particularly focus on the advances made in studies on retroviruses, DNA viruses, and influenza viruses over the last decades and which have not recently been exhaustively reviewed. Further, this review presents how the research on some of these viruses has contributed to our understanding of the cellular pathways of the nuclear export of mRNAs.

Nucleocytoplasmic Transport of Macromolecules

The precise evolutionary mechanisms behind the transition from prokaryotes to eukaryotes are still debated. However, it is widely acknowledged that this transition was a gradual change requiring the coevolution of numerous cellular components and mechanisms. One of the more obvious novelties is the NPC, the only gateway connecting the nucleus to the cytoplasm, which allows the selective transition of macromolecules between the two cell compartments [9,10].

The NPC is formed of multiple copies of 30 distinct proteins called nucleoporins (Nups) and reaches a size of over 110 MDa in humans [11]. Scaffolding Nups form the structural basis of the NPC. At the junction of the inner and outer nuclear membranes sits the inner pore ring forming a continuous channel with a diameter of around 40 nm (Figure 1, zoomed rectangle) [12,13,14]. This inner pore ring connects to a nuclear ring and a cytoplasmic ring from which the nuclear basket and the cytoplasmic filaments respectively emanate [15,16]. Classified separately from the scaffolding Nups are the FG-repeat-containing Nups (FG-Nups), which are named due to their intrinsically disordered domains that are rich in Phenylalanine and

Glycine residues [17]. These FG-Nups are anchored to the inner pore ring with the disordered regions protruding into the channel [14,18,19]. The FG-repeats form a dynamic mesh-like structure [20,21] that limits passive diffusion across the pores to molecules of less than 40 kDa or $\approx$ 5 nm in size [22,23,24]. The transport of macromolecules, such as messenger RNAs (mRNAs), or larger proteins requires the cargo to bind a nuclear transport receptor (NTR). The most common NTRs all possess multiple FG-repeat-binding sites allowing for multivalent but low-affinity interactions with FG-Nups inside the central channel of the NPC. The ultra-fast binding and unbinding dynamics between NTRs and FG-repeats allow the NTR–cargo complex to transition through the NPC in less than 5 ms [25]. How FG-repeats can efficiently plug the pores while still allowing for such fast and specific transport remains a highly intriguing subject that is still widely debated. The most likely model suggests that FG-domains behave like ideal polymers by balancing cohesive and repulsive interactions [26].

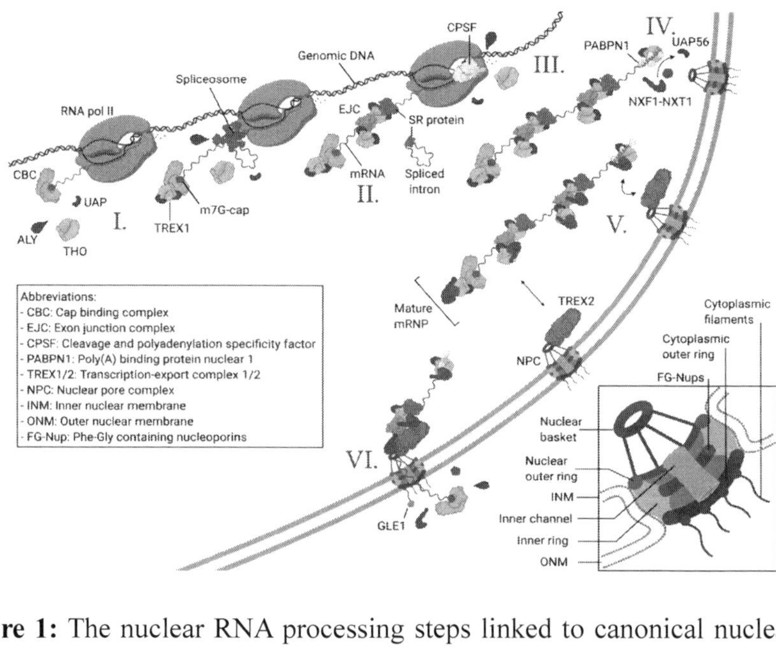

Figure 1: The nuclear RNA processing steps linked to canonical nuclear export. I. RNA polymerase II initiates the transcription of genomic DNA (red and blue helix) and produces a capped transcript (purple). The cap is recognized by the cap-binding complex (CBC). The TREX core complex THO and the adapters ALY (Ally of AML-1 and LEF-1) and UAP56 assemble at the CBC. II. The spliceosome assembles at the splicing sites of nascent mRNA and promotes the intron excision. The exon junction complex (EJC) is deposited 23 nt upstream of the splicing site where the Serine-Arginine-rich (SR) pro-

teins remain attached to the mRNA. TREX1 interacts with the EJC and SR proteins. III. The cleavage and polyadenylation specificity factor complex (CPSF) cleaves the nascent mRNA at the termination sites and promotes the synthesis of the poly(A) tail. IV. Nuclear RNA export factor 1 (NXF1)–NTF2-related export protein 1 (NXT1) exportins are recruited by the TREX1 complex starting at the 3′ end of the mature mRNP. V. NXF1–NXT1 interacts with TREX2 to direct the mature mRNP towards the nuclear pore complex (NPC) where the mRNP is exported through the interaction of NXF1–NXT1 with FG-Nups inside the NPC pore VI. During export, many protein components of the mRNP are removed, while others dissociate from the mRNA in the cytoplasm and are recycled to the nucleus. The square in the bottom right corner shows a more detailed representation of the NPC.

The most prominent NTRs belong to the karyopherin family (reviewed in [27]). Within this family importin-α and importin-β are responsible for the nuclear import of proteins (reviewed in [28]), while exportin 1, also known as Chromosome region maintenance 1 (CRM1), allows the export of proteins but also the export of ribosomal subunits and small RNAs [29]. Together, these karyopherins transport over 1000 different cargoes to and from the nucleus [30,31]. The cargo molecules are recognized through specific transport signals, such as the nuclear localization signal (NLS) [32,33], bound either directly by importin-β [34,35] or indirectly by importin-α [36,37], or the nuclear export signal (NES) recognized by CRM1 [38]. The association between karyopherins and their cargoes is modulated by the small GTPase Ran and depends on the phosphorylation state of the bound guanosine molecule. Ran-GTP is concentrated within the nucleus where it stimulates the binding of karyopherin exportins and their NES-containing cargoes [39]. In the cytoplasm, Ran-GDP is predominant and promotes the association between importins and their cargoes [40]. Thus, the gradient of Ran-GTP/GDP across the nuclear envelope is responsible for the directionality of karyopherin-dependent nucleocytoplasmic transport (reviewed in [41]).

The nuclear export of mRNA is somewhat more complicated and tightly connected to the processing of the pre-mRNA. Eukaryotic mRNAs originate as pre-mRNAs that undergo 5′ capping, splicing, and 3′ polyadenylation. During each of these steps, specific proteins remain associated with the mRNA, forming heterogenous messenger ribonucleoprotein complexes (mRNPs) (reviewed in [42]). The export of bulk mRNA is Ran-independent [43], and the implicated NTRs, Nuclear RNA export factor 1 (NXF1, also known as tip-associated protein (TAP)) and its co-factor NTF2-related export

such as adenine methylation, could also be influential [78,79]. It remains, however, unclear if this influence is due to the direct involvement of the chemical modification itself, the proteins left on the RNA after the modification or the role of m6A in splicing decisions. This topic was reviewed and discussed extensively in [80]. Interestingly, recent work has also suggested that different types of RNAs show different dependencies on NXF1 and TREX components for their export, underlining the complex dynamics within RNA export machinery and suggesting different degrees of redundancy [81].

Although most mRNA is exported through this canonical NXF1–NXT1 pathway, we must consider some alternative cellular mRNA export pathways that can be exploited by viruses. One of these pathways is the alternative RNA export (ALREX) pathway, which is mainly used for the export of intronless mRNAs [82,83]. In ALREX, the export of these mRNAs also relies on NXF1–NXT1 export receptors but requires a reduced number of TREX1 adapters, such as ALY or UAP56, which are mainly recruited during the capping process by the CBC, without the direct involvement of the splicing machinery [84].

Finally, while, the bulk of mRNA exports are karyopherin independent, some specific stress-induced messengers undergo export via the CRM1 pathway (reviewed in [85,86]). Initially thought to be only responsible for the nuclear export of proteins, ribosomal subunits and non-coding RNAs, CRM1 can also bind mRNAs through different specific NES-containing adapter proteins [86]. Among the mRNAs exported by the CRM1 pathway are mRNAs coding proto-oncogenes, cytokines, and lymphokines [85], many of which contain sequences rich in adenine and uracil residues (also called ARE elements) within their 3′ untranslated regions (UTRs) [87] (reviewed in [88]). These sequence motifs are recognized by Human-antigen R (HuR), a ubiquitous adapter protein, which, together with its cofactors acidic leucine-rich nuclear phosphoprotein 31 A and B (ANP32A and ANP32B), can recruit CRM1 to a subset of ARE containing mRNAs [87,89]. HuR and both its cofactors possess different NES motifs, while HuR also has an ARE-binding motif [90]. However, not all transcripts that are exported by CRM1 rely on HuR: The ARE containing mRNA encoding human interferon-gamma (IFN- γ) is CRM1-dependent but HuR-independent [91].

CRM1 also plays an important role in the Eukaryotic translation initiation factor 4E (eIF4E) dependent export of mRNAs, which encode proteins involved in cell proliferation and survival, as well as the formation of cancer metastasis [92]. In addition to its cytoplasmic localization, significant amounts of the translation initiation factor eIF4E are present in the cell nucleus [93]. Here, eIF4E allows the export mRNAs that possess specific secondary structures in their 3′ UTR, called the eIF4E-sensitivity element (4E-SE). 4E-SE is approximately 50 nt in length and is bound by the Leucine-rich PPR motif-containing (LRPPRC) [94] adapter protein, which interacts with both eIF4E and CRM1 [95]. In this export mechanism, different components of the NXF1–NXT1 pathway, such as UAP56, DDX3, and the heterogenous nuclear RNP A1 (hnRNPA1), also play a role, while the CBC does not seem to be implicated [94].

NUCLEAR EXPORT OF VIRAL MRNAS

Gaining access to the nucleus allows viruses to exploit the nuclear cellular transcription and replication machinery for their own multiplication. Even though the RNAs of these *nuclear viruses* are, apart from influenza viruses, mainly synthesized by the cellular RNA polymerase II, they often harbor features that could be considered aberrant in cellular mRNAs. Contrary to most cellular mRNAs, viruses often make use of partially spliced and/or unspliced mRNAs to increase the coding capacity of their genomes or to ensure their genome replication. In addition to intron retention, viral mRNAs often contain long non-translated regions and premature termination codons. These features are usually considered to be contradictory to the efficient nuclear export of mRNA and detrimental to mRNA stability, inducing their degradation via the nuclear RNA degradation pathway (reviewed in [96]). How viruses avoid the targeting of their RNAs by this surveillance mechanism is beyond the scope of this article, but interesting reviews on this subject can be found in the Viruses Special Issue, "Viral Interactions with Host RNA Decay Pathway" [97,98,99]. To still be able to export all their mRNAs, despite their beforementioned differences to cellular mRNAs, *nuclear viruses* have evolved numerous and diverse mechanisms to exploit the existing cellular export pathways. In the following paragraphs, we present a number of examples from different well-known pathogenic viruses to provide a general overview of the mechanisms used while also illustrating how viruses have contributed to the discovery of the major cellular export mechanisms.

The Success of Retroviruses

Retroviruses are among the most prominent *nuclear viruses*. Named after the mechanism of reverse transcription, which allows them to synthesize a DNA copy of their single-stranded RNA genome, retroviruses insert this DNA copy into the genome of the host cell [100,101,102]. This integration allows the virus to undergo latent infection, during which the proviral insert stays transcriptionally silent [103,104]. During an active infection, however, viral genes from the integrated provirus are transcribed by cellular RNA polymerase II. Simple retroviruses only code the three precursor proteins, Gag, Pol, and Env [105]. Gag is processed into matrix, capsid, and nucleocapsid structural proteins; Pol harbors protease, reverse transcriptase, RNAse H, and integrase functions; and Env produces the envelope proteins responsible for viral tropism [106]. While Env is expressed from a partially spliced mRNA, Gag and Gag-Pol are expressed as polyproteins from unspliced genomic RNA (gRNA) spanning the entire length of the proviral DNA [107]. In addition to the three genes, complex retroviruses code auxiliary proteins harboring important regulatory functions essential for the viral replication cycle [108].

With the discovery of Human Immunodeficiency Virus 1 (HIV-1) as the causative agent of Acquired Immunodeficiency Syndrome (AIDS) in 1983 [109,110], research began on this newly discovered virus family. For a critical review on the discovery of retroviruses and HIV-1, see [111]. While the precise cellular mRNA export pathways were still unknown at the time, it was already demonstrated that intron retention inhibits the nuclear export of mRNAs [112,113]. Thus, an important challenge was to understand how HIV-1 could efficiently export its partially spliced and unspliced mRNAs, which are essential for the expression of its envelope, structural, and enzymatic proteins [106]. Relevant research accelerated in the late 1980s, continued throughout the 1990s, and presently remains ongoing. In the following paragraph, we discuss the breakthroughs in HIV-1 research that not only allowed us to determine how the virus exports its mRNAs but also led to the elucidation of the two major cellular mRNA export pathways.

HIV-1 Research Paved the Way for Nuclear Export Pathway Discovery

HIV-1 is a complex retrovirus that codes six auxiliary proteins expressed from more than 20 different RNA isoforms [114]. The first exported messengers code the auxiliary proteins Tat (Transactivating regulatory protein), Rev

(Regulator of expression of viral proteins), and Nef (Negative factor) and are very similar to cellular mRNAs since they are entirely spliced, capped, and polyadenylated. They are thus competent for canonical mRNA export [115].

In 1989, Hadzopoulou-Cladaras et al. first observed that the presence of Rev, one of the newly synthesized auxiliary proteins, is necessary for the efficient production of intron-containing mRNAs and the accumulation of structural viral proteins. For its function, Rev requires a *cis*-acting RNA element named the Rev responsive element (RRE), which is localized within the Env coding region [116]. Later that year, the same team determined that Rev increases the stability of the unspliced viral RNAs and contributes to their efficient nuclear export, without any effect on entirely spliced RNAs [117]. The RRE is a 234 nt long and highly structured region, which contains four stem-loops and one branched stem-loop [118]. The in vitro binding of Rev to the RRE was also first described in 1989 [119], and the cooperative binding of multiple Rev proteins to the same RRE was suggested in 1991 [120]. The uncovering of the HIV-1 Rev nuclear export signal (NES) [121] was followed in 1995 by the discovery of the first potential export factor, an FG-repeat-containing protein aptly named human Rev interacting protein (hRIP), which interacts with Rev [122]. However, Rev was also found to interact with numerous other FG-repeat-containing proteins in the yeast two-hybrid system [123]. This and the fact that yeast homolog Rip1p was determined to be a *bona fide* NPC Nup led to the hypothesis that another—at the time undiscovered—protein links Rev to the FG-repeat-containing Nups, with hRIP acting as a carrier to deliver the Rev-RNA-containing complex to the NPC [124]. In 1997, Wolff et al. showed that Rev-dependent nuclear export could be inhibited by Leptomycin B (LMB) treatment [125]. A previously described target of LMB was the yeast Chromosome Region Maintenance Protein 1 (Crm1p) [126]. With this information, Neville et al. demonstrated that Crm1p bridges the interactions between Rev and the FG-Nups of the NPC [127]. Independently, Stade et al. proved that mammalian CRM1, also known as Exportin 1, is an important nuclear export factor [29]. This not only demonstrated that HIV-1 partially spliced and unspliced mRNAs are exported through RRE–Rev–CRM1 interactions but also that CRM1 is the major exportin of NES-containing proteins. CRM1 was later also described as being the major export factor for ribosomal subunits [128]. Interestingly, it was later established that hRIP still plays an important role in HIV-1 RNA nuclear export and is necessary for the correct release of exported HIV-1 mRNA from the exit of the NPC [129].

Thanks to these early advances, the precise mechanism of the Rev-dependent export of intron-containing HIV-1 RNAs is relatively clear today. As soon as the RRE is transcribed and correctly folded, Rev is recruited and oligomerizes co-transcriptionally with the help of the helicase cofactor DDX1 [130]. The oligomerization of Rev exposes multiple NESs that recruit CRM1—this time together with another helicase co-factor, DDX3 [131,132] (Figure 2a). Once assembled, the complex migrates towards the NPC where it is exported in a Ran-GTP dependent manner [133]. In this way, CRM1 is indirectly recruited to the secondary mRNA structure, in this case, RRE, through the intermediary of a viral adapter protein, Rev.

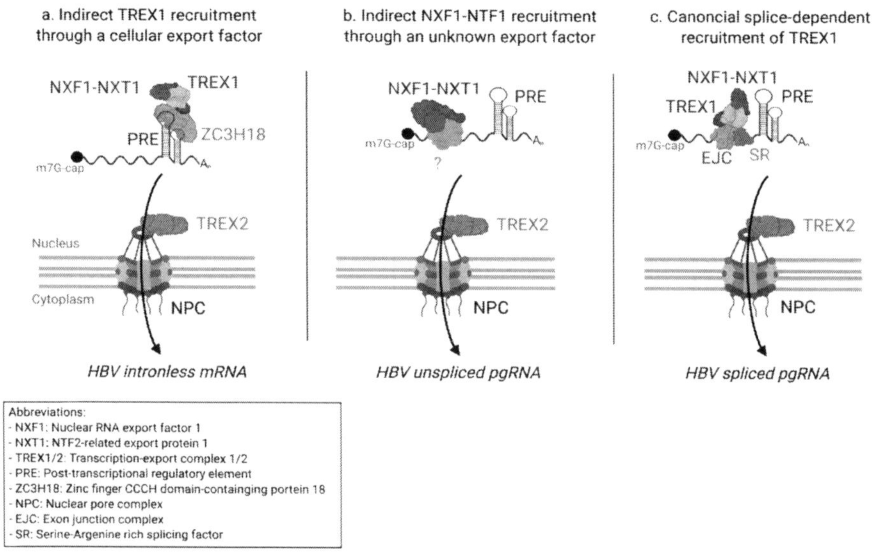

Figure 2: Examples of the mRNA nuclear export mechanisms of retroviruses. (a) The export of HIV-1 intron-containing mRNAs occurs via the indirect recruitment of Chromosome region maintenance 1 (CRM1) to the structured Rev responsive element (RRE) RNA element through the multimerized viral Rev protein. (b) Mason–Pfizer Monkey Virus intron-containing RNAs are exported through the direct recruitment of cellular nuclear export receptors NXF1–NXT1 to the constitutive transport elements (CTE) secondary RNA structure. For the canonical cellular mRNA export mechanism, NXF1–NXT1 allows the anchorage of the mRNP to the TREX2 complex at the NPC, facilitating its consequent export. (c) Intron-containing mRNAs of the Prototype Foamy Virus are exported through indirect recruitment of cellular CRM1 mediated by the

cellular adapter protein HuR, which is usually implicated in the nuclear export of ARE-containing cellular RNA.

It was recently suggested that the NES of Rev is not the only domain responsible for its interactions with CRM1 [134]. Additional work also demonstrated that different cellular CRM1-associated proteins, such as RNA binding protein 14 (RBM14) [135] and Phosphofurin acidic cluster sorting protein 1 (PACS1) [136], act as Rev co-factors and function in the Rev-mediated nuclear export of viral RNAs. Other cellular proteins, like eIF5a [137] and Src-associated in mitosis protein 68 kDa (Sam68) [138], are also considered to be Rev cofactors and play a role in Rev/RRE functions and HIV-1 replication (reviewed in [139]).

Finally, very recently, Hulver et al. discovered that cellular Tat-specific factor 1 (Tat-SF1) can interact independently of the viral auxiliary protein Tat with the 5′ transactivating response region (TAR) structure and also with additional 5′ regions of the viral genome [140]. These interactions are responsible for the selective nuclear export of viral RNAs, thereby promoting the export of unspliced genomic RNAs but not that of single-spliced RNA isoforms [140].

From HIV-1 to the Simple Mason–Pfizer Monkey Virus

While an important effort was made to uncover the mechanism behind the export of intron retaining mRNA using the HIV-1 Rev/RRE system, knowledge on the canonical export pathway of the bulk cellular mRNA was somewhat lacking. While the importance of splicing was established early on, the nuclear transport receptor remained long undiscovered. In 1994, Bray et al. showed that the insertion of a small element from the Mason–Pfizer Monkey virus (MPMV) genome into unspliced HIV-1 RNAs allows their export and translation in a Rev- and RRE-independent manner [141]. The same observations were also made for a similar RNA element of Simian Retrovirus type 1 (SRV1) [142]. Both MPMV and SRV1 were later defined as being serologically distinct strains of the same Mason–Pfizer Monkey Virus species [143]. As a simple retrovirus, MPMV only encodes the essential Gag, Pol, and Env precursor proteins with no Rev-like protein that could enable the export of the unspliced viral mRNAs. A different mechanism for mRNA export, independent of a viral protein, was thus suspected to exist for simple retroviruses. After the discovery of these RNA elements named constitutive transport elements (CTEs), it was rapidly hypothesized that CTEs could tap into a new RNA nuclear export pathway, potentially the one responsible for the export of cellular mRNA. The essential structure for Rev and RRE substitution is between 150 and 175 nt long, depending on the serotype, and forms a long stem-loop containing two internal loops [144,145,146]. Saavedra et al. showed that when microinjected into *Xenopus* oocyte nuclei, RNAs containing the SRV1 CTE can outcompete the export of cellular mRNA, thus suggesting that retroviral CTE-dependent mRNA export uses the same pathway as the bulk of cellular mRNA [147]. In the same year, Pasquinelli et al. described similar findings for MPMV CTE containing RNAs [148]. Using MPMV CTE RNA as bait for protein capture on an affinity column, Grüter et al. discovered a strongly interacting protein, which they identified as human NXF1 [149], a homolog of the yeast mRNA export factor 67 (Mex67) protein shown previously to be involved in mRNA export [150]. The importance of NXF1 in cellular mRNA export was further uncovered in yeast, where Mex67 was reported to form an export complex with mRNA transport regulator 2 (Mtr2) [151]. In 2000, competition experiments confirmed the role of the homologous NXF1–NXT1 transporter dimer in human cells [152].

Thus, simple retrovirus MPMV, like HIV-1, uses a structural RNA element to export its intron-containing mRNAs. Because it is a simple retrovirus, however, MPMV does not code a viral export protein. Instead, the constitutive transport element directly recruits the cellular export receptor NXF1–NXT1, which allows binding to TREX2 and access to the nuclear pore complex (Figure 2b).

Diversity within the Retroviridae Family

Numerous complex retroviruses use mechanisms very similar to those of HIV-1. Mouse Mammary Tumor Virus (MMTV), Jaagsiekte Sheep Retrovirus 1 (JSRV), and Human Endogenous retrovirus K (HERV-K) all code Rev like auxiliary proteins that bind *cis*-acting secondary RNA structures to direct intron-containing RNA towards the CRM1 export pathway [153,154,155,156]. Interestingly, human T-cell leukemia virus type 1 (HTLV-1) also uses a similar strategy by coding the Rex protein [157,158]. However, the Rex responsive element (RxRE) is located on the 3′ LTR of all viral transcripts [159]. It has been suggested that RxRE plays a role in HTLV-1 mRNA polyadenylation, where the poly(A) signal and actual poly(A) site are unusually far apart [160,161]. Rex has also been shown to interact with viral p30 protein to balance viral gene expression and infection latency [162] (reviewed in [163]).

Rous Sarcoma Virus (RSV) lies between simple and complex retroviruses as it includes a single auxiliary gene *src* of cellular origin providing its oncogenic properties [164,165]. RSV is often considered to be a simple retrovirus since *src* does not have a direct regulatory effect on viral gene expression. Considering the nuclear export of unspliced viral mRNAs, RSV uses a similar strategy to other simple retroviruses. The direct repeat (DR) sequences flanking the *src* gene bind NXF1 and DDX5 [166,167]. Interestingly, however, the RSV Gag protein displays partial nuclear localization and also contributes to the export of intron-containing viral mRNAs in a CRM1-dependant manner [168]. A recent study also succeeded in visualizing the active nuclear export of the viral RNA–Gag complex through the NPC, further highlighting the active role of Gag in unspliced viral mRNA export [169]. Other studies suggest, however, that Gag does not promote the export of DR-containing reporter constructs [167]. Despite this discrepancy, it is possible that RSV exports intron-containing mRNAs via two distinct pathways to direct them towards either Env protein translation or encapsidation. The Gag proteins of several other retroviruses, including HIV-1 and MPMV, also display partial nuclear localization [170,171]. It is, however, still debated how this contributes to the nuclear export of intron-containing mRNAs and their subsequent translation or encapsidation (reviewed in [172]).

Another interesting example is the Murine leukemia virus (MLV), which is often characterized as one of the simplest retroviruses. However, MLV's strategy for the nuclear export of unspliced mRNAs cannot be considered simple by any means. Contrary to other simple retroviruses, such as MPMV, which use a single *cis*-acting RNA element, MLV has five such elements that are involved in RNA export and are scattered along the entire viral RNA sequence. Some of these elements are found on both spliced and unspliced RNAs (reviewed in [173]). One of these five elements, γ-CTE, which harbors a conserved AAGACA sequence in its stem-loop, directly recruits the NXF1 exportin [174]. The 3′ located cytoplasmic accumulation element (CAE) likely interacts with multiple components of the THO core complex [175]. None of the *cis*-elements identified to date are located within the *env* gene, as is the case with other simple retroviruses. The presence of such numerous distinct elements suggests that MLV mRNA nuclear export is highly regulated and tightly linked to other viral or cellular mechanisms [173]. Moreover, it has been shown that this nuclear export is dependent on some particular components of the TREX1 complex: While spliced MLV mRNAs require the UAP56 helicase for export, unspliced genomic mRNAs instead depend on THOC5, THOC7, and the splice factor SRp20 [175,176]. Recent work showed that MLV RNAs are exported through both the CRM1 and NXF1 pathways, with the CRM1 pathway priming exported unspliced RNAs for packaging [177]. This further proves that nuclear export mechanisms can influence the downstream fate of viral RNAs.

Bridging the gap between retroviruses and the related pararetroviruses, which will be discussed in the following paragraph, is the Prototype foamy virus (PFV) of the *Spumavirinae* subfamily. Foamy viruses (FVs) are considered retroviruses because of their genomic organization and integration of proviral DNA into the host genome. However, they present several significant differences to other retroviruses. For example, FVs' Pol enzymatic functions are not expressed in a Gag–Pol fusion protein but as separate Pol precursor proteins from a spliced mRNA, and an internal promoter within the *env* gene allows the expression of the *tas* auxiliary gene [178,179]. Another feature unique to FVs is that for 5 to 10 % of FV particles, reverse transcription of the packaged genomic viral RNA already starts in the preassembling capsids [180]. This means that the released viral particles contain both genomic RNA and genomic DNA [181]. Furthermore, the budding of preassembled FV capsids takes place not only at the plasma membrane but also in the intracellular compartments (endoplasmic reticulum and Golgi) [182]. When it comes to the export of mRNA, FVs are still faced with the same challenges of the nuclear retention of unspliced mRNAs. Here, FVs seem to combine the two major pathways used by other retroviruses. FV unspliced mRNA export is viral-protein independent but relies on the cellular CRM1 export pathway [183]. Unlike other retroviruses, CRM1 is thus not recruited to viral RNA by a viral protein but rather by the host adapter HuR, which is also implicated in the export of cellular ARE-containing RNAs [90,183] (Figure 2c). It has been suggested that other adapter proteins (ANP32A and ANP32B) from the same pathway are also involved, bridging HuR to CRM1 [183]. Foamy viruses are often referred to as the oldest retroviruses [184], suggesting that the HuR export pathway might have paved the way for the evolution of Rev-like and CTE-dependent strategies. For other retroviruses, the FV Gag protein is partially located within the nucleus, and correct viral egress is dependent on this nuclear localization [185], further suggesting that Gag is involved in the sorting of unspliced mRNA for protein synthesis or virion assembly [186].

mRNA Nuclear Export of Pararetroviruses

Pararetroviruses, like retroviruses, replicate their genomes by reverse transcription but do not integrate their genomic DNA into the host genome. Like for foamy viruses, reverse transcription occurs late in viral replication, leading to the encapsidation of a partially double-stranded DNA, also called relaxed circular DNA (rcDNA) [187].

The only pararetrovirus known to infect humans is the Hepatitis B virus (HBV), a member of the *Hepadnaviridae* family of strictly hepatotropic viruses. Only 3.2 kbp in size, HBV has the smallest DNA genome of all known mammalian viruses [188]. Upon infection, rcDNA is transported to the nucleus where the host machinery converts it into covalently closed circular DNA (cccDNA) [189]. This cccDNA then associates with cellular histones to form a mini-chromosome, which persists within the host nucleus [190]. Due to its small size, the HBV genome has a very high coding density with four overlapping genes. To efficiently express each of these genes, HBV relies on four promoters, two enhancers, and a single termination and polyadenylation site leading to the synthesis of six mRNAs (reviewed in [191]). Aside from being used as a template for reverse transcription, pregenomic RNA (pgRNA), which is slightly longer than the genome, is also a messenger for polymerase and core protein expression. The precore mRNA (preC RNA) facilitates the expression of the Hepatitis B e antigen (HBeAg) [192]. The subgenomic mRNAs PreS1 and PreS2, which are 2.3 kb and 2.1 kb in length, respectively, code the three different variants of the surface antigens L, M, and S [193]. The last subgenomic mRNA is rather small (0.7 kb) and codes the X regulatory protein [194]. The long pg and preC mRNAs can be spliced, resulting in mRNAs coding the HBV splice-generated protein (HBSP) [195], which has been linked to HBV pathogenesis [196]. Overall, HBV transcripts can be conceptually divided into three groups: unspliced pg and preC mRNAs, spliced pg and preC mRNAs, and intronless subgenomic mRNAs. Each of these isoforms are exported to the cytoplasm either for translation or for encapsidation and reverse transcription [197].

All HBV transcripts share a 365 nt long *cis*-acting RNA element, the posttranscriptional regulatory element (PRE), which overlaps the X coding sequence [198]. After PRE's discovery in 1993, it was rapidly proposed that PRE might play an important role in mRNA nuclear export, as is the case for HIV-1 RRE and MPMV CTE [199]. Interestingly, however, the PRE-linked export mechanism is CRM1-independent and also functions differently from MPMV CTE-dependent export, as the PRE is unable to directly bind the NXF1–NXT1 export receptors [200]. The PRE can be divided into two export relevant regions, the sub-elements PRE 1 and 2 (SEP1 and SEP2), each forming multiple stem-loop secondary structures [201]. Several PRE interacting proteins have been identified and proposed to play a role in the nuclear export of HBV mRNAs. These proteins include glyceraldehyde-3-phosphate dehydrogenase (GADPH), Polypyrimidine tract-binding (PTB) protein, and the La protein [202,203,204]. The precise role of each of these proteins in HBV mRNA export remains unclear [205]. A much more promising candidate is the zinc finger CCCH domain-containing protein 18 (ZC3H18), which has been found to bind SEP1 and to recruit several components of the TREX complex, including THOC2 [206] (Figure 3a). Once the TREX1 complex is recruited, intronless subgenomic mRNAs are presumably exported through the NXF1–NXT1 pathway.

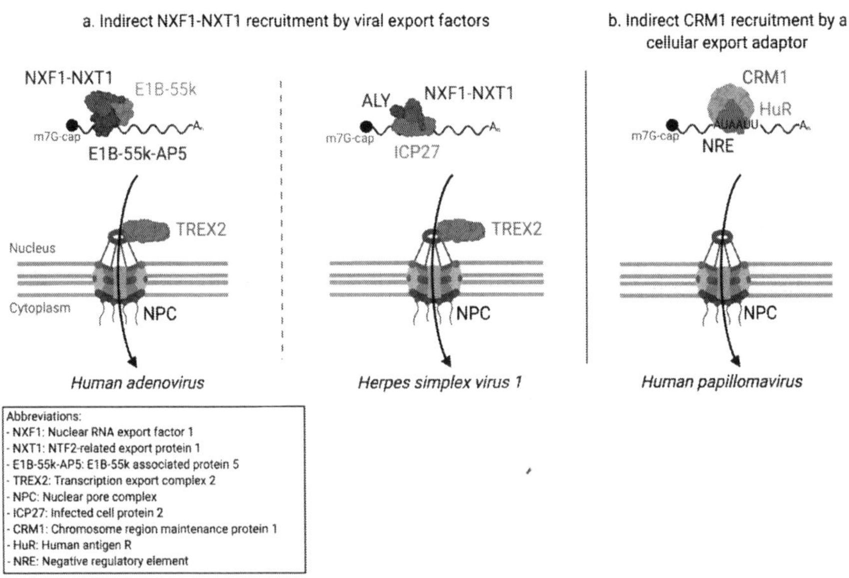

Figure 3: Hepatitis B Virus mRNA nuclear export by the NXF1–NTF1 pathway. (**a**) Hepatitis B virus intronless subgenomic mRNAs are exported through the association of the TREX1 complex to the PRE element via the cellular adapting protein ZC3H18. As is the case for the canonical mRNA export pathway, TREX1 recruits NXF1–NXT1 nuclear transport receptors allowing export through the NPC after interaction with the TREX2 complex. (b) Unspliced pgRNA is exported through the recruitment of NXF1–NXT1 in a PRE- and Core-independent manner. (c) Spliced Hepatitis B virus (HBV) pgRNAs are exported in a splicing-dependent manner, most likely through the canonical cellular mRNA export pathway.

Unspliced variants of HBV pgRNA and preC mRNAs are exported in a PRE-independent manner, suggesting that HBV might, like HIV-1, use different pathways to export different mRNA isoforms [199]. It has been shown that the HBV core protein displays nucleo-cytoplasmic shuttling properties [207] and interacts in the nucleus with pgRNAs forming pgRNPs [208]. Core also interacts with TREX components and NXF1–NXT1 export receptors [209]. The nuclear export of pgRNA is, however, independent of the core protein [209] (Figure 3b). Recent research identified a new HBV RNA binding protein involved in RNA nuclear export, the TAR DNA-binding protein (TARDBP). TARDBP has been suggested to play a passive role in mRNA nuclear export by inhibiting the splicing of pgRNA, thus favoring the export of unspliced pgRNA isoforms [210]. Hu et al. also recently determined that UAP56 interacts with the HBV X protein, potentializing the export of single-exon HBV mRNAs in cell cultures [211]. This supports the hypothesis that the export of spliced HBV mRNAs is linked to splicing even if HBx does not interact directly with NXF1 or NXT1 (Figure 3c). In summary, HBV likely exports all its different mRNA isoforms through an NXF1–NXT1-dependant mechanism. However, only the export of intronless subgenomic mRNAs depends on the PRE *cis*-element. Recent research suggests that pregenomic and precore RNA export is partially regulated by splicing factors. Although the export of HBV transcripts has been studied for nearly 30 years, important details on its precise mechanisms remain elusive. Understanding the export mechanisms used by HBV could lead to the development of new and specific antiviral components needed in the pursuit of Hepatitis B control and prevention (reviewed in [212]).

Nuclear Replicating DNA Viruses

Apart from the previously discussed pararetroviruses and some other exceptions, such as *Poxiridae* and *Megaviridae*, most DNA viruses replicate their genomes within the nucleus of the host cell. Most such viruses use complex mechanisms to regulate their gene expression from an early phase, which provides the proteins needed for genome replication, to the late expression of the structural proteins needed to package DNA and form virions (reviewed in [213]). Although most nuclear DNA viruses undergo splicing for most of their transcripts, some mRNAs retain at least one intron that hampers their export via the canonical cellular mRNA export pathway. Thus, even though they do not require the export of a genomic or pregenomic

RNA like retroviruses and pararetroviruses, DNA viruses have also had to evolve mechanisms to guarantee the efficient export of their intron-containing mRNAs. After providing an overview of the mechanisms used by two large DNA virus families, *Adenoviridae* and *Herpesviridae*, for which experimental data on this topic are plentiful, we will also discuss the historical and current findings concerning human oncogenic *Papilloma* and *Polyomaviridae*.

Adenovirus mRNA Nuclear Export

Adenoviruses (AdVs) are non-enveloped viruses with a linear double-stranded DNA genome of about 36 kbp. They were first discovered in adenoid cell culture, hence the name (*adénos*, gland) [214]. Adenoviruses are ubiquitous, and infection is usually asymptomatic or mild and self-limiting in immunocompetent hosts. However, in immunocompromised individuals, adenoviral infection can lead to more severe illness [215].

The human Adenovirus genome is rather dense, with over 40 genes [216]. To assure efficient viral replication, these genes are expressed in an orderly manner to reprogram the host cell towards virus replication without alerting the immune system or inducing cell damage, leading to apoptosis [217]. The first gene expressed is the early gene 1A (E1A) [218], which drives the expression of other early proteins originating from the E2, E3, and E4 transcription units [219]. Later during infection, the expression profile switches gradually towards the major late promoter (MLP), inducing the synthesis of the late gene products necessary for the assembly of the viral progeny [219]. Overall the transcriptional program of AdV is highly regulated by different mechanisms, one of which is alternative splicing, leaving individual transcripts either partially or entirely unspliced [220]. Some viral transcripts are also specifically retained in the nucleus until the appropriate time for their expression [219].

The study of AdV mRNA export mechanisms has, to date, mainly focused on late messengers, and the situation seems rather complex. A potential role of the viral E1B-55k protein in the export of late AdV transcripts was first observed in 1986 [221]. However, its role in export was suggested to be rather indirect and upstream of the actual export event [222]. E1B-55k is, however, the chief orchestrator of the events leading to late AdV mRNA export by first reorganizing, together with the viral protein E4orf3, promyelocytic leukemia (PML) nuclear bodies at the periphery of which viral transcription and replication occur [223,224]. At later stages of infection, another viral protein, E4orf6, enters into competition with E4orf3 and induces the relocalization of E1B-55k towards the periphery of the nuclear speckles, a storage place for splicing factors and mature, export-competent, viral, and cellular transcripts [224]. All these sequential virus-induced modifications of nuclear structures inhibit the maturation and export of cellular mRNAs while promoting the export of intron-containing viral transcripts [225]. Furthermore, E1B-55k, E4orf6, and some cellular components forming a ubiquitin-3-ligase complex are required for the efficient transport of late mRNAs towards the cytoplasm [226]. The relevant mechanism relies on the inhibition of the cellular components essential for the degradation of intron-containing mRNAs [227]. For the many other roles of E1B-55k in the regulation of Adenovirus gene expression, see [228].

The current model of export for AdV late mRNAs suggests the participation of the cellular E1B-55k-associated protein (E1B-55k-AP5), which interacts with both the viral mRNAs and NXF1 [152,229,230] (Figure 4a). In this model, E1B-55k could play a role in the release of late mRNAs from their sites of transcription, but the precise molecular mechanisms by which E1B-55k and E1B-55k-AP5 lead to the selective export of AdV mRNAs remain unclear.

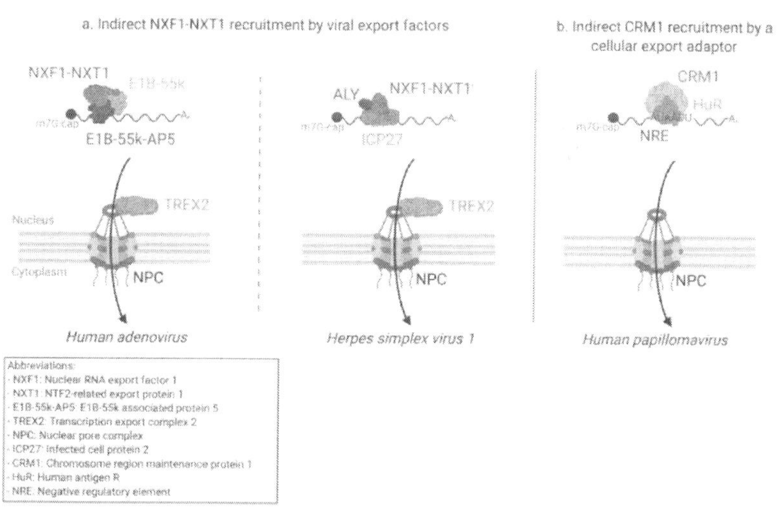

Figure 4: Examples of the mRNA nuclear export of DNA viruses. (a) Both Adenoviruses and Herpesviruses recruit NXF1–NXT1 exportins to their mRNAs indirectly through viral adapter proteins. The Adenovirus early protein E1B-55k and its associated cellular partner, E1B-55k-AP5, bind to viral mRNAs and recruit NXF1–NXT1 which promotes nuclear export of the mRNP complex

through the TREX2 NPC interaction mechanism. Herpes simplex virus 1 codes the ICP27 export protein, which is well conserved within *Herpesviridae*. ICP27 recruits NXF1–NXT1 export receptors through the cellular ALY protein, inducing transport through the NPC by interacting with the TREX2 complex. (b) Human Papillomavirus late mRNAs presumably recruit cellular CRM1 through the cellular adapter protein HuR, which binds to the negative regulatory element (NRE) RNA element.

While the roles of Adenoviruses in human oncogenesis are still debated, there have been numerous attempts at using viruses to target cancer cells [231,232,233] (reviewed in [234]). The most challenging part of this approach is to modify a virus in a way that it becomes specific to cancer cells while not affecting normal cells. In many types of cancer, the export mechanisms of mRNAs are highly deregulated, and the retention of intron-containing mRNAs is no longer efficient [235]. Thus, using modified viruses, which are unable to export their mRNAs from the nucleus under normal conditions, is a promising approach. Indeed, E1B-55k-deficient Adenoviruses have a selective oncolytic effect in combination with cisplatin, when compared to the use of the classic anti-cancer drug alone [236]. It was later demonstrated that the selectivity of the virus could be attributed to inefficient viral mRNA export in healthy cells [235].

Herpesvirus mRNA Export—A Shared Strategy with Different Tools

Herpesviridae are a large family of viruses that induce numerous pathologies in a wide range of hosts. They have a sizeable linear dsDNA genome enclosed in an icosahedral capsid surrounded by a proteinaceous tegument layer and a lipidic membrane [237]. Herpesviruses are unusual as they can undergo latent infection and reactivate at a later stage [238]. The nine different herpesviruses infecting humans are classified into three different subfamilies, *Alphaherpesvidinae*, *Betaherpesvirinae*, and *Gammaherpesvirinae* [239]. Herpesviruses are ubiquitous and present in around 80% of the human population [240]. Despite their large genomes, they possess very few intronic sequences, with up to 98 % of their genes being intronless. Although new splice sites have recently been mapped to the genome of Herpes simplex virus 1 (HSV1 or Human alphaherpesvirus 1), their functions remain to be confirmed [241]. Herpesviruses thus need splicing independent mRNA export mechanisms. A viral mRNA export factor has been discovered within each of the three subfamilies, and most experimental data on the precise mechanism by which they function are available for HSV1 and its infected cell protein 27 (ICP27).

The role of ICP27 as an export adapter for viral mRNAs was first suggested after the discovery of its nuclear localization, RNA binding, and nucleo-cytoplasmic shuttling [242,243,244,245,246,247]. ICP27, which itself is expressed from a spliced mRNA, gives HSV1 intronless transcripts access to the canonical mRNA export pathway by interacting with ALY, NFX1, and cargo mRNAs [248,249,250] (Figure 4a). Although it has been suggested that ALY is dispensable for HSV1 export [250], its presence significantly improves efficiency, likely by adding specificity to ICP27s' interactions with the target mRNAs [251,252,253]. While the RNA binding region of ICP27 was mapped to its RGG motif early on [243], no RNA sequence specificity for this interaction was found, and the RNA elements necessary for ICP27-dependent export remain thus undiscovered. Several other proteins have been found to interact with ICP27, including the splicing factor UIF and nucleoporin Nup62 [254,255]. While the latter interaction could compete with cellular export, their precise roles in RNA export are not currently known. Finally, ICP27 has also been suggested to be involved in splicing and polyadenylation and could be linked to the switch from a latent to a lytic infection [241].

The viral export factors within the three herpesvirus subfamilies share an overall amino acid homology of 20 to 30 % and a C-terminal homology of 40 to 60 % with ICP27 [256,257]. They also all share many other common features, such as nucleocytoplasmic shuttling, RNA binding activity, and the requirement for phosphorylation by cellular kinases. However, the kinases responsible for this phosphorylation vary from one subfamily of *Herpesviridae* to another, as does the nature and position of the RNA binding domain (reviewed in [258]). These small but significant differences are exemplified by the fact that neither HSV1 ICP27 nor the Human cytomegalovirus's unique long 69 gene product (UL69) can entirely complement the mutant Epstein–Barr virus EB2 protein, which is essential for viral mRNA export and virus production [259]. Overall, all the herpesvirus export protein homologs share a common mechanism allowing the export of viral mRNAs by forming ribonucleoprotein complexes with components of the TREX complex and the cellular export factor NXF1.

Small DNA Virus mRNA Export

Papilloma- and Polyomaviruses are both small dsDNA viruses that historically belonged to the same *Papoviviridae* family. Although they were split into two distinct families, the International Committee on the Taxonomy of Viruses (ICTV) recently included these two families within the same *Papovaviricetes* class due to their numerous similarities and phylogenetic proximity [260]. Both families include viruses that have oncogenic potential, some of which are associated with human cancers, particularly with cervical, skin, prostate, and colorectal cancer [261]. Viral gene expression is finely regulated during an early and a late expression phase, and, in both cases, alternative splicing is an essential regulatory mechanism [262,263,264,265] (reviewed in [266] and [267]). These alternative splicing events lead to the production of intron-retention or entirely unspliced transcripts that need to overcome nuclear retention to be exported to the cytoplasm for translation.

The most common human polyomaviruses are the BK polyomavirus (BKPyV) and JC polyomavirus (JCPyV), bearing the names of the patients from whom they were first isolated [268,269]. Both these viruses are ubiquitous among the human population without any clinical significance in healthy individuals. However, in immunocompromised settings, both can lead to severe disease. Further, their clinical significance has risen in recent years, starting with the global AIDS epidemic and continuing today with a growing number of immunocompromised organ receivers and gene therapy patients [270,271,272,273]. The model species of polyomaviruses is simian virus 40 (SV40). Interestingly, SV40 also circulated in the human population after contaminated oral polio vaccines were administrated to millions of people between 1954 and 1961 [274]. Polyomaviruses are highly cancerogenic in non-permissive hosts. However, the implications of SV40 in human cancers are still debated to date [275,276,277,278,279] (reviewed in [280]). Little is known about the precise mRNA export mechanisms used by human polyomaviruses. It was, however, due to experiments with SV40 that the close relationships between splicing and mRNA export were first suggested, which are now widely acknowledged [281]. Similar information on polyomavirus mRNA export came from studies on the murine polyomavirus (MuPyV) performed in the 1990s, showing that the 5′ splice site, which is retained in the unspliced late viral protein 2 (VP2) mRNA, influences its nuclear export [282]. Interestingly, it was also demonstrated that while the strength of the splice site is inversely proportional to export efficiency, splicing itself does not increase the amount of exported mRNAs

[283]. It is probable that splicing factors bind this suboptimal splice site, and, even though they do not lead to splicing, they might interact with some components of the export machinery and thus slightly promote the inefficient and leaky export of this unspliced mRNA.

The knowledge about the role of Papillomaviruses in human tumors is well established. HPV is most commonly known as the causative agent of cervical cancer in women but can also cause, among others, penile and anal cancer, as well as head and neck cancer, alongside more benign genital and cutaneous warts. According to the centers for disease control and prevention (CDC), HPV infection is the most common sexually transmitted infection today [284]. Early and late HPV gene expression is not only regulated over time but is also highly dependent on the differentiation process of the infected epithelial cells [285,286,287,288]. Initial infection occurs at the basal layer of the epithelium [289], where the early phase of gene expression is centered on viral genome maintenance, the stimulation of cell division, and apoptosis inhibition, which all allow viral genomes to spread among the dividing cells [290]. Infected daughter cells migrate upwards through the various epithelial layers while completing keratinocyte differentiation [291]. Cellular differentiation leads to tightly regulated changes in viral gene expression, each of them linked to a particular epithelial layer [292,293,294]. Late viral genes are only expressed in the upper layers of the epithelium, where they lead to the production of structural proteins and the assembly and release of new viral particles [295,296,297]. The regulation of gene expression is orchestrated by polyadenylation and alternative splicing [298] (reviewed in [299]). Recent advances also suggest an essential role of cellular DNA damage response in the regulation of late HPV gene expression [300]. Interestingly, even if late transcripts are first detected in mid-layer epithelial cells, fully processed and exported mRNAs are only present in the uppermost layers [301,302,303,304]. This observation suggests that, although they are transcribed earlier on, the processing and export of late mRNAs also depends on the differentiation of keratinocytes.

The regulation of late gene expression is further linked to an AU-rich late negative regulatory element (NRE), present on all but one late transcript [305]. As the name implies, the NRE has an inhibitory effect on late HPV gene expression [306]. Reporter assays have shown that the inhibitory effect of this regulatory element can be overcome via the addition of HIV-1 RRE to the reporter transcript while providing the HIV-1 Rev protein *in trans* [307]. This suggests the role of mRNA nuclear export or retention in the regulation of late gene expression. Several cellular proteins were found to bind the

NER [308], and one of these proteins was later identified as HuR [309], which is essential for the export of ARE-containing mRNAs [90] (Figure 4b). Further underlining the potential role of HuR is the fact that upon the viral-induced proliferation of the epithelial cells, HuR is relocated from the nucleus to the cytoplasm [310]. The expression of late genes L1 and L2 is also correlated to the overexpression of HuR in cell cultures [311]. If the HPV late gene expression is indeed linked to viral mRNA export by HuR through the ARE-RNA export pathway, one might expect LMB treatment to abolish this effect by inhibiting the CRM1 exportin. HPV infected cells have been treated with LMB, but this treatment rapidly induced apoptosis linked to the expression of viral protein E7 [312]. The effect of this treatment on HPV late gene expression and mRNA export was, therefore, not tested. Thus, the role of HuR and the ARE containing RNA export pathway in the nuclear export of late HPV transcripts still needs to be clarified.

The One Exception—The Nuclear Export of Influenza Virus RNAs

We have seen that all reverse transcribing—and most DNA—viruses access the cell nucleus to exploit cellular transcription and replication machinery. Conventional RNA viruses code their own RNA-dependent RNA polymerase (RdRp) and are thus able to replicate their genomes within the cytoplasm without requiring direct access to the nucleus. However, while many RNA viruses only interact with the nucleus through the import of viral proteins that affect host gene expression, there is one known exception to this rule: influenza viruses. Influenza viruses have a segmented negative single-stranded RNA genome (with eight segments for influenza A viruses (IAV)), which is replicated with high mutation rates and has the potential for the re-assortment of the segmented genomes between different viral strains [313,314]. This genomic variability leads to high antigenic variability requiring annual vaccinations. Sporadically, the emergence of a new IAV strain due to an antigenic shift can lead to pandemics and high mortality due to a lack of immunity conferred by vaccination or prior infections [314,315,316]; for reviews, see [317,318].

Influenza virus RdRps lack methyltransferase activities and are unable to cap the 5′ ends of viral mRNAs, which is required for their proper processing and function. Thus, during a process called cap-snatching, the influenza virus RdRp binds to the C-terminal domain of cellular RNA polymerase

II and cleaves the newly synthesized cellular mRNA 10–13 nucleotides downstream of the 5′ cap [319,320,321] (reviewed in [322]).

Among the eight genomic segments of IAV, the viral RdRp transcribes segments 1 to 6 into six intronless mRNAs, while segments 7 and 8 are transcribed into mRNAs that can undergo alternative splicing [323,324,325]. Viral RdRp is also responsible for the replication of the viral genomic segments that also takes place within the host cell nucleus [326,327]. Thus, four different types of viral RNAs are synthesized during the IAV replication cycle: intronless mRNAs, spliced isoforms of intron-containing mRNAs, unspliced intron-containing mRNAs, and newly produced viral genomic segments, which are associated with the viral polymerase complex and are encapsidated by the viral nucleoprotein (NP) forming the eight viral ribonucleoproteins (vRNPs) [328,329,330].

Export of Influenza A Virus mRNAs

Empirically, one might expect all intronless viral mRNAs to be exported through the ALREX pathway. However, the export of the different intronless viral mRNAs is differentially affected by the depletion of NXF1, the major export receptor involved in the ALREX export pathway [331]. Surprisingly, the export of early transcripts coding the three polymerase subunits PA, PB1, and PB2 is entirely insensitive to NXF1 depletion but also to LMB inhibition of the CRM1 pathway [331]. These findings suggest that early influenza transcripts use a unique nuclear export mechanism independent of NXF1 and CRM1. Unlike intronless polymerase subunit transcripts, intronless mRNAs transcribed later during the infection cycle of IAV display, at least for some of them, sensitivity towards NXF1 depletion [331]. Under this condition, the export of mRNAs coding neuraminidase (NA) and haemagglutinin (HA) is compromised, whereas the NP coding mRNA remains unaffected [331]. Larsen et al. showed that the effect of NXF1 depletion on influenza mRNA export was different from one cell type to another and that NP mRNA export was, indeed, NXF1-dependent in human epithelial lung cells (A549) [332]. Since these cells reflect in vivo IAV infection more accurately, it is reasonable to consider NP mRNA export as comparable to the NXF1-dependent export of HA and NA mRNAs.

It is important to highlight that despite the fact that all IAV mRNAs undergo capping and polyadenylation, two processes that are involved in the recruitment of the NXF1–NXT1 nuclear export machinery, these export receptors are used only by some of the IAV mRNAs. This observation can be explained by the fact that the mechanisms by which IAV mRNAs are processed significantly differs from the canonical cellular mechanisms. Although NCBP1 has been reported to be involved in IAV mRNA export, it is still unclear whether the CBC might be disturbed during cap-snatching, thus interfering with its canonical role in RNA export [51,333]. Furthermore, the poly(A) tail is not synthesized by the PAP but rather by viral RdRp stuttering on the poly(U) stretch before its dissociation from the template RNA [334]. This implies that CPSF and PABP cannot play their usual roles in the recruitment and activation of the TREX1 complex. It has, however, been shown that the efficient export of viral mRNA requires ongoing Polymerase II activity [335], suggesting that viral transcription likely takes place in an environment allowing for at least some of the interactions that usually link mRNA export to processing to occur. Further complicating our current understanding of IAV intronless mRNA export is the fact that a poly(A) tail is required for the efficient export of cellular intronless mRNA by the ALREX pathway [83]. Whether or not the atypical polyadenylation of IAV mRNAs can contribute equally towards this mechanism remains unknown.

Since the mRNAs transcribed from segments 7 and 8 can be spliced to produce the non-structural proteins NS2 and NS3 and matrix proteins M2, M3, and M4, respectively [323,324,325], one might expect these spliced mRNAs to be exported by the splicing-dependent NXF1–NXT1 export receptors. This is indeed the case, but, surprisingly, the unspliced mRNAs coding the NS1 and M1 proteins seem to follow a similar pathway [331]. Recent studies have shown that the precise mechanism might deviate from the canonical pathway, at least for the export of M coding mRNAs as the splicing of the M1 mRNA has been shown to occur post-transcriptionally at the nuclear speckles [336]. Viral NS1 protein and its cellular partner, the NS1 binding protein (NS1-BP) [336], are also involved in the splicing of M1 mRNA, as well as in the export of both the spliced and unspliced forms of M mRNAs [337,338]. This suggests that the nuclear accumulation of NS1 is responsible for the M1 transcripts' relocalization towards the nuclear speckles where they undergo splicing before being exported towards the cytoplasm, leading to the subsequent accumulation of the M2 protein later during infection. This temporal regulation is essential since high levels of M2 protein are cytotoxic and only required during the later phases of the IAV replication cycle [339]. Another critical component identified for M mRNA nuclear export is UAP56 [331], which co-localizes with NS1, further corroborating its importance in the mechanism [340].

Interestingly, recent work showed that the knockdown of cellular NS1-BP leads to a significant reduction in M mRNA export without altering cellular bulk mRNA export [336]. This suggests that IAV M mRNA speckle-dependent nuclear export is unique and thus a highly attractive target for the development of new antiviral agents [341]. Surprisingly, like for the export of intronless mRNAs, the mechanism of M mRNA nuclear export is not the same as that of the NS mRNAs. It has been previously shown that NS2 mRNA export is insensitive to UAP56 depletion [331] and that there are no temporal differences in the accumulation of NS1 and NS2 proteins. In addition, the splicing and export of NS mRNA are independent of the NS1 protein [342]. Further emphasizing the differences in the nuclear export of M and NS mRNAs is the fact that a chemical component that inhibits M mRNA export only has mild effects on NS mRNA export [341]. Contradicting the presumably distinct splicing and export mechanisms of the M and NS mRNAs is the involvement of the RNA helicase eIF4A3 in the splicing and export of both RNAs [343]. Other cellular RNA helicases are also involved in IAV mRNA export. For instance, it has been shown that DDX19 is associated with viral polymerase and is involved in the export of all viral mRNAs. Moreover, its involvement can be bypassed during the later phases of infection [344]. This suggests that IAVs have evolved some redundancies in their viral mRNA nuclear export, which might explain the difficulties in proposing a single model for the mechanisms involved.

Export of Influenza A Virus vRNPs

As mentioned before, influenza viruses export their replicated genomic segments into the cytoplasm as vRNPs, in which negative viral RNAs are encapsidated by the NP protein and associated with the polymerase complex (reviewed in [345]). The export of these vRNPs depends on the NS2 viral protein [346], also known as the Nuclear Export Protein (NEP), and M1 [347,348]. Both NEP and M1 are associated with purified virions [349,350]. While a daisy chain model of interactions was first proposed where M1 binds NP and NEP binds M1, the current model suggests that M1 and NEP interact with each other and that both also interact with NP. NEP possesses a canonical and an atypical leucine rich NES, which are both involved in the CRM1-dependent export of vRNPs [351,352,353]. Yeast and mammalian two-hybrid experiments recently showed that NEP interacts with Nup98 and Nup214, two nucleoporins that are essential for IAV nuclear export and replication [354,355]. Interestingly, NP itself also possesses several NES motifs and displays CRM1-dependent nucleo-cytoplasmic shuttling [356,357]. NP also interacts with NXT1, co-factor of the major mRNA export receptor, NXF1 [358]. Surprisingly, interaction with NXT1 stimulates the CRM1-dependent nuclear export of NP (and probably also that of vRNPs) [358]. While these findings suggest a novel redundancy in the IAV replication cycle, here centered on the nuclear export of vRNPs and the different viral proteins that are able to recruit the export machinery, the precise role of NP in nuclear export remains unclear. A recent study also suggested the influence of cell polarity on influenza vRNP nuclear export [359].

The nuclear export of both IAV mRNAs and vRNPs has been explored as a potential target for anti-influenza treatment strategies [341]. Overall, the precise nuclear export mechanisms of all types of IAV RNAs remain unclear due to many contradictory results, which might be due to the use of different cell types or differences between the viral strains. Recent research suggests that the impaired nuclear export of vRNPs might be responsible for host restrictions between avian and human influenza [360], further underlining the need for a better understanding of these mechanisms.

CONCLUDING COMMENTS

Viruses that require access to the nucleus of their eukaryotic host cells are faced with numerous challenges. One of these is the requirement for the nuclear export of their different mRNA isoforms, which are essential for the viral replication cycle. The cellular mechanism for mRNA export is highly dependent on proper pre-mRNA processing, including 5′ capping, 3′ polyadenylation, and, most importantly, intron splicing. However, many viral transcripts retain introns and contain other features often considered contradictory to efficient RNA export, such as long non-translated regions harboring premature stop codons.

To overcome these obstacles, viruses have evolved different strategies to exploit the cellular pathways of mRNA nuclear export. The precise mechanisms of export are highly variable between different viral families, and, in some instances, even within the same family. A common characteristic among them, however, is their dependency on the formation of ribonucleoprotein complexes, including nuclear export receptors. These complexes can be formed in many different ways. In some cases, like MPMV, nuclear transport receptors are recruited directly to viral mRNAs through a secondary RNA structure. In other cases, an adapter protein of cellular or viral origin, provides the link between the viral mRNA and the nuclear export receptor. The resulting ribonucleoprotein complexes can be relatively simple, but, more often, they are complex and very dynamic.

The CRM1 export pathway is exploited to export partially spliced and unspliced mRNAs, particularly in complex retroviruses that code for an adapter protein recruiting CRM1 on the secondary structures of viral mRNAs. The nucleolar localization of these adapter proteins is probably related to that of CRM1 but also to the escape of viral intron-containing mRNAs from the host surveillance machinery. It is important to highlight that CRM1 is also the major exporter of proteins (cellular and viral) and is systematically over-expressed and deregulated in cancer cells, leading, for example, to the abnormal accumulation of tumor suppressor proteins in the cytoplasm and their functional inactivation (for a review, see [361]). For this reason, research on, and the identification of, effective and reversible CRM1 inhibitors, called SINE (selective inhibitor of nuclear export), has been very active for many years, and several molecules have already been tested and selected after phase I, II, and III clinical trials. One of them, selexinor, was approved in the United States in 2019 by the Food and Drug Administration (FDA) [362]. The specific inhibition of CRM1 is, of course, also deleterious to viral replication and could, therefore, represent an effective antiviral treatment. As we have seen, CRM1 inhibition not only considerably reduces the replication of several influenza viruses (H1N1, H3N2, H5N1, and H7N9) but also that of HIV-1 by blocking the export of its genomic RNA and the production of new viral particles [351,363,364].

Finally, we should keep in mind that virology has greatly contributed to our current understanding of the nuclear export mechanisms of mRNAs. The discovery of NESs and their importance in CRM1-mediated export and of CTEs and the NXF1–NXT1 export receptor marked a milestone in our understanding of the mechanisms by which mRNAs are recruited for export into the cytoplasm.

Much knowledge has been accumulated on the nuclear export of viral mRNAs, showing that viruses, by evolving and adapting to their hosts, have developed very different strategies to hijack the cellular export machinery and ensure the efficient expression of their own proteins. Some of these strategies, which are not completely understood because they are highly complex and vary according to the species of viral RNA, continue to attract the interest of researchers. Ongoing and future studies should provide a complete picture of the export of mRNAs of viruses, such as hepatitis B virus, influenza viruses, and polyoma- and papillomaviruses. Virology will thus continue to contribute to our understanding of the processes that regulate the post-transcriptional fate of viral and cellular mRNAs.

ACKNOWLEDGMENTS

The authors gratefully acknowledge Mario Keller (IBMP, Strasbourg) for his stimulating discussions and Ólöf Garðarsdottir (University of Iceland, Reykjavík) for her critical reading of the manuscript. All figures are original and were created with BioRender.com.

REFERENCES

1. Callan, H.G.; Tomlin, S.G. Experimental studies on amphibian oocyte nuclei I. Investigation of the structure of the nuclear membrane by means of the electron microscope. Proc. R. Soc. Lond. Ser. B Biol. Sci. 1950, 137, 367–378.
2. Gall, J. Observations on the nuclear membrane with the electron microscope. Exp. Cell Res. 1954, 7, 197–200.
3. Lin, D.H.; Hoelz, A. The Structure of the Nuclear Pore Complex (An Update). Annu. Rev. Biochem. 2019, 88, 725–783.
4. Edwards, R.A.; Rohwer, F. Viral metagenomics. Nat. Rev. Genet. 2005, 3, 504–510.
5. Koonin, E.V.; Dolja, V.V.; Krupovic, M. Origins and evolution of viruses of eukaryotes: The ultimate modularity. Virology 2015, 479, 2–25.
6. Kobiler, O.; Drayman, N.; Butin-Israeli, V.; Oppenheim, A. Virus strategies for passing the nuclear envelope barrier. Nucleus 2012, 3, 526–539.
7. Fay, N.; Panté, N. Nuclear entry of DNA viruses. Front. Microbiol. 2015, 6, 467.
8. Tessier, T.M.; Dodge, M.J.; Prusinkiewicz, M.A.; Mymryk, J.S. Viral Appropriation: Laying Claim to Host Nuclear Transport Machinery. Cells 2019, 8, 559.
9. Cavalier-Smith, T. Origin of the cell nucleus, mitosis and sex: Roles of intracellular coevolution. Biol. Direct 2010, 5, 7.
10. Ribbeck, K.; Görlich, D. Kinetic analysis of translocation through nuclear pore complexes. EMBO J. 2001, 20, 1320–1330.
11. Fabergé, A.C. Direct demonstration of eight-fold symmetry in nuclear pores. Zeitschrift für Zellforschung und Mikroskopische Anatomie 1973, 136, 183–190.
12. Akey, C.W.; Radermacher, M. Architecture of the Xenopus nuclear pore complex revealed by three-dimensional cryo-electron microscopy. J. Cell Biol. 1993, 122, 1–19.
13. Beck, M.; Förster, F.; Ecke, M.; Plitzko, J.M.; Melchior, F.; Gerisch, G.; Baumeister, W.; Medalia, O. Nuclear Pore Complex Structure and Dynamics Revealed by Cryoelectron Tomography. Science 2004, 306, 1387–1390.

14. Elad, N.; Maimon, T.; Frenkiel-Krispin, D.; Lim, R.Y.; Medalia, O. Structural analysis of the nuclear pore complex by integrated approaches. Curr. Opin. Struct. Biol. 2009, 19, 226–232.
15. Goldberg, M.W.; Allen, T.D. High resolution scanning electron microscopy of the nuclear envelope: Demonstration of a new, regular, fibrous lattice attached to the baskets of the nucleoplasmic face of the nuclear pores. J. Cell Biol. 1992, 119, 1429–1440.
16. Jarnik, M.; Aebi, U. Toward a more complete 3-D structure of the nuclear pore complex. J. Struct. Biol. 1991, 107, 291–308.
17. Denning, D.P.; Patel, S.S.; Uversky, V.; Fink, A.L.; Rexach, M. Disorder in the nuclear pore complex: The FG repeat regions of nucleoporins are natively unfolded. Proc. Natl. Acad. Sci. USA 2003, 100, 2450–2455.
18. Lim, R.Y.; Fahrenkrog, B.; Köser, J.; Schwarz-Herion, K.; Deng, J.; Aebi, U. Nanomechanical Basis of Selective Gating by the Nuclear Pore Complex. Science 2007, 318, 640–643.
19. Lim, R.Y.; Köser, J.; Huang, N.-P.; Schwarz-Herion, K.; Aebi, U. Nanomechanical interactions of phenylalanine–glycine nucleoporins studied by single molecule force–volume spectroscopy. J. Struct. Biol. 2007, 159, 277–289.
20. Frey, S.; Görlich, D. A Saturated FG-Repeat Hydrogel Can Reproduce the Permeability Properties of Nuclear Pore Complexes. Cell 2007, 130, 512–523.
21. Frey, S.; Görlich, D. FG/FxFG as well as GLFG repeats form a selective permeability barrier with self-healing properties. EMBO J. 2009, 28, 2554–2567.
22. Paine, P.L.; Moore, L.C.; Horowitz, S.B. Nuclear envelope permeability. Nature 1975, 254, 109–114.
23. Bonner, W.M. Protein migration into nuclei. I. Frog oocyte nuclei in vivo accumulate microinjected histones, allow entry to small proteins, and exclude large proteins. J. Cell Biol. 1975, 64, 421–430.
24. Keminer, O.; Peters, R. Permeability of Single Nuclear Pores. Biophys. J. 1999, 77, 217–228.
25. Milles, S.; Mercadante, D.; Aramburu, I.V.; Jensen, M.R.; Banterle, N.; Koehler, C.; Tyagi, S.; Clarke, J.; Shammas, S.L.; Blackledge, M.; et al. Plasticity of an ultrafast interaction between nucleoporins and nuclear transport receptors. Cell 2015, 163, 734–745.

26. Davis, L.K.; Ford, I.J.; Šarić, A.; Hoogenboom, B.W. Intrinsically disordered nuclear pore proteins show ideal-polymer morphologies and dynamics. Phys. Rev. E 2020, 101, 022420.
27. Pemberton, L.F.; Paschal, B.M. Mechanisms of Receptor-Mediated Nuclear Import and Nuclear Export. Traffic 2005, 6, 187–198.
28. Stewart, M. Molecular mechanism of the nuclear protein import cycle. Nat. Rev. Mol. Cell Biol. 2007, 8, 195–208.
29. Stade, K.; Ford, C.S.; Guthrie, C.; Weis, K. Exportin 1 (Crm1p) Is an Essential Nuclear Export Factor. Cell 1997, 90, 1041–1050.
30. Cokol, M.; Nair, R.; Rost, B. Finding nuclear localization signals. EMBO Rep. 2000, 1, 411–415.
31. Kırlı, K.; Karaca, S.; Dehne, H.J.; Samwer, M.; Pan, K.-T.; Lenz, C.; Urlaub, H.; Görlich, D. A deep proteomics perspective on CRM1-mediated nuclear export and nucleocytoplasmic partitioning. Elife 2015, 4, e11466.
32. Tinland, B.; Koukolíková-Nicola, Z.; Hall, M.N.; Hohn, B. The T-DNA-linked VirD2 protein contains two distinct functional nuclear localization signals. Proc. Natl. Acad. Sci. USA 1992, 89, 7442–7446.
33. Moede, T.; Leibiger, B.; Pour, H.G.; Berggren, P.-O.; Leibiger, I.B. Identification of a nuclear localization signal, RRMKWKK, in the homeodomain transcription factor PDX-1. FEBS Lett. 1999, 461, 229–234.
34. Imamoto, N.; Shimamoto, T.; Kose, S.; Takao, T.; Tachibana, T.; Matsubae, M.; Sekimoto, T.; Shimonishi, Y.; Yoneda, Y.; Naoko, I.; et al. The nuclear pore-targeting complex binds to nuclear pores after association with a karyophile. FEBS Lett. 1995, 368, 415–419.
35. Radu, A.; Blobel, G.; Moore, M.S. Identification of a protein complex that is required for nuclear protein import and mediates docking of import substrate to distinct nucleoporins. Proc. Natl. Acad. Sci. USA 1995, 92, 1769–1773.
36. Görlich, D.; Henklein, P.; Laskey, R.A.; Hartmann, E. A 41 amino acid motif in importin-alpha confers binding to importin-beta and hence transit into the nucleus. EMBO J. 1996, 15, 1810–1817.
37. Görlich, D.; Kostka, S.; Kraft, R.; Dingwall, C.; Laskey, R.A.; Hartmann, E.; Prehn, S. Two different subunits of importin cooperate to recognize nuclear localization signals and bind them to the nuclear envelope. Curr. Biol. 1995, 5, 383–392.

38. Fukuda, M.; Asano, S.; Nakamura, T.; Adachi, M.; Yoshida, M.; Yanagida, M.; Nishida, E. CRM1 is responsible for intracellular transport mediated by the nuclear export signal. Nature 1997, 390, 308–311.
39. Arnaoutov, A.; Azuma, Y.; Ribbeck, K.; Joseph, J.; Boyarchuk, Y.; Karpova, T.S.; McNally, J.; Dasso, M. Crm1 is a mitotic effector of Ran-GTP in somatic cells. Nat. Cell Biol. 2005, 7, 626–632.
40. Bischoff, F.; Görlich, D. RanBP1 is crucial for the release of RanGTP from importin β-related nuclear transport factors. FEBS Lett. 1997, 419, 249–254.
41. Kim, Y.-H.; Han, M.-E.; Oh, S.-O. The molecular mechanism for nuclear transport and its application. Anat. Cell Biol. 2017, 50, 77–85.
42. Björk, P.; Wieslander, L. Integration of mRNP formation and export. Cell Mol. Life Sci. 2017, 74, 2875–2897.
43. Clouse, K.N.; Luo, M.-J.; Zhou, Z.; Reed, R. A Ran-independent pathway for export of spliced mRNA. Nature 2001, 3, 97–99.
44. Viphakone, N.; Hautbergue, G.M.; Walsh, M.; Chang, C.-T.; Holland, A.; Folco, E.G.; Reed, R.; Wilson, S.A. TREX exposes the RNA-binding domain of Nxf1 to enable mRNA export. Nat. Commun. 2012, 3, 1–14.
45. Sträßer, K.; Masuda, S.; Mason, P.; Pfannstiel, J.; Oppizzi, M.; Rodríguez-Navarro, S.; Rondón, A.G.; Aguilera, A.; Struhl, K.; Reed, R.; et al. TREX is a conserved complex coupling transcription with messenger RNA export. Nature 2002, 417, 304–308.
46. Gatfield, D.; Le Hir, H.; Schmitt, C.; Braun, I.C.; Köcher, T.; Wilm, M.; Izaurralde, E.; Köcher, T. The DExH/D box protein HEL/UAP56 is essential for mRNA nuclear export in Drosophila. Curr. Biol. 2001, 11, 1716–1721.
47. Hautbergue, G.M.; Hung, M.-L.; Walsh, M.J.; Snijders, A.P.; Chang, C.-T.; Jones, R.; Ponting, C.P.; Dickman, M.J.; Wilson, S.A. UIF, a New mRNA Export Adaptor that Works Together with REF/ALY, Requires FACT for Recruitment to mRNA. Curr. Biol. 2009, 19, 1918–1924.
48. Viphakone, N.; Sudbery, I.; Griffith, L.; Heath, C.G.; Sims, D.; Wilson, H.L. Co-transcriptional Loading of RNA Export Factors Shapes the Human Transcriptome. Mol. Cell 2019, 75, 310–323.e8.
49. Ghosh, A.; Lima, C.D. Enzymology of RNA cap synthesis. Wiley Interdiscip. Rev. RNA 2010, 1, 152–172.

50. Rozen, F.; Sonenberg, N. Identification of nuclear cap specific proteins in HeLa cells. Nucleic Acids Res. 1987, 15, 6489–6500.
51. Cheng, H.; Dufu, K.; Lee, C.-S.; Hsu, J.L.; Dias, A.; Reed, R. Human mRNA Export Machinery Recruited to the 5′ End of mRNA. Cell 2006, 127, 1389–1400.
52. Sen, R.; Barman, P.; Kaja, A.; Ferdoush, J.; Lahudkar, S.; Roy, A.; Bhaumik, S.R. Distinct Functions of the Cap-Binding Complex in Stimulation of Nuclear mRNA Export. Mol. Cell. Biol. 2019, 39, e00540-18.
53. McCloskey, A.; Taniguchi, I.; Shinmyozu, K.; Ohno, M. hnRNP C Tetramer Measures RNA Length to Classify RNA Polymerase II Transcripts for Export. Science 2012, 335, 1643–1646.
54. Lee, Y.; Rio, D.C. Mechanisms and Regulation of Alternative Pre-mRNA Splicing. Annu. Rev. Biochem. 2015, 84, 291–323.
55. Huang, Y.; Steitz, J.A. Splicing Factors SRp20 and 9G8 Promote the Nucleocytoplasmic Export of mRNA. Mol. Cell 2001, 7, 899–905.
56. Müller-McNicoll, M.; Botti, V.; Domingues, A.M.; Brandl, H.; Schwich, O.D.; Steiner, M.C.; Curk, T.; Poser, I.; Zarnack, K.; Neugebauer, K.M. SR proteins are NXF1 adaptors that link alternative RNA processing to mRNA export. Genes Dev. 2016, 30, 553–566.
57. Shen, H. UAP56—A key player with surprisingly diverse roles in pre-mRNA splicing and nuclear export. BMB Rep. 2009, 42, 185–188.
58. Edmonds, M.; Abrams, R. Polynucleotide biosynthesis: Formation of a sequence of adenylate units from adenosine triphosphate by an enzyme from thymus nuclei. J. Biol. Chem. 1960, 235, 1142–1149.
59. Dantonel, J.-C.; Murthy, K.G.; Manley, J.L.; Tora, L. Transcription factor TFIID recruits factor CPSF for formation of 3′ end of mRNA. Nature 1997, 389, 399–402.
60. Elkon, R.; Ugalde, A.P.; Agami, R. Alternative cleavage and polyadenylation: Extent, regulation and function. Nat. Rev. Genet. 2013, 14, 496–506.
61. Johnson, S.A.; Cubberley, G.; Bentley, D.L. Cotranscriptional Recruitment of the mRNA Export Factor Yra1 by Direct Interaction with the 3′ End Processing Factor Pcf11. Mol. Cell 2009, 33, 215–226.
62. Qu, X.; Lykke-Andersen, S.; Nasser, T.; Saguez, C.; Bertrand, E.; Jensen, T.H.; Moore, C.L. Assembly of an Export-Competent mRNP Is Needed for Efficient Release of the 3′-End Processing Complex after

Polyadenylation—IGMM. Mol. Cell. Biol. 2009, 29, 5327–5338.

63. Wahle, E. Poly(A) Tail Length Control Is Caused by Termination of Processive Synthesis. J. Biol. Chem. 1995, 270, 2800–2808.
64. Iglesias, N.; Tutucci, E.; Gwizdek, C.; Vinciguerra, P.; Von Dach, E.; Corbett, A.H.; Dargemont, C.; Stutz, F. Ubiquitin-mediated mRNP dynamics and surveillance prior to budding yeast mRNA export. Genes Dev. 2010, 24, 1927–1938.
65. Hautbergue, G.M.; Hung, M.-L.; Golovanov, A.P.; Lian, L.-Y.; Wilson, S.A. Mutually exclusive interactions drive handover of mRNA from export adaptors to TAP. Proc. Natl. Acad. Sci. USA 2008, 105, 5154–5159.
66. Mor, A.; Suliman, S.; Ben-Yishay, R.; Yunger, S.; Brody, Y.; Shav-Tal, Y. Dynamics of single mRNP nucleocytoplasmic transport and export through the nuclear pore in living cells. Nat. Cell Biol. 2010, 12, 543–552.
67. Dilworth, D.J.; Tackett, A.J.; Rogers, R.S.; Yi, E.C.; Christmas, R.H.; Smith, J.J.; Siegel, A.F.; Chait, B.T.; Wozniak, R.W.; Aitchison, J.D. The mobile nucleoporin Nup2p and chromatin-bound Prp20p function in endogenous NPC-mediated transcriptional control. J. Cell Biol. 2005, 171, 955–965.
68. Ishii, K.; Arib, G.; Lin, C.; Van Houwe, G.; Laemmli, U.K. Chromatin Boundaries in Budding Yeast: The nuclear pore connection. Cell 2002, 109, 551–562.
69. Burns, L.T.; Wente, S.R. From Hypothesis to Mechanism: Uncovering Nuclear Pore Complex Links to Gene Expression. Mol. Cell. Biol. 2014, 34, 2114–2120.
70. Jani, D.; Valkov, E.; Stewart, M. Structural basis for binding the TREX2 complex to nuclear pores, GAL1 localisation and mRNA export. Nucleic Acids Res. 2014, 42, 6686–6697.
71. Fischer, T.; Strässer, K.; Rácz, A.; Rodríguez-Navarro, S.; Oppizzi, M.; Ihrig, P.; Lechner, J.; Hurt, E. The mRNA export machinery requires the novel Sac3p–Thp1p complex to dock at the nucleoplasmic entrance of the nuclear pores. EMBO J. 2002, 21, 5843–5852.
72. Kurshakova, M.M.; Krasnov, A.N.; Kopytova, D.V.; Shidlovskii, Y.V.; Nikolenko, J.V.; Nabirochkina, E.N.; Spehner, D.; Schultz, P.; Tora, L.; Georgieva, S.G. SAGA and a novel Drosophila export complex anchor efficient transcription and mRNA export to NPC. EMBO J. 2007, 26, 4956–4965.

73. Jani, D.; Lutz, S.; Marshall, N.J.; Fischer, T.; Köhler, A.; Ellisdon, A.M.; Hurt, E.; Stewart, M. Sus1, Cdc31, and the Sac3 CID Region Form a Conserved Interaction Platform that Promotes Nuclear Pore Association and mRNA Export. Mol. Cell 2009, 33, 727–737.
74. Wickramasinghe, V.O.; McMurtrie, P.I.; Mills, A.D.; Takei, Y.; Penrhyn-Lowe, S.; Amagase, Y.; Main, S.; Marr, J.; Stewart, M.; Laskey, R.A. mRNA Export from Mammalian Cell Nuclei Is Dependent on GANP. Curr. Biol. 2010, 20, 25–31.
75. Alcázar-Román, A.R.; Tran, E.J.; Guo, S.; Wente, S.R. Inositol hexakisphosphate and Gle1 activate the DEAD-box protein Dbp5 for nuclear mRNA export. Nat. Cell Biol. 2006, 8, 711–716.
76. Lin, D.H.; Correia, A.R.; Cai, S.W.; Huber, F.M.; Jette, C.A.; Hoelz, A. Structural and functional analysis of mRNA export regulation by the nuclear pore complex. Nat. Commun. 2018, 9, 2319.
77. Folkmann, A.W.; Noble, K.N.; Cole, C.N.; Wente, S.R. Dbp5, Gle1-IP6 and Nup. Nucleus 2011, 2, 540–548.
78. Yang, X.; Yang, Y.; Sun, B.-F.; Chen, Y.-S.; Xu, J.-W.; Lai, W.-Y.; Li, A.; Wang, X.; Bhattarai, D.P.; Xiao, W.; et al. 5-methylcytosine promotes mRNA export-NSUN2 as the methyltransferase and ALYREF as an m5C reader. Cell Res. 2017, 27, 606–625.
79. Roundtree, I.A.; Luo, G.; Zhang, Z.; Wang, X.; Zhou, T.; Cui, Y.; Sha, J.-H.; Huang, X.; Guerrero, L.; Xie, P.; et al. YTHDC1 mediates nuclear export of N6-methyladenosine methylated mRNAs. Elife 2017, 6, e31311.
80. Lesbirel, S.; Wilson, S.A. The m6A-methylase complex and mRNA export. Biochim. Biophys. Acta Gene Regul. Mech. 2019, 1862, 319–328.
81. Zuckerman, B.; Ron, M.; Mikl, M.; Segal, E.; Ulitsky, I. Gene Architecture and Sequence Composition Underpin Selective Dependency of Nuclear Export of Long RNAs on NXF1 and the TREX Complex. Mol. Cell 2020, 79, 251–267.e6.
82. Palazzo, A.F.; Springer, M.; Shibata, Y.; Lee, C.-S.; Dias, A.P.; Rapoport, T.A. The signal sequence coding region promotes nuclear export of mRNA. PLoS Biol. 2007, 5, e322.
83. Akef, A.; Zhang, H.; Masuda, S.; Palazzo, A.F. Trafficking of mRNAs containing ALREX-promoting elements through nuclear speckles. Nucleus 2013, 4, 326–340.

84. Lei, H.; Dias, A.P.; Reed, R. Export and stability of naturally intronless mRNAs require specific coding region sequences and the TREX mRNA export complex. Proc. Natl. Acad. Sci. USA 2011, 108, 17985–17990.
85. Hutten, S.; Kehlenbach, R.H. CRM1-mediated nuclear export: To the pore and beyond. Trends Cell Biol. 2007, 17, 193–201.
86. Delaleau, M.; Borden, K.L. Multiple Export Mechanisms for mRNAs. Cells 2015, 4, 452–473.
87. Brennan, C.M.; Gallouzi, I.-E.; Steitz, J.A. Protein Ligands to Hur Modulate Its Interaction with Target Mrnas in Vivo. J. Cell Biol. 2000, 151, 1–14.
88. Barreau, C.; Paillard, L.; Osborne, H.B. AU-rich elements and associated factors: Are there unifying principles? Nucleic Acids Res. 2005, 33, 7138–7150.
89. Gallouzi, I.-E.; Steitz, J.A. Delineation of mRNA Export Pathways by the Use of Cell-Permeable Peptides. Science 2001, 294, 1895–1901.
90. Fries, B.; Heukeshoven, J.; Hauber, I.; Grüttner, C.; Stocking, C.; Kehlenbach, R.H.; Hauber, J.; Chemnitz, J. Analysis of Nucleocytoplasmic Trafficking of the HuR Ligand APRIL and Its Influence on CD83 Expression. J. Biol. Chem. 2007, 282, 4504–4515.
91. Kimura, T.; Fujisawa, J.-I.; Hashimoto, I.; Nagase, T. CRM1-dependent, but not ARE-mediated, nuclear export of IFN-alpha1 mRNA. J. Cell Sci. 2004, 117, 2259–2270.
92. Culjkovic, B.; Topisirovic, I.; Skrabanek, L.; Ruiz-Gutierrez, M.; Borden, K.L. eIF4E is a central node of an RNA regulon that governs cellular proliferation. J. Cell Biol. 2006, 175, 415–426.
93. Rong, L.; Livingstone, M.; Sukarieh, R.; Petroulakis, E.; Gingras, A.-C.; Crosby, K.; Smith, B.; Polakiewicz, R.D.; Pelletier, J.; Ferraiuolo, M.A.; et al. Control of eIF4E cellular localization by eIF4E-binding proteins, 4E-BPs. RNA 2008, 14, 1318–1327.
94. Topisirovic, I.; Siddiqui, N.; Lapointe, V.L.; Trost, M.; Thibault, P.; Bangeranye, C.; Piñol-Roma, S.; Borden, K.L. Molecular dissection of the eukaryotic initiation factor 4E (eIF4E) export-competent RNP. EMBO J. 2009, 28, 1087–1098.
95. Volpon, L.; Culjkovic-Kraljacic, B.; Sohn, H.S.; Blanchet-Cohen, A.; Osborne, M.J.; Borden, K.L. A biochemical framework for eIF4E-dependent mRNA export and nuclear recycling of the export machinery. RNA 2017, 23, 927–937.

96. Bresson, S.; Tollervey, D. Surveillance-ready transcription: Nuclear RNA decay as a default fate. Open Biol. 2018, 8, 170270.
97. Toro-Ascuy, D.; Rojas-Araya, B.; Valiente-Echeverría, F.; Rifo, R.S. Interactions between the HIV-1 Unspliced mRNA and Host mRNA Decay Machineries. Viruses 2016, 8, 320.
98. Balagopal, V.; Beemon, K.L. Rous Sarcoma Virus RNA Stability Element Inhibits Deadenylation of mRNAs with Long 3′UTRs. Viruses 2017, 9, 204.
99. Balistreri, G.; Bognanni, C.; Mühlemann, O. Virus Escape and Manipulation of Cellular Nonsense-Mediated mRNA Decay. Viruses 2017, 9, 24.
100. Temin, H.M. Nature of the provirus of Rous sarcoma. Nat. Cancer Inst. Monogr. 1964, 17, 557–570.
101. Baltimore, D. Viral RNA-dependent DNA Polymerase: RNA-dependent DNA Polymerase in Virions of RNA Tumour Viruses. Nature 1970, 226, 1209–1211.
102. Temin, H.M.; Mizutani, S. Viral RNA-dependent DNA Polymerase: RNA-dependent DNA Polymerase in Virions of Rous Sarcoma Virus. Nature 1970, 226, 1211–1213.
103. Brahic, M.; Stowring, L.; Ventura, P.; Haase, A.T. Gene expression in visna virus infection in sheep. Nature 1981, 292, 240–242.
104. Folks, T.; Powell, D.; Lightfoote, M.; Benn, S.; Martin, M.; Fauci, A. Induction of HTLV-III/LAV from a nonvirus-producing T-cell line: Implications for latency. Science 1986, 231, 600–602.
105. Coffin, J.M.; Hughes, S.H.; Varmus, H.E. Genetic Organization; Cold Spring Harbor Laboratory Press: New York, NY, USA, 1997; ISBN 978-0-87969-497-5.
106. Swanstrom, R.; Wills, J.W. Synthesis, Assembly, and Processing of Viral Proteins. In Retroviruses; Coffin, J.M., Hughes, S.H., Varmus, H.E., Eds.; Cold Spring Harbor Laboratory Press: New York, NY, USA, 1997; ISBN 978-0-87969-571-2.
107. Rabson, A.B.; Graves, B.J. Synthesis and Processing of Viral RNA. In Retroviruses; Coffin, J.M., Hughes, S.H., Varmus, H.E., Eds.; Cold Spring Harbor Laboratory Press: New York, NY, USA, 1997; ISBN 978-0-87969-571-2.

108. Coffin, J.M.; Hughes, S.H.; Varmus, H.E. Retroviral "Lifestyles": Simple Versus Complex; Cold Spring Harbor Laboratory Press: New York, NY, USA, 1997; ISBN 978-0-87969-497-5.
109. Barré-Sinoussi, F.; Chermann, J.; Rey, F.; Nugeyre, M.; Chamaret, S.; Gruest, J.; Dauguet, C.; Axler-Blin, C.; Vezinet-Brun, F.; Rouzioux, C.; et al. Isolation of a T-lymphotropic retrovirus from a patient at risk for acquired immune deficiency syndrome (AIDS). Science 1983, 220, 868–871.
110. Vilmer, E.; Barré-Sinoussi, F.; Rouzioux, C.; Gazengel, C.; Brun, F.V.; Dauguet, C.; Fischer, A.; Manigne, P.; Chermann, J.; Griscelli, C.; et al. Isolation of new lymphotropic retrovirus from two siblings with haemophilia B, one with AIDS. Lancet 1984, 323, 753–757.
111. Vahlne, A. A historical reflection on the discovery of human retroviruses. Retrovirology 2009, 6, 40.
112. Lai, C.J.; Khoury, G. Deletion mutants of simian virus 40 defective in biosynthesis of late viral mRNA. Proc. Natl. Acad. Sci. USA 1979, 76, 71–75.
113. Valencia, P.; Dias, A.P.; Reed, R. Splicing promotes rapid and efficient mRNA export in mammalian cells. Proc. Natl. Acad. Sci. USA 2008, 105, 3386–3391.
114. Ocwieja, K.E.; Sherrill-Mix, S.; Mukherjee, R.; Custers-Allen, R.; David, P.; Brown, M.; Wang, S.; Link, D.R.; Olson, J.; Travers, K.; et al. Dynamic regulation of HIV-1 mRNA populations analyzed by single-molecule enrichment and long-read sequencing. Nucleic Acids Res. 2012, 40, 10345–10355.
115. Mahiet, C.; Swanson, C.M. Control of HIV-1 gene expression by SR proteins. Biochem. Soc. Trans. 2016, 44, 1417–1425.
116. Hadzopoulou-Cladaras, M.; Felber, B.K.; Cladaras, C.; Athanassopoulos, A.; Tse, A.; Pavlakis, G.N. The rev (trs/art) protein of human immunodeficiency virus type 1 affects viral mRNA and protein expression via a cis-acting sequence in the env region. J. Virol. 1989, 63, 1265–1274.
117. Felber, B.K.; Hadzopoulou-Cladaras, M.; Cladaras, C.; Copeland, T.; Pavlakis, G.N. Rev protein of human immunodeficiency virus type 1 affects the stability and transport of the viral mRNA. Proc. Natl. Acad. Sci. USA 1989, 86, 1495–1499.

118. Malim, M.H.; Hauber, J.; Le, S.-Y.; Maizel, J.V.; Cullen, B.R. The HIV-1 rev trans-activator acts through a structured target sequence to activate nuclear export of unspliced viral mRNA. Nature 1989, 338, 254–257.
119. Daly, T.J.; Cook, K.S.; Gray, G.S.; Maione, T.E.; Rusche, J.R. Specific binding of HIV-1 recombinant Rev protein to the Rev-responsive element in vitro. Nature 1989, 342, 816–819.
120. Kjems, J.; Brown, M.; Chang, D.D.; Sharp, P.A. Structural analysis of the interaction between the human immunodeficiency virus Rev protein and the Rev response element. Proc. Natl. Acad. Sci. USA 1991, 88, 683–687.
121. Wen, W.; Meinkotht, J.L.; Tsien, R.Y.; Taylor, S.S. Identification of a signal for rapid export of proteins from the nucleus. Cell 1995, 82, 463–473.
122. Fritz, C.C.; Zapp, M.L.; Green, M.R. A human nucleoporin-like protein that specifically interacts with HIV Rev. Nature 1995, 376, 530–533.
123. Dicks, M.D.J.; Betancor, G.; Jimenez-Guardeño, J.M.; Pessel-Vivares, L.; Apolonia, L.; Goujon, C.; Malim, M.H. Multiple components of the nuclear pore complex interact with the amino-terminus of MX2 to facilitate HIV-1 restriction. PLoS Pathog. 2018, 14, e1007408.
124. Fritz, C.C.; Green, M.R. HIV Rev uses a conserved cellular protein export pathway for the nucleocytoplasmic transport of viral RNAs. Curr. Biol. 1996, 6, 848–854.
125. Wolff, B.; Sanglier, J.-J.; Wang, Y. Leptomycin B is an inhibitor of nuclear export: Inhibition of nucleo-cytoplasmic translocation of the human immunodeficiency virus type 1 (HIV-1) Rev protein and Rev-dependent mRNA. Chem. Biol. 1997, 4, 139–147.
126. Nishi, K.; Yoshida, M.; Fujiwara, D.; Nishikawa, M.; Horinouchi, S.; Beppu, T. Leptomycin B targets a regulatory cascade of crm1, a fission yeast nuclear protein, involved in control of higher order chromosome structure and gene expression. J. Biol. Chem. 1994, 269, 6320–6324.
127. Neville, M.; Stutz, F.; Lee, L.; Davis, L.I.; Rosbash, M. The importin-beta family member Crm1p bridges the interaction between Rev and the nuclear pore complex during nuclear export. Curr. Biol. 1997, 7, 767–775.
128. Ho, J.H.-N.; Kallstrom, G.; Johnson, A.W. Nmd3p Is a Crm1p-Dependent Adapter Protein for Nuclear Export of the Large Ribosomal Subunit. J. Cell Biol. 2000, 151, 1057–1066.

129. Sánchez-Velar, N.; Udofia, E.B.; Yu, Z.; Zapp, M.L. hRIP, a cellular cofactor for Rev function, promotes release of HIV RNAs from the perinuclear region. Genes Dev. 2004, 18, 23–34.
130. Edgcomb, S.P.; Carmel, A.B.; Naji, S.; Ambrus-Aikelin, G.; Reyes, J.R.; Saphire, A.C.; Gerace, L.; Williamson, J.R. DDX1 Is an RNA-Dependent ATPase Involved in HIV-1 Rev Function and Virus Replication. J. Mol. Biol. 2012, 415, 61–74.
131. Yedavalli, V.S.; Neuveut, C.; Chi, Y.-H.; Kleiman, L.; Jeang, K.-T. Requirement of DDX3 DEAD Box RNA Helicase for HIV-1 Rev-RRE Export Function. Cell 2004, 119, 381–392.
132. Garbelli, A.; Beermann, S.; Di Cicco, G.; Dietrich, U.; Maga, G. A Motif Unique to the Human Dead-Box Protein DDX3 Is Important for Nucleic Acid Binding, ATP Hydrolysis, RNA/DNA Unwinding and HIV-1 Replication. PLoS ONE 2011, 6, e19810.
133. Askjaer, P.; Jensen, T.H.; Nilsson, J.; Englmeier, L.; Kjems, J. The Specificity of the CRM1-Rev Nuclear Export Signal Interaction Is Mediated by RanGTP. J. Biol. Chem. 1998, 273, 33414–33422.
134. Yue, Y.; Coskun, A.K.; Jawanda, N.; Auer, J.; Sutton, R.E. Differential interaction between human and murine Crm1 and lentiviral Rev proteins. Virology 2018, 513, 1–10.
135. Budhiraja, S.; Liu, H.; Couturier, J.; Malovannaya, A.; Qin, J.; Lewis, D.E.; Rice, A.P. Mining the Human Complexome Database Identifies RBM14 as an XPO1-Associated Protein Involved in HIV-1 Rev Function. J. Virol. 2015, 89, 3557–3567.
136. Liu, H.; Hu, P.-W.; Budhiraja, S.; Misra, A.; Couturier, J.; Lloyd, R.E.; Lewis, D.E.; Kimata, J.T.; Rice, A.P. PACS1 is an HIV-1 cofactor that functions in Rev-mediated nuclear export of viral RNA. Virology 2019, 540, 88–96.
137. Bevec, D.; Jaksche, H.; Oft, M.; Wöhl, T.; Himmelspach, M.; Pacher, A.; Schebesta, M.; Koettnitz, K.; Dobrovnik, M.; Csonga, R.; et al. Inhibition of HIV-1 Replication in Lymphocytes by Mutants of the Rev Cofactor eIF-5A. Science 1996, 271, 1858–1860.
138. Modem, S.; Badri, K.R.; Holland, T.C.; Reddy, T.R. Sam68 is absolutely required for Rev function and HIV-1 production. Nucleic Acids Res. 2005, 33, 873–879.
139. Suhasini, M.; Reddy, T.R. Cellular proteins and HIV-1 Rev function. Curr. HIV Res. 2009, 7, 91–100.

140. Hulver, M.J.; Trautman, J.P.; Goodwin, A.P.; Roszczenko, S.K.; Fogarty, K.H.; Miller, H.B. Human Tat-specific factor 1 binds the HIV-1 genome and selectively transports HIV-1 RNAs. Mol. Biol. Rep. 2020, 47, 1759–1772.

141. Bray, M.; Prasad, S.; Dubay, J.W.; Hunter, E.; Jeang, K.T.; Rekosh, D.; Hammarskjold, M.L. A small element from the Mason-Pfizer monkey virus genome makes human immunodeficiency virus type 1 expression and replication Rev-independent. Proc. Natl. Acad. Sci. USA 1994, 91, 1256–1260.

142. Zolotukhin, A.S.; Valentin, A.; Pavlakis, G.N.; Felber, B.K. Continuous propagation of RRE(-) and Rev(-)RRE(-) human immunodeficiency virus type 1 molecular clones containing a cis-acting element of simian retrovirus type 1 in human peripheral blood lymphocytes. J. Virol. 1994, 68, 7944–7952.

143. Van Regenmortel, M.; Fauquet, C.; Bishop, D.; Carstens, E.; Estes, M.; Lemon, S.; Maniloff, J.; Mayo, M.; McGeoch, D.; Pringle, C.; et al. Virus Taxonomy: Classification and Nomenclature of Viruses. Seventh Report of the International Committee on Taxonomy of Viruses; London Academic Press: London, UK, 2000; ISBN 0-12-370200-3.

144. Tabernero, C.; Zolotukhin, A.S.; Valentin, A.; Pavlakis, G.N.; Felber, B.K. The posttranscriptional control element of the simian retrovirus type 1 forms an extensive RNA secondary structure necessary for its function. J. Virol. 1996, 70, 5998–6011.

145. Ernst, R.K.; Bray, M.; Rekosh, D.; Hammarskjold, M.L. Secondary structure and mutational analysis of the Mason-Pfizer monkey virus RNA constitutive transport element. RNA 1997, 3, 210–222.

146. Ernst, R.K.; Bray, M.; Rekosh, D.; Hammarskjöld, M.L. A structured retroviral RNA element that mediates nucleocytoplasmic export of intron-containing RNA. Mol. Cell. Biol. 1997, 17, 135–144.

147. Saavedra, C.; Felber, B.; Izaurralde, E. The simian retrovirus-1 constitutive transport element, unlike the HIV-1 RRE, uses factors required for cellular mRNA export. Curr. Biol. 1997, 7, 619–628.

148. Pasquinelli, A.E.; Ernst, R.K.; Lund, E.; Grimm, C.; Zapp, M.L.; Rekosh, D.; Hammarskjöld, M.; Dahlberg, J.E. The constitutive transport element (CTE) of Mason-Pfizer monkey virus (MPMV) accesses a cellular mRNA export pathway. EMBO J. 1997, 16, 7500–7510.

149. Grüter, P.; Tabernero, C.; Von Kobbe, C.; Schmitt, C.; Saavedra, C.; Bachi, A.; Wilm, M.; Felber, B.K.; Izaurralde, E. TAP, the Human Homolog of Mex67p, Mediates CTE-Dependent RNA Export from the Nucleus. Mol. Cell 1998, 1, 649–659.

150. Segref, A.; Sharma, K.; Doye, V.; Hellwig, A.; Huber, J.; Lührmann, R.; Hurt, E. Mex67p, a novel factor for nuclear mRNA export, binds to both poly(A)+ RNA and nuclear pores. EMBO J. 1997, 16, 3256–3271.

151. Santos-Rosa, H.; Moreno, H.; Simos, G.; Segref, A.; Fahrenkrog, B.; Panté, N.; Hurt, E. Nuclear mRNA Export Requires Complex Formation between Mex67p and Mtr2p at the Nuclear Pores. Mol. Cell. Biol. 1998, 18, 6826–6838.

152. Bachi, A.; Braun, I.C.; Rodrigues, J.P.; Panté, N.; Ribbeck, K.; Von Kobbe, C.; Kutay, U.; Wilm, M.; Görlich, D.; Carmo-Fonseca, M.; et al. The C-terminal domain of TAP interacts with the nuclear pore complex and promotes export of specific CTE-bearing RNA substrates. RNA 2000, 6, 136–158.

153. Mertz, J.A.; Simper, M.S.; Lozano, M.M.; Payne, S.M.; Dudley, J.P. Mouse Mammary Tumor Virus Encodes a Self-Regulatory RNA Export Protein and Is a Complex Retrovirus. J. Virol. 2005, 79, 14737–14747.

154. Müllner, M.; Salmons, B.; Günzburg, W.H.; Indik, S. Identification of the Rem-responsive element of mouse mammary tumor virus. Nucleic Acids Res. 2008, 36, 6284–6294.

155. Hofacre, A.; Nitta, T.; Fan, H. Jaagsiekte Sheep Retrovirus Encodes a Regulatory Factor, Rej, Required for Synthesis of Gag Protein. J. Virol. 2009, 83, 12483–12498.

156. Yang, J.; Bogerd, H.P.; Peng, S.; Wiegand, H.; Truant, R.; Cullen, B.R. An ancient family of human endogenous retroviruses encodes a functional homolog of the HIV-1 Rev protein. Proc. Natl. Acad. Sci. USA 1999, 96, 13404–13408.

157. Hidaka, M.; Inoue, J.; Yoshida, M.; Seiki, M. Post-transcriptional regulator (rex) of HTLV-1 initiates expression of viral structural proteins but suppresses expression of regulatory proteins. EMBO J. 1988, 7, 519–523.

158. Hanly, S.M.; Rimsky, L.T.; Malim, M.H.; Kim, J.H.; Hauber, J.; Dodon, M.D.; Le, S.Y.; Maizel, J.V.; Cullen, B.R.; Greene, W.C. Comparative analysis of the HTLV-I Rex and HIV-1 Rev trans-regulatory proteins and their RNA response elements. Genes Dev. 1989, 3, 1534–1544.

159. Seiki, M.; Inoue, J.; Hidaka, M.; Yoshida, M. Two cis-acting elements responsible for posttranscriptional trans-regulation of gene expression of human T cell leukemia virus type I. Proc. Natl. Acad. Sci. USA 1988, 85, 7124–7128.
160. Seiki, M.; Hattori, S.; Hirayama, Y.; Yoshida, M. Human adult T-cell leukemia virus: Complete nucleotide sequence of the provirus genome integrated in leukemia cell DNA. Proc. Natl. Acad. Sci. USA 1983, 80, 3618–3622.
161. Ahmed, Y.F.; Gilmartin, G.M.; Hanly, S.M.; Nevins, J.R.; Greene, W.C. The HTLV-I rex response element mediates a novel form of mRNA polyadenylation. Cell 1991, 64, 727–737.
162. Sinha-Datta, U.; Datta, A.; Ghorbel, S.; Dodon, M.D.; Nicot, C. Human T-cell Lymphotrophic Virus Type I Rex and p30 Interactions Govern the Switch between Virus Latency and Replication. J. Biol. Chem. 2007, 282, 14608–14615.
163. Moles, R.; Sarkis, S.; Galli, V.; Omsland, M.; Purcell, D.F.J.; Yurick, D.; Khoury, G.; Pise-Masison, C.A.; Franchini, G. p30 protein: A critical regulator of HTLV-1 viral latency and host immunity. Retrovirology 2019, 16, 1–15.
164. Martin, G.S. Rous Sarcoma Virus: A Function required for the Maintenance of the Transformed State. Nature 1970, 227, 1021–1023.
165. Bernstein, A.; MacCormick, R.; Martin, G. Transformation-defective mutants of avian sarcoma viruses: The genetic relationship between conditional and nonconditional mutants. Virology 1976, 70, 206–209.
166. Paca, R.E.; Ogert, R.A.; Hibbert, C.S.; Izaurralde, E.; Beemon, K.L. Rous Sarcoma Virus DR Posttranscriptional Elements Use a Novel RNA Export Pathway. J. Virol. 2000, 74, 9507–9514.
167. Leblanc, J.J.; Uddowla, S.; Abraham, B.; Clatterbuck, S.; Beemon, K.L. Tap and Dbp5, but not Gag, are involved in DR-mediated nuclear export of unspliced Rous sarcoma virus RNA. Virology 2007, 363, 376–386.
168. Scheifele, L.Z.; Garbitt, R.A.; Rhoads, J.D.; Parent, L.J. Nuclear entry and CRM1-dependent nuclear export of the Rous sarcoma virus Gag polyprotein. Proc. Natl. Acad. Sci. USA 2002, 99, 3944–3949.
169. Maldonado, R.J.K.; Rice, B.; Chen, E.C.; Tuffy, K.M.; Chiari, E.F.; Fahrbach, K.M.; Hope, T.J.; Parent, L.J. Visualizing Association of the Retroviral Gag Protein with Unspliced Viral RNA in the Nucleus. Mbio 2020, 11, e00524-20.

170. Yu, K.L.; Lee, S.H.; Lee, E.S.; You, J.C. HIV-1 nucleocapsid protein localizes efficiently to the nucleus and nucleolus. Virology 2016, 492, 204–212.
171. Weldon, R.A.; Sarkar, P.; Brown, S.M.; Weldon, S.K. Mason–Pfizer monkey virus Gag proteins interact with the human sumo conjugating enzyme, hUbc. Virology 2003, 314, 62–73.
172. Maldonado, R.K.; Parent, L.J. Orchestrating the Selection and Packaging of Genomic RNA by Retroviruses: An Ensemble of Viral and Host Factors. Viruses 2016, 8, 257.
173. Pessel-Vivares, L.; Houzet, L.; Lainé, S.; Mougel, M. Insights into the nuclear export of murine leukemia virus intron-containing RNA. RNA Biol. 2015, 12, 942–949.
174. Pessel-Vivares, L.; Ferrer, M.; Laine, S.; Mougel, M. MLV requires Tap/NXF1-dependent pathway to export its unspliced RNA to the cytoplasm and to express both spliced and unspliced RNAs. Retrovirology 2014, 11, 21.
175. Sakuma, T.; Tonne, J.M.; Ikeda, Y. Murine Leukemia Virus Uses TREX Components for Efficient Nuclear Export of Unspliced Viral Transcripts. Viruses 2014, 6, 1135–1148.
176. Bartels, H.; Luban, J. Gammaretroviral pol sequences act in cis to direct polysome loading and NXF1/NXT-dependent protein production by gag-encoded RNA. Retrovirology 2014, 11, 73.
177. Mougel, M.; Akkawi, C.; Chamontin, C.; Feuillard, J.; Pessel-Vivares, L.; Socol, M.; Laine, S. NXF1 and CRM1 nuclear export pathways orchestrate nuclear export, translation and packaging of murine leukaemia retrovirus unspliced RNA. RNA Biol. 2020, 17, 528–538.
178. Yu, S.F.; Baldwin, D.N.; Gwynn, S.R.; Yendapalli, S.; Linial, M.L. Human Foamy Virus Replication: A Pathway Distinct from That of Retroviruses and Hepadnaviruses. Science 1996, 271, 1579–1582.
179. Löchelt, M.; Muranyi, W.; Flügel, R.M. Human foamy virus genome possesses an internal, Bel-1-dependent and functional promoter. Proc. Natl. Acad. Sci. USA 1993, 90, 7317–7321.
180. Moebes, A.; Enssle, J.; Bieniasz, P.D.; Heinkelein, M.; Lindemann, D.; Bock, M.; McClure, M.O.; Rethwilm, A. Human foamy virus reverse transcription that occurs late in the viral replication cycle. J. Virol. 1997, 71, 7305–7311.

181. Yu, S.F.; Sullivan, M.D.; Linial, M.L. Evidence that the Human Foamy Virus Genome Is DNA. J. Virol. 1999, 73, 1565–1572.

182. Goepfert, P.A.; Shaw, K.L.; Ritter, G.D.; Mulligan, M.J. A sorting motif localizes the foamy virus glycoprotein to the endoplasmic reticulum. J. Virol. 1997, 71, 778–784.

183. Bodem, J.; Schied, T.; Gabriel, R.; Rammling, M.; Rethwilm, A. Foamy Virus Nuclear RNA Export Is Distinct from That of Other Retroviruses. J. Virol. 2010, 85, 2333–2341.

184. Rethwilm, A.; Bodem, J. Evolution of Foamy Viruses: The Most Ancient of All Retroviruses. Viruses 2013, 5, 2349–2374.

185. Schliephake, A.W.; Rethwilm, A. Nuclear localization of foamy virus Gag precursor protein. J. Virol. 1994, 68, 4946–4954.

186. Renault, N.; Tobaly-Tapiero, J.; Paris, J.; Giron, M.-L.; Coiffic, A.; Roingeard, P.; Saïb, A. A nuclear export signal within the structural Gag protein is required for prototype foamy virus replication. Retrovirology 2011, 8, 6.

187. Summers, J.; Mason, W.S. Replication of the genome of a hepatitis B-like virus by reverse transcription of an RNA intermediate. Cell 1982, 29, 403–415.

188. Chen, Y.; Michitaka, K.; Matsubara, H.; Yamamoto, K.; Horiike, N.; Onji, M. Complete genome sequence of hepatitis B virus (HBV) from a patient with fulminant hepatitis without precore and core promoter mutations: Comparison with HBV from a patient with acute hepatitis infected from the same infectious source. J. Hepatol. 2003, 38, 84–90.

189. Kitamura, K.; Que, L.; Shimadu, M.; Koura, M.; Ishihara, Y.; Wakae, K.; Nakamura, T.; Watashi, K.; Wakita, T.; Muramatsu, M. Flap endonuclease 1 is involved in cccDNA formation in the hepatitis B virus. PLoS Pathog. 2018, 14, e1007124.

190. Werle–Lapostolle, B.; Bowden, S.; Locarnini, S.; Wursthorn, K.; Petersen, J.; Lau, G.; Trepo, C.; Marcellin, P.; Goodman, Z.; Delaney, W.E.; et al. Persistence of cccDNA during the natural history of chronic hepatitis B and decline during adefovir dipivoxil therapy. Gastroenterology 2004, 126, 1750–1758.

191. Quasdorff, M.; Protzer, U. Control of hepatitis B virus at the level of transcription. J. Viral Hepat. 2010, 17, 527–536.

192. Takahashi, K.; Aoyama, K.; Ohno, N.; Iwata, K.; Akahane, Y.; Baba, K.; Yoshizawa, H.; Mishiro, S. The precore/core promoter mutant

(T1762A1764) of hepatitis B virus: Clinical significance and an easy method for detection. J. Gen. Virol. 1995, 76, 3159–3164.

193. Cattaneo, R.; Will, H.; Schaller, H. Hepatitis B virus transcription in the infected liver. EMBO J. 1984, 3, 2191–2196.
194. Lucifora, J.; Arzberger, S.; Durantel, D.; Belloni, L.; Strubin, M.; Levrero, M.; Zoulim, F.; Hantz, O.; Protzer, U. Hepatitis B virus X protein is essential to initiate and maintain virus replication after infection. J. Hepatol. 2011, 55, 996–1003.
195. Soussan, P.; Tuveri, R.; Nalpas, B.; Garreau, F.; Zavala, F.; Masson, A.; Pol, S.; Brechot, C.; Kremsdorf, D. The expression of hepatitis B spliced protein (HBSP) encoded by a spliced hepatitis B virus RNA is associated with viral replication and liver fibrosis. J. Hepatol. 2003, 38, 343–348.
196. Chen, J.-Y.; Chen, W.-N.; Liu, L.-L.; Lin, W.-S.; Jiao, B.-Y.; Wu, Y.-L.; Lin, J.-Y.; Lin, X. Hepatitis B spliced protein (HBSP) generated by a spliced hepatitis B virus RNA participates in abnormality of fibrin formation and functions by binding to fibrinogen γ chain. J. Med. Virol. 2010, 82, 2019–2026.
197. Wang, J.C.-Y.; Nickens, D.G.; Lentz, T.B.; Loeb, D.D.; Zlotnick, A. Encapsidated hepatitis B virus reverse transcriptase is poised on an ordered RNA lattice. Proc. Natl. Acad. Sci. USA 2014, 111, 11329–11334.
198. Huang, J.; Liang, T.J. A novel hepatitis B virus (HBV) genetic element with Rev response element-like properties that is essential for expression of HBV gene products. Mol. Cell. Biol. 1993, 13, 7476–7486.
199. Huang, Z.M.; Yen, T.S. Role of the hepatitis B virus posttranscriptional regulatory element in export of intronless transcripts. Mol. Cell. Biol. 1995, 15, 3864–3869.
200. Zang, W.-Q.; Yen, T.B. Distinct Export Pathway Utilized by the Hepatitis B Virus Posttranscriptional Regulatory Element. Virology 1999, 259, 299–304.
201. Smith, G.J.; Donello, J.E.; Hope, T.J.; Lück, R.; Steger, G. The hepatitis B virus post-transcriptional regulatory element contains two conserved RNA stem-loops which are required for function. Nucleic Acids Res. 1998, 26, 4818–4827.

202. Zang, W.-Q.; Li, B.; Huang, P.-Y.; Lai, M.M.C.; Yen, T.B. Role of Polypyrimidine Tract Binding Protein in the Function of the Hepatitis B Virus Posttranscriptional Regulatory Element. J. Virol. 2001, 75, 10779–10786.

203. Li, Y.; Huang, T.; Zhang, X.; Wan, T.; Hu, J.; Huang, A.; Tang, H. Role of glyceraldehyde-3-phosphate dehydrogenase binding to hepatitis B virus posttranscriptional regulatory element in regulating expression of HBV surface antigen. Arch. Virol. 2009, 154, 519–524.

204. Horke, S.; Reumann, K.; Rang, A.; Heise, T. Molecular Characterization of the Human La Protein·Hepatitis B Virus RNA.B Interactionin Vitro. J. Biol. Chem. 2002, 277, 34949–34958.

205. Kamath, R.V.; Leary, D.J.; Huang, S. Nucleocytoplasmic Shuttling of Polypyrimidine Tract-binding Protein Is Uncoupled from RNA Export. Mol. Biol. Cell 2001, 12, 3808–3820.

206. Chi, B.; Wang, K.; Du, Y.; Gui, B.; Chang, X.; Wang, L.; Fan, J.; Chen, S.; Wu, X.; Li, G.; et al. A Sub-Element in PRE enhances nuclear export of intronless mRNAs by recruiting the TREX complex via ZC3H18. Nucleic Acids Res. 2014, 42, 7305–7318.

207. Li, H.-C.; Huang, E.-Y.; Su, P.-Y.; Wu, S.-Y.; Yang, C.-C.; Lin, Y.-S.; Chang, W.-C.; Shih, C. Nuclear Export and Import of Human Hepatitis B Virus Capsid Protein and Particles. PLoS Pathog. 2010, 6, e1001162.

208. Porterfield, J.Z.; Dhason, M.S.; Loeb, D.D.; Nassal, M.; Stray, S.J.; Zlotnick, A. Full-Length Hepatitis B Virus Core Protein Packages Viral and Heterologous RNA with Similarly High Levels of Cooperativity. J. Virol. 2010, 84, 7174–7184.

209. Yang, C.-C.; Huang, E.-Y.; Li, H.-C.; Su, P.-Y.; Shih, C. Nuclear Export of Human Hepatitis B Virus Core Protein and Pregenomic RNA Depends on the Cellular NXF1-p15 Machinery. PLoS ONE 2014, 9, e106683.

210. Makokha, G.N.; Abe-Chayama, H.; Chowdhury, S.; Hayes, C.N.; Tsuge, M.; Yoshima, T.; Ishida, Y.; Zhang, Y.; Uchida, T.; Tateno, C.; et al. Regulation of the Hepatitis B virus replication and gene expression by the multi-functional protein TARDBP. Sci. Rep. 2019, 9, 1–18.

211. Hu, B.; Yu, L.; Zhu, N.-S.; Xie, J. Cellular UAP56 interacts with the HBx protein of the hepatitis B virus and is involved in viral RNA nuclear export in hepatocytes. Exp. Cell Res. 2020, 390, 111929.

212. Zoulim, F. Hepatitis B virus resistance to antiviral drugs: Where are we going? Liver Int. 2011, 31, 111–116.
213. Gale, M.; Tan, S.-L.; Katze, M.G. Translational Control of Viral Gene Expression in Eukaryotes. Microbiol. Mol. Biol. Rev. 2000, 64, 239–280.
214. Rowe, W.P.; Huebner, R.J.; Gilmore, L.K.; Parrott, R.H.; Ward, T.G. Isolation of a Cytopathogenic Agent from Human Adenoids Undergoing Spontaneous Degeneration in Tissue Culture. Proc. Soc. Exp. Biol. Med. 1953, 84, 570–573.
215. Lion, T. Adenovirus Infections in Immunocompetent and Immunocompromised Patients. Clin. Microbiol. Rev. 2014, 27, 441–462.
216. Sprengel, J.; Schmitz, B.; Heuss-Neitzel, D.; Zock, C.; Doerfler, W. Nucleotide sequence of human adenovirus type 12 DNA: Comparative functional analysis. J. Virol. 1994, 68, 379–389.
217. Berk, A.J. Adenoviridae. In Fields Virology; Knipe, D.M., Howley, P.M., Eds.; Wolters Kluwer Health/Lippincott Williams & Wilkins: Philadelphia, PA, USA, 2013; ISBN 978-1-4511-0563-6.
218. Crisostomo, L.; Soriano, A.M.; Mendez, M.; Graves, D.; Pelka, P. Temporal dynamics of adenovirus 5 gene expression in normal human cells. PLoS ONE 2019, 14, e0211192.
219. Fessler, S.P.; Young, C.S.H. Control of Adenovirus Early Gene Expression during the Late Phase of Infection. J. Virol. 1998, 72, 4049–4056.
220. Biasiotto, R.; Akusjärvi, G. Regulation of Human Adenovirus Alternative RNA Splicing by the Adenoviral L4-33K and L4-22K Proteins. Int. J. Mol. Sci. 2015, 16, 2893–2912.
221. Pilder, S.; Moore, M.; Logan, J.; Shenk, T. The adenovirus E1B-55K transforming polypeptide modulates transport or cytoplasmic stabilization of viral and host cell mRNAs. Mol. Cell. Biol. 1986, 6, 470–476.
222. Leppard, K.N.; Shenk, T. The adenovirus E1B 55 kd protein influences mRNA transport via an intranuclear effect on RNA metabolism. EMBO J. 1989, 8, 2329–2336.
223. Leppard, K.N.; Everett, R.D. The adenovirus type 5 E1b 55K and E4 Orf3 proteins associate in infected cells and affect ND10 components. J. Gen. Virol. 1999, 80, 997–1008.

224. Ornelles, D.A.; Shenk, T. Localization of the adenovirus early region 1B 55-kilodalton protein during lytic infection: Association with nuclear viral inclusions requires the early region 4 34-kilodalton protein. J. Virol. 1991, 65, 424–429.

225. Beltz, G.A.; Flint, S.J. Inhibition of HeLa cell protein synthesis during adenovirus infection: Restriction of cellular messenger RNA sequences to the nucleus. J. Mol. Biol. 1979, 131, 353–373.

226. Blanchette, P.; Kindsmüller, K.; Groitl, P.; Dallaire, F.; Speiseder, T.; Branton, P.E.; Dobner, T. Control of mRNA Export by Adenovirus E4orf6 and E1B55K Proteins during Productive Infection Requires E4orf6 Ubiquitin Ligase Activity. J. Virol. 2008, 82, 2642–2651.

227. Woo, J.L.; Berk, A.J. Adenovirus Ubiquitin-Protein Ligase Stimulates Viral Late mRNA NuclearExport. J. Virol. 2007, 81, 575–587.

228. Hidalgo, P.; Ip, W.H.; Dobner, T.; Gonzalez, R.A. The biology of the adenovirus E1B 55K protein. FEBS Lett. 2019, 593, 3504–3517.

229. Gabler, S.; Schütt, H.; Groitl, P.; Wolf, H.; Shenk, T.; Dobner, T. E1B 55-Kilodalton-Associated Protein: A Cellular Protein with RNA-Binding Activity Implicated in Nucleocytoplasmic Transport of Adenovirus and Cellular mRNAs. J. Virol. 1998, 72, 7960–7971.

230. Yatherajam, G.; Huang, W.; Flint, S.J. Export of Adenoviral Late mRNA from the Nucleus Requires the Nxf1/Tap Export Receptor. J. Virol. 2011, 85, 1429–1438.

231. Muik, A.; Stubbert, L.J.; Jahedi, R.Z.; Geiβ, Y.; Kimpel, J.; Dold, C.; Tober, R.; Volk, A.; Klein, S.; Dietrich, U.; et al. Re-engineering Vesicular Stomatitis Virus to Abrogate Neurotoxicity, Circumvent Humoral Immunity, and Enhance Oncolytic Potency. Cancer Res. 2014, 74, 3567–3578.

232. Heo, J.; Reid, T.; Ruo, L.; Breitbach, C.J.; Rose, S.; Bloomston, M.; Cho, M.; Lim, H.Y.; Chung, H.C.; Kim, C.W.; et al. Randomized dose-finding clinical trial of oncolytic immunotherapeutic vaccinia JX-594 in liver cancer. Nat. Med. 2013, 19, 329–336.

233. Sauthoff, H.; Pipiya, T.; Heitner, S.; Chen, S.; Bleck, B.; Reibman, J.; Chang, W.; Norman, R.G.; Rom, W.N.; Hay, J.G. Impact of E1a Modifications on Tumor-Selective Adenoviral Replication and Toxicity. Mol. Ther. 2004, 10, 749–757.

234. Zheng, M.; Huang, J.; Tong, A.; Yang, H. Oncolytic Viruses for Cancer Therapy: Barriers and Recent Advances. Mol. Ther. Oncolytics 2019, 15, 234–247.

235. O'Shea, C.C.; Johnson, L.; Bagus, B.; Choi, S.; Nicholas, C.; Shen, A.; Boyle, L.; Pandey, K.; Soria, C.; Kunich, J.; et al. Late viral RNA export, rather than p53 inactivation, determines ONYX-015 tumor selectivity. Cancer Cell 2004, 6, 611–623.
236. Khuri, F.R.; Nemunaitis, J.; Ganly, I.; Arseneau, J.; Tannock, I.F.; Romel, L.; Gore, M.; Ironside, J.; MacDougall, R.; Heise, C.; et al. A controlled trial of intratumoral ONYX-015, a selectively-replicating adenovirus, in combination with cisplatin and 5-fluorouracil in patients with recurrent head and neck cancer. Nat. Med. 2000, 6, 879–885.
237. Liu, F.; Zhou, Z.H. Comparative virion structures of human herpesviruses. In Human Herpesviruses; Cambridge University Press: Cambridge, UK, 2010; pp. 27–43.
238. Singh, N.; Tscharke, D.C. Herpes Simplex Virus Latency Is Noisier the Closer We Look. J. Virol. 2020, 94, e01701-19.
239. McGeoch, D.J.; Dolan, A.; Ralph, A.C. Toward a Comprehensive Phylogeny for Mammalian and Avian Herpesviruses. J. Virol. 2000, 74, 10401–10406.
240. Wald, A.; Corey, L. HSV: Persistence in the population: Epidemiology, transmission. In Human Herpesviruses; Cambridge University Press: Cambridge, UK, 2010; pp. 656–672.
241. Tang, S.; Patel, A.; Krause, P.R. Hidden regulation of herpes simplex virus 1 pre-mRNA splicing and polyadenylation by virally encoded immediate early gene ICP27. PLoS Pathog. 2019, 15, e1007884.
242. Mears, W.E.; Lam, V.; Rice, S.A. Identification of nuclear and nucleolar localization signals in the herpes simplex virus regulatory protein ICP27. J. Virol. 1995, 69, 935–947.
243. Mears, W.E.; Rice, S.A. The RGG box motif of the herpes simplex virus ICP27 protein mediates an RNA-binding activity and determines in vivo methylation. J. Virol. 1996, 70, 7445–7453.
244. Mears, W.E.; Rice, S.A. The Herpes Simplex Virus Immediate-Early Protein ICP27 Shuttles between Nucleus and Cytoplasm. Virology 1998, 242, 128–137.
245. Phelan, A.; Clements, J.B. Herpes simplex virus type 1 immediate early protein IE63 shuttles between nuclear compartments and the cytoplasm. J. Gen. Virol. 1997, 78, 3327–3331.

246. Sandri-Goldin, R.M. ICP27 mediates HSV RNA export by shuttling through a leucine-rich nuclear export signal and binding viral intronless RNAs through an RGG motif. Genes Dev. 1998, 12, 868–879.

247. Koffa, M.D.; Clements, J.B.; Izaurralde, E.; Wadd, S.; Wilson, S.A.; Mattaj, I.W.; Kuersten, S. Herpes simplex virus ICP27 protein provides viral mRNAs with access to the cellular mRNA export pathway. EMBO J. 2001, 20, 5769–5778.

248. Chen, I.-H.B.; Sciabica, K.S.; Sandri-Goldin, R.M. ICP27 Interacts with the RNA Export Factor Aly/REF To Direct Herpes Simplex Virus Type 1 Intronless mRNAs to the TAP Export Pathway. J. Virol. 2002, 76, 12877–12889.

249. Hernandez, F.P.; Sandri-Goldin, R.M. Head-to-Tail Intramolecular Interaction of Herpes Simplex Virus Type 1 Regulatory Protein ICP27 Is Important for Its Interaction with Cellular mRNA Export Receptor TAP/NXF1. Mbio 2010, 1, e00268-10.

250. Johnson, L.A.; Sandri-Goldin, R.M. Efficient Nuclear Export of Herpes Simplex Virus 1 Transcripts Requires both RNA Binding by ICP27 and ICP27 Interaction with TAP/NXF1. J. Virol. 2009, 83, 1184–1192.

251. Tunnicliffe, R.B.; Hautbergue, G.M.; Kalra, P.; Jackson, B.R.; Whitehouse, A.; Wilson, S.A.; Golovanov, A.P. Structural Basis for the Recognition of Cellular mRNA Export Factor REF by Herpes Viral Proteins HSV-1 ICP27 and HVS ORF. PLoS Pathog. 2011, 7, e1001244.

252. Tian, X.; Devi-Rao, G.; Golovanov, A.P.; Sandri-Goldin, R.M. The Interaction of the Cellular Export Adaptor Protein Aly/REF with ICP27 Contributes to the Efficiency of Herpes Simplex Virus 1 mRNA Export. J. Virol. 2013, 87, 7210–7217.

253. Tunnicliffe, R.B.; Hautbergue, G.M.; Wilson, S.A.; Kalra, P.; Golovanov, A.P. Competitive and Cooperative Interactions Mediate RNA Transfer from Herpesvirus Saimiri ORF57 to the Mammalian Export Adaptor ALYREF. PLoS Pathog. 2014, 10, e1003907.

254. Malik, P.; Tabarraei, A.; Kehlenbach, R.H.; Korfali, N.; Iwasawa, R.; Graham, S.V.; Schirmer, E.C. Herpes Simplex Virus ICP27 Protein Directly Interacts with the Nuclear Pore Complex through Nup62, Inhibiting Host Nucleocytoplasmic Transport Pathways. J. Biol. Chem. 2012, 287, 12277–12292.

255. Tunnicliffe, R.B.; Tian, X.; Storer, J.; Sandri-Goldin, R.M.; Golovanov, A.P. Overlapping motifs on the herpes viral proteins ICP27 and ORF57 mediate interactions with the mRNA export adaptors ALYREF and UIF. Sci. Rep. 2018, 8, 15005.
256. Davison, A.J.; Scott, J.E. The Complete DNA Sequence of Varicella-Zoster Virus. J. Gen. Virol. 1986, 67, 1759–1816.
257. Winkler, M.; Rice, S.A.; Stamminger, T. UL69 of human cytomegalovirus, an open reading frame with homology to ICP27 of herpes simplex virus, encodes a transactivator of gene expression. J. Virol. 1994, 68, 3943–3954.
258. Ote, I.; Piette, J.; Sadzot-Delvaux, C. The Varicella-Zoster virus IE4 protein: A conserved member of the herpesviral mRNA export factors family and a potential alternative target in antiherpetic therapies. Biochem. Pharmacol. 2010, 80, 1973–1980.
259. Gruffat, H.; Batisse, J.; Pich, D.; Neuhierl, B.; Manet, E.; Hammerschmidt, W.; Sergeant, A. Epstein-Barr Virus mRNA Export Factor EB2 Is Essential for Production of Infectious Virus. J. Virol. 2002, 76, 9635–9644.
260. Koonin, E.V.; Dolja, V.V.; Krupovic, M.; Varsani, A.; Wolf, Y.I.; Yutin, N.; Zerbini, M.; Kuhn, J.H. Create a Megataxonomic Framework, Filling All Principal Taxonomic Ranks, for ssDNA Viruses; ICTV TaxoProp: Berlin, Germany, 2019.
261. De Martel, C.; Georges, D.; Bray, F.; Ferlay, J.; Clifford, G.M. Global burden of cancer attributable to infections in 2018: A worldwide incidence analysis. Lancet Glob. Health 2020, 8, e180–e190.
262. Baker, C.; Calef, C. Maps of Papillomavirus mRNA Transcripts. In Human Papillomaviruses 1996: A Compilation and Analysis of Nucleic Acid and Amino Acid Sequences; Theoretical Biology and Biophysics: Los Alamos, NM, USA, 1996; pp. 1–20.
263. Van Der Meijden, E.; Kazem, S.; Dargel, C.A.; Van Vuren, N.; Hensbergen, P.J.; Feltkamp, M.C. Characterization of T Antigens, Including Middle T and Alternative T, Expressed by the Human Polyomavirus Associated with Trichodysplasia Spinulosa. J. Virol. 2015, 89, 9427–9439.
264. Abend, J.R.; Joseph, A.E.; Das, D.; Campbell-Cecen, D.B.; Imperiale, M.J. A truncated T antigen expressed from an alternatively spliced BK virus early mRNA. J. Gen. Virol. 2009, 90, 1238–1245.

265. Trowbridge, P.W.; Frisque, R.J. Identification of three new JC virus proteins generated by alternative splicing of the early viral mRNA. J. Neurovirol. 1995, 1, 195–206.
266. Graham, S.V.; Faizo, A.A.A. Control of human papillomavirus gene expression by alternative splicing. Virus Res. 2017, 231, 83–95.
267. Huang, Y.; Carmichael, G.G. RNA processing in the polyoma virus life cycle. Front. Biosci. 2009, 14, 4968–4977.
268. Gardner, S.D.; Field, A.M.; Coleman, D.V.; Hulme, B. New human papovavirus (B.K.) isolated from urine after renal transplantation. Lancet 1971, 297, 1253–1257.
269. Padgett, B.L.; Walker, D.L.; ZuRhein, G.M.; Eckroade, R.J.; Dessel, B.H. Cultivation of papova-like virus from human brain with progressive multifocal leucoencephalopathy. Lancet 1971, 1, 1257–1260.
270. Miller, M.A.; Weibel, C.; Ferguson, D.; Landry, M.L.; Kahn, J.S. WU Polyomavirus in Patients Infected with HIV or Hepatitis C Virus, Connecticut, USA, 2007. Emerg. Infect. Dis. 2009, 15, 1095–1097.
271. Cedeno-Laurent, F.; Penalva de Oliveira, A.C.; Vidal, J.E.; Trujillo, J.R. Human Polyomavirus-Associated Cerebral Disorders in the Post-HAART Era. Available online: https://www.hindawi.com/journals/pri/2011/562427/ (accessed on 19 August 2020).
272. Di Taranto, C.; Pietropaolo, V.; Orsi, G.; Jin, L.; Sinibaldi, L.; Degener, A.; Lin, L. Detection of BK polyomavirus genotypes in healthy and HIV-positive children. Eur. J. Epidemiol. 1997, 13, 653–657.
273. Maglennon, G.A.; McIntosh, P.B.; Doorbar, J. Immunosuppression Facilitates the Reactivation of Latent Papillomavirus Infections. J. Virol. 2014, 88, 710–716.
274. Cutrone, R.; Lednicky, J.; Dunn, G.; Rizzo, P.; Bocchetta, M.; Chumakov, K.; Minor, P.; Carbone, M. Some Oral Poliovirus Vaccines Were Contaminated with Infectious SV40 after 1961. Cancer Res. 2005, 65, 10273–10279.
275. Testa, J.R.; Carbone, M.; Hirvonen, A.; Khalili, K.; Krynska, B.; Linnainmaa, K.; Pooley, F.D.; Rizzo, P.; Rusch, V.; Xiao, G.H. A multi-institutional study confirms the presence and expression of simian virus 40 in human malignant mesotheliomas. Cancer Res. 1998, 58, 4505–4509.

276. Lopez-Rios, F.; Illei, P.B.; Rusch, V.W.; Ladanyi, M. Evidence against a role for SV40 infection in human mesotheliomas and high risk of false-positive PCR results owing to presence of SV40 sequences in common laboratory plasmids. Lancet 2004, 364, 1157–1166.
277. Strickler, H.D. A multicenter evaluation of assays for detection of SV40 DNA and results in masked mesothelioma specimens. Cancer Epidemiol. Biomarker. Prev. 2001, 10, 523–532.
278. Engels, E.A.; Sarkar, C.; Daniel, R.W.; Gravitt, P.E.; Verma, K.; Quezado, M.; Shah, K.V. Absence of simian virus 40 in human brain tumors from northern India. Int. J. Cancer 2002, 101, 348–352.
279. De Rienzo, A.; Tor, M.; Sterman, D.H.; Aksoy, F.; Albelda, S.M.; Testa, J.R. Detection of SV40 DNA sequences in malignant mesothelioma specimens from the United States, but not from Turkey. J. Cell. Biochem. 2002, 84, 455–459.
280. Rotondo, J.C.; Mazzoni, E.; Bononi, I.; Tognon, M.; Martini, F. Association between Simian Virus 40 and Human Tumors. Front. Oncol. 2019, 9, 670.
281. Ryu, W.-S.; Mertz, J.E. Simian virus 40 late transcripts lacking excisable intervening sequences are defective in both stability in the nucleus and transport to the cytoplasm. J. Virol. 1989, 63, 4386–4394.
282. Barrett, N.L.; Carmichael, G.G.; Luo, Y. Splice site requirement for the efficient accumulation of polyoma virus late mRNAs. Nucleic Acids Res. 1991, 19, 3011–3017.
283. Huang, Y.; Carmichael, G.G. A suboptimal 5′ splice site is a cis-acting determinant of nuclear export of polyomavirus late mRNAs. Mol. Cell. Biol. 1996, 16, 6046–6054.
284. Centers for Disease Control and Prevention. Genital HPV Infection—CDC Fact Sheet; Centers for Disease Control and Prevention: Atlanta, GA, USA, 2017.
285. Ozbun, M.A.; Meyers, C. Temporal Usage of Multiple Promoters during the Life Cycle of Human Papillomavirus Type 31b. J. Virol. 1998, 72, 2715–2722.
286. Geisen, C.; Kahn, T. Promoter Activity of Sequences Located Upstream of the Human Papillomavirus Types 16 and 18 Late Regions. J. Gen. Virol. 1996, 77, 2193–2200.
287. Braunstein, T.H.; Madsen, B.S.; Gavnholt, B.; Rosenstierne, M.W.; Johnsen, C.K.; Norrild, B. Identification of a new promoter in the early

region of the human papillomavirus type 16 genome. J. Gen. Virol. 1999, 80, 3241–3250.

288. Pycon, D.; Pearce, S.M.; Lank, S.M.; Ahlquist, P.; Lambert, P.F. Establishment of Human Papillomavirus Infection Requires Cell Cycle Progression. PLoS Pathog. 2009, 5, e1000318.

289. Shafti-Keramat, S.; Handisurya, A.; Kriehuber, E.; Meneguzzi, G.; Slupetzky, K.; Kirnbauer, R. Different Heparan Sulfate Proteoglycans Serve asCellular Receptors for HumanPapillomaviruses. J. Virol. 2003, 77, 13125–13135.

290. Ozbun, M.A. Human Papillomavirus Type 31b Infection of Human Keratinocytes and the Onset of Early Transcription. J. Virol. 2002, 76, 11291–11300.

291. Nakahara, T.; Nishimura, A.; Tanaka, M.; Ueno, T.; Ishimoto, A.; Sakai, H. Modulation of the Cell Division Cycle by Human Papillomavirus Type 18 E4. J. Virol. 2002, 76, 10914–10920.

292. Hansen, C.N.; Nielsen, L.; Norrild, B. Activities of E7 promoters in the human papillomavirus type 16 genome during cell differentiation. Virus Res. 2010, 150, 34–42.

293. Ozbun, M.A.; Meyers, C. Two Novel Promoters in the Upstream Regulatory Region of Human Papillomavirus Type 31b Are Negatively Regulated by Epithelial Differentiation. J. Virol. 1999, 73, 3505–3510.

294. Hummel, M.; Hudson, J.B.; Laimins, L.A. Differentiation-induced and constitutive transcription of human papillomavirus type 31b in cell lines containing viral episomes. J. Virol. 1992, 66, 6070–6080.

295. Wilson, R.; Ryan, G.B.; Knight, G.L.; Laimins, L.A.; Roberts, S. The full-length E1∧E4 protein of human papillomavirus type 18 modulates differentiation-dependent viral DNA amplification and late gene expression. Virology 2007, 362, 453–460.

296. Genther, S.M.; Sterling, S.; Duensing, S.; Munger, K.; Sattler, C.; Lambert, P.F. Quantitative Role of the Human Papillomavirus Type 16 E5 Gene during the Productive Stage of the Viral Life Cycle. J. Virol. 2003, 77, 2832–2842.

297. Bryan, J.T.; Brown, D.R. Association of the Human Papillomavirus Type 11 E1∧E4 Protein with Cornified Cell Envelopes Derived from Infected Genital Epithelium. Virology 2000, 277, 262–269.

298. Wang, X.; Meyers, C.; Wang, H.-K.; Chow, L.T.; Zheng, Z.-M. Construction of a Full Transcription Map of Human Papillomavirus

Type 18 during Productive Viral Infection. J. Virol. 2011, 85, 8080–8092.

299. Graham, S.V. Keratinocyte Differentiation-Dependent Human Papillomavirus Gene Regulation. Viruses 2017, 9, 245.
300. Nilsson, K.; Wu, C.; Schwartz, S. Role of the DNA Damage Response in Human Papillomavirus RNA Splicing and Polyadenylation. Int. J. Mol. Sci. 2018, 19, 1735.
301. Crum, C.P.; Nuovo, G.; Friedman, D.; Silverstein, S.J. Accumulation of RNA homologous to human papillomavirus type 16 open reading frames in genital precancers. J. Virol. 1988, 62, 84–90.
302. Stoler, M.H.; Wolinsky, S.; Whitbeck, A.; Broker, T.R.; Chow, L.T. Differentiation-linked human papillomavirus types 6 and 11 transcription in genital condylomata revealed by in situ hybridization with message-specific RNA probes. Virology 1989, 172, 331–340.
303. Beyer-Finkler, E.; Stoler, M.; Girardi, F.; Pfister, H. Cell differentiation-related gene expression of human papillomavirus. Med. Microbiol. Immunol. 1990, 179, 185–192.
304. Stoler, M.H.; Rhodes, C.R.; Whitbeck, A.; Wolinsky, S.; Chow, L.T.; Broker, T.R. Human papillomavirus type 16 and 18 gene expression in cervical neoplasias. Hum. Pathol. 1992, 23, 117–128.
305. Kennedy, I.M.; Haddow, J.K.; Clements, J.B. Analysis of human papillomavirus type 16 late mRNA 3′ processing signals in vitro and in vivo. J. Virol. 1990, 64, 1825–1829.
306. Kennedy, I.M.; Haddow, J.K.; Clements, J.B. A negative regulatory element in the human papillomavirus type 16 genome acts at the level of late mRNA stability. J. Virol. 1991, 65, 2093–2097.
307. Sokolowski, M.; Zhao, C.; Tan, W.; Schwartz, S. AU-rich mRNA instability elements on human papillomavirus type 1 late mRNAs and c-fos mRNAs interact with the same cellular factors. Oncogene 1997, 15, 2303–2319.
308. Zhao, C.; Sokolowski, M.; Tan, W.; Schwartz, S. Characterisation and partial purification of cellular factors interacting with a negative element on human papillomavirus type 1 late mRNAs. Virus Res. 1998, 55, 1–13.
309. Sokolowski, M.; Furneaux, H.; Schwartz, S. The inhibitory activity of the AU-rich RNA element in the human papillomavirus type 1 late

3′ untranslated region correlates with its affinity for the elav-like HuR protein. J. Virol. 1999, 73, 1080–1091.

310. Koffa, M.D.; Graham, S.V.; Takagaki, Y.; Manley, J.L.; Clements, J.B. The human papillomavirus type 16 negative regulatory RNA element interacts with three proteins that act at different posttranscriptional levels. Proc. Natl. Acad. Sci. USA 2000, 97, 4677–4682.

311. Cumming, S.; Chuen-Im, T.; Zhang, J.; Graham, S.V. The RNA stability regulator HuR regulates L1 protein expression in vivo in differentiating cervical epithelial cells. Virology 2009, 383, 142–149.

312. Jolly, C.E.; Gray, L.J.; Parish, J.L.; Lain, S.; Herrington, C.S. Leptomycin B induces apoptosis in cells containing the whole HPV 16 genome. Int. J. Oncol. 2009, 35, 649–656.

313. Steel, J.; Lowen, A.C. Influenza A Virus Reassortment. In Influenza Pathogenesis and Control-Volume I.; Springer: Berlin/Heidelberg, Germany, 2014; pp. 377–401.

314. Smith, G.J.D.; Bahl, J.; Vijaykrishna, D.; Zhang, A.J.X.; Poon, L.L.M.; Chen, H.; Webster, R.; Peiris, J.S.M.; Guan, Y. Dating the emergence of pandemic influenza viruses. Proc. Natl. Acad. Sci. USA 2009, 106, 11709–11712.

315. Lam, T.T.-Y.; Zhu, H.; Wang, J.; Smith, D.K.; Holmes, E.C.; Webster, R.G.; Webby, R.; Peiris, J.M.; Guan, Y. Reassortment Events among Swine Influenza A Viruses in China: Implications for the Origin of the 2009 Influenza Pandemic. J. Virol. 2011, 85, 10279–10285.

316. Fang, R.; Jou, W.M.; Huylebroeck, D.; Devos, R.; Fiers, W. Complete structure of A/duck/Ukraine/63 influenza hemagglutinin gene: Animal virus as progenitor of human H3 Hong Kong 1968 influenza hemagglutinin. Cell 1981, 25, 315–323.

317. Saunders-Hastings, P.; Krewski, D. Reviewing the History of Pandemic Influenza: Understanding Patterns of Emergence and Transmission. Pathogens 2016, 5, 66.

318. Nickol, M.E.; Kindrachuk, J. A year of terror and a century of reflection: Perspectives on the great influenza pandemic of 1918–1919. BMC Infect. Dis. 2019, 19, 1–10.

319. Krug, R.M.; Broni, B.A.; Bouloy, M. Are the 5′ ends of influenza viral mrnas synthesized in vivo donated by host mRNAs? Cell 1979, 18, 329–334.

320. Fodor, E.; Pritlove, D.C.; Brownlee, G.G. The influenza virus panhandle is involved in the initiation of transcription. J. Virol. 1994, 68, 4092–4096.
321. Engelhardt, O.G.; Smith, M.; Fodor, E. Association of the Influenza A Virus RNA-Dependent RNA Polymerase with Cellular RNA Polymerase II. J. Virol. 2005, 79, 5812–5818.
322. De Vlugt, C.; Sikora, D.; Pelchat, M. Insight into Influenza: A Virus Cap-Snatching. Viruses 2018, 10, 641.
323. Inglis, S.C.; Brown, C.M. Spliced and unspliced RNAs encoded by virion RNA segment 7 of influenza virus. Nucleic Acids Res. 1981, 9, 2727–2740.
324. Lamb, R.A.; Lai, C.-J. Spliced and unspliced messenger RNAs synthesized from cloned influenza virus M DNA in an SV40 vector: Expression of the influenza virus membrane protein (M1). Virology 1982, 123, 237–256.
325. Lamb, R.A.; Choppin, P.W.; Chanock, R.M.; Lai, C.J. Mapping of the two overlapping genes for polypeptides NS1 and NS2 on RNA segment 8 of influenza virus genome. Proc. Natl. Acad. Sci. USA 1980, 77, 1857–1861.
326. Jackson, D.A.; Caton, A.J.; McCready, S.J.; Cook, P.R. Influenza virus RNA is synthesized at fixed sites in the nucleus. Nature 1982, 296, 366–368.
327. Dou, D.; Revol, R.; Östbye, H.; Wang, H.; Daniels, R. Influenza A Virus Cell Entry, Replication, Virion Assembly and Movement. Front. Immunol. 2018, 9, 1581.
328. Turrell, L.; Lyall, J.W.; Tiley, L.S.; Fodor, E.; Vreede, F.T. The role and assembly mechanism of nucleoprotein in influenza A virus ribonucleoprotein complexes. Nat. Commun. 2013, 4, 1591.
329. Resa-Infante, P.; Recuero-Checa, M.A.; Zamarreño, N.; Llorca, O.; Ortín, J. Structural and Functional Characterization of an Influenza Virus RNA Polymerase-Genomic RNA Complex. J. Virol. 2010, 84, 10477–10487.
330. Arranz, R.; Coloma, R.; Chichon, F.J.; Conesa, J.J.; Carrascosa, J.L.; Valpuesta, J.M.; Ortín, J.; Martín-Benito, J.; Ruiz-Garcia, D. The Structure of Native Influenza Virion Ribonucleoproteins. Science 2012, 338, 1634–1637.

331. Read, E.K.C.; Digard, P. Individual influenza a virus mRNAs show differential dependence on cellular NXF1/TAP for their nuclear export. J. Gen. Virol. 2010, 91, 1290–1301.

332. Larsen, S.; Bui, S.; Pérez, V.; Mohammad, A.; Medina-Ramirez, H.; Newcomb, L.L. Influenza polymerase encoding mRNAs utilize atypical mRNA nuclear export. Virol. J. 2014, 11, 1–11.

333. Bier, K.; York, A.; Fodor, E. Cellular cap-binding proteins associate with influenza virus mRNAs. J. Gen. Virol. 2011, 92, 1627–1634.

334. Poon, L.L.M.; Pritlove, D.C.; Fodor, E.; Brownlee, G.G. Direct Evidence that the Poly(A) Tail of Influenza A Virus mRNA Is Synthesized by Reiterative Copying of a U Track in the Virion RNA Template. J. Virol. 1999, 73, 3473–3476.

335. Amorim, M.J.; Read, E.K.; Dalton, R.M.; Medcalf, L.; Digard, P. Nuclear Export of Influenza a Virus mRNAs Requires Ongoing RNA Polymerase II Activity. Traffic 2007, 8, 1–11.

336. Mor, A.; White, A.; Zhang, K.; Thompson, M.; Esparza, M.; Muñoz-Moreno, R.; Koide, K.; Lynch, K.W.; García-Sastre, A.; Fontoura, B.M.A. Influenza virus mRNA trafficking through host nuclear speckles. Nat. Microbiol. 2016, 1, 16069.

337. Pereira, C.F.; Read, E.K.C.; Wise, H.M.; Amorim, M.J.; Digard, P. Influenza A Virus NS1 Protein Promotes Efficient Nuclear Export of Unspliced Viral M1 mRNA. J. Virol. 2017, 91, e00528-17.

338. Pereira, C.F.; Wise, H.M.; Kurian, D.; Pinto, R.M.; Amorim, M.J.; Gill, A.C.; Digard, P. Effects of mutations in the effector domain of influenza A virus NS1 protein. BMC Res. Notes 2018, 11, 673.

339. Manzoor, R.; Igarashi, M.; Takada, A. Influenza a Virus M2 Protein: Roles from Ingress to Egress. Int. J. Mol. Sci. 2017, 18, 2649.

340. Chiba, S.; Hill-Batorski, L.; Neumann, G.; Kawaoka, Y. The Cellular DExD/H-Box RNA Helicase UAP56 Co-localizes With the Influenza A Virus NS1 Protein. Front. Microbiol. 2018, 9, 2192.

341. Esparza, M.; Mor, A.; Niederstrasser, H.; White, K.M.; White, A.; Zhang, K.; Gao, S.; Wang, J.; Liang, J.; Sho, S.; et al. Chemical intervention of influenza virus mRNA nuclear export. PLoS Pathog. 2020, 16, e1008407.

342. Robb, N.C.; Jackson, D.; Vreede, F.T.; Fodor, E. Splicing of influenza A virus NS1 mRNA is independent of the viral NS1 protein. J. Gen. Virol. 2010, 91, 2331–2340.

343. Ren, X.; Yu, Y.; Li, H.; Huang, J.; Zhou, A.; Liu, S.; Hu, P.; Li, B.; Qi, W.; Liao, M. Avian Influenza A Virus Polymerase Recruits Cellular RNA Helicase eIF4A3 to Promote Viral mRNA Splicing and Spliced mRNA Nuclear Export. Front. Microbiol. 2019, 10, 1625.

344. Diot, C.; Fournier, G.; Dos Santos, M.; Magnus, J.; Komarova, A.; Van Der Werf, S.; Munier, S.; Naffakh, N. Influenza A Virus Polymerase Recruits the RNA Helicase DDX19 to Promote the Nuclear Export of Viral mRNAs. Sci. Rep. 2016, 6, 33763.

345. Eisfeld, A.J.; Neumann, G.; Kawaoka, Y. At the centre: Influenza A virus ribonucleoproteins. Nat. Rev. Microbiol. 2015, 13, 28–41.

346. O'Neill, R.E.; Talon, J.; Palese, P. The influenza virus NEP (NS2 protein) mediates the nuclear export of viral ribonucleoproteins. EMBO J. 1998, 17, 288–296.

347. Bui, M.; Wills, E.G.; Helenius, A.; Whittaker, G.R. Role of the Influenza Virus M1 Protein in Nuclear Export of Viral Ribonucleoproteins. J. Virol. 2000, 74, 1781–1786.

348. Cao, S.; Liu, X.; Yu, M.; Li, J.; Jia, X.; Bi, Y.; Sun, L.; Gao, G.F.; Liu, W. A Nuclear Export Signal in the Matrix Protein of Influenza A Virus Is Required for Efficient Virus Replication. J. Virol. 2012, 86, 4883–4891.

349. Yasuda, J.; Nakada, S.; Kato, A.; Toyoda, T.; Ishihama, A. Molecular Assembly of Influenza Virus: Association of the NS2 Protein with Virion Matrix. Virology 1993, 196, 249–255.

350. Richardson, J.C.; Akkina, R.K. NS2 protein of influenza virus is found in purified virus and phosphorylated in infected cells. Arch. Virol. 1991, 116, 69–80.

351. Neumann, G.; Hughes, M.T.; Kawaoka, Y. Influenza A virus NS2 protein mediates vRNP nuclear export through NES-independent interaction with hCRM1. EMBO J. 2000, 19, 6751–6758.

352. Iwatsuki-Horimoto, K.; Horimoto, T.; Fujii, Y.; Kawaoka, Y. Generation of Influenza A Virus NS2 (NEP) Mutants with an Altered Nuclear Export Signal Sequence. J. Virol. 2004, 78, 10149–10155.

353. Yu, M.; Liu, X.; Cao, S.; Zhao, Z.; Zhang, K.; Xie, Q.; Chen, C.; Gao, S.; Bi, Y.; Sun, L.; et al. Identification and Characterization of Three Novel Nuclear Export Signals in the Influenza A Virus Nucleoprotein. J. Virol. 2012, 86, 4970–4980.

354. Chen, J.; Huang, S.; Chen, Z. Human cellular protein nucleoporin hNup98 interacts with influenza A virus NS2/nuclear export protein and overexpression of its GLFG repeat domain can inhibit virus propagation. J. Gen. Virol. 2010, 91, 2474–2484.

355. Akyazi, B.Ş.; PİRİNÇAL, A.; Kawaguchi, A.; Nagata, K.; Turan, K. Interaction of influenza A virus NS2/NEP protein with the amino-terminal part of Nup214. Turk. J. Biol. 2020, 44, 82–92.

356. Huang, S.; Chen, J.; Chen, Q.; Wang, H.; Yao, Y.; Chen, Z.; Chen, J.; Chen, J. A Second CRM1-Dependent Nuclear Export Signal in the Influenza A Virus NS2 Protein Contributes to the Nuclear Export of Viral Ribonucleoproteins. J. Virol. 2013, 87, 767–778.

357. Li, J.; Yu, M.; Zheng, W.; Liu, W. Nucleocytoplasmic Shuttling of Influenza A Virus Proteins. Viruses 2015, 7, 2668–2682.

358. Chutiwitoonchai, N.; Aida, Y. NXT1, a Novel Influenza A NP Binding Protein, Promotes the Nuclear Export of NP via a CRM1-Dependent Pathway. Viruses 2016, 8, 209.

359. Liu, J.; Wang, H.; Fang, M.; Chen, X.; Zeng, X. A human cell polarity protein Lgl2 regulates influenza A virus nucleoprotein exportation from nucleus in MDCK cells. J. Biosci. 2020, 45, 1–9.

360. Kumar, S.; Yeo, D.; Muralidharan, N.H.; Lai, S.K.; Tong, C.; Tan, B.H.; Sugrue, R. Impaired Nuclear Export of the Ribonucleoprotein Complex and Virus-Induced Cytotoxicity Combine to Restrict Propagation of the A/Duck/Malaysia/02/2001 (H9N2) Virus in Human Airway Cells. Cells 2020, 9, 355.

361. Sun, Q.; Chen, X.; Zhou, Q.; Burstein, E.; Yang, S.; Jia, D. Inhibiting cancer cell hallmark features through nuclear export inhibition. Signal Transduct. Target. Ther. 2016, 1, 16010.

362. Chari, A.; Vogl, D.T.; Gavriatopoulou, M.; Nooka, A.K.; Yee, A.J.; Huff, C.A.; Moreau, P.; Dingli, D.; Cole, C.; Lonial, S.; et al. Oral Selinexor-Dexamethasone for Triple-Class Refractory Multiple Myeloma. N. Engl. J. Med. 2019, 381, 727–738.

363. Perwitasari, O.; Johnson, S.; Yan, X.; Howerth, E.; Shacham, S.; Landesman, Y.; Baloglu, E.; McCauley, D.; Tamir, S.; Tompkins, S.M.; et al. Verdinexor, a Novel Selective Inhibitor of Nuclear Export, Reduces Influenza A Virus Replication In Vitro and In Vivo. J. Virol. 2014, 88, 10228–10243.

364. Perwitasari, O.; Johnson, S.; Yan, X.; Register, E.; Crabtree, J.; Gabbard, J.; Howerth, E.; Shacham, S.; Carlson, R.; Tamir, S.; et al. Antiviral Efficacy of Verdinexor In Vivo in Two Animal Models of Influenza A Virus Infection. PLoS ONE 2016, 11, e0167221.

CHAPTER 10

NUCLEOPHAGY—IMPLICATIONS FOR MICROAUTOPHAGY AND HEALTH

Florian Bo Otto and Michael Thumm

Department of Cellular Biochemistry, University Medical Center Göttingen, Humboldtallee 23, D-37073 Göttingen, Germany

ABSTRACT

Nucleophagy, the selective subtype of autophagy that targets nuclear material for autophagic degradation, was not only shown to be a model system for the study of selective macroautophagy, but also for elucidating the role of the core autophagic machinery within microautophagy. Nucleophagy also emerged as a system associated with a variety of disease conditions including cancer, neurodegeneration and ageing. Nucleophagic processes are part of natural cell development, but also act as a response to various stress conditions. Upon releasing small portions of nuclear material, micronuclei,

Citation: Bo Otto, F.; Thumm, M. Nucleophagy—Implications for Microautophagy and Health. Int. J. Mol. Sci. 2020, 21, 4506.

the autophagic machinery transfers these micronuclei to the vacuole for subsequent degradation. Despite sharing many cargos and requiring the core autophagic machinery, recent investigations revealed the aspects that set macro- and micronucleophagy apart. Central to the discrepancies found between macro- and micronucleophagy is the nucleus vacuole junction, a large membrane contact site formed between nucleus and vacuole. Exclusion of nuclear pore complexes from the junction and its exclusive degradation by micronucleophagy reveal compositional differences in cargo. Regarding their shared reliance on the core autophagic machinery, micronucleophagy does not involve normal autophagosome biogenesis observed for macronucleophagy, but instead maintains a unique role in overall microautophagy, with the autophagic machinery accumulating at the neck of budding vesicles.

Keywords: nucleophagy; piecemeal microautophagy of the nucleus (PMN); nucleus vacuole junction; microautophagy; Atg39; ER-phagy; nuclear pore complexes; endosomal sorting complexes required for transport (ESCRT)

INTRODUCTION

Autophagy is a highly conserved process that facilitates the degradation and recycling of intracellular components and plays a central role in the maintenance of cellular homeostasis. Involved in a wide range of human pathologies, autophagy has garnered attention with respect to novel therapeutics. The autophagic subtype that targets nuclear material is termed nucleophagy. As a principal cellular organelle, regulation of the different nuclear aspects by nucleophagy has emerged as a system with increasing links to ageing, neurodegenerative diseases, as well as cancer [1,2,3].

Nucleophagic processes occur under various conditions. Starvation, rapamycin induced TORC1 inactivation, genotoxic stress, expansion of the nucleus vacuole junction (NVJ), as well as defects in the nuclear envelope and lamina lead to enhanced rates of nucleophagy [3,4,5,6,7]. Detailed investigations towards their initiation showed that macro- and micronucleophagy occur dependent on the Nem1/Spo7–Pah1 axis, a downstream process of TORC1 inactivation [8]. The Nem1/Spo7–Pah1 axis which is part of lipid metabolism, is also involved in other autophagic processes, including endosomal sorting complexes required for transport (ESCRT)-dependent and autophagy independent micro-ER-phagy [9].

Nucleophagy occurs in both a macro- and micronucleophagic manner in the budding yeast *Saccharomyces cerevisiae* [4,5]. Designation of nuclear material for autophagic degradation is mediated by autophagic cargo receptor Atg39, independent of the nucleophagic mode [10]. Nucleophagy involves the formation of micronuclei, small parts of nuclear material that bud off of the main nucleus, which are then degraded by delivery to the vacuole.

Macronucleophagy is a classic example of selective macroautophagy involving the core macroautophagic machinery (hereafter the core autophagic machinery), with the addition of factors conveying selectivity. Macroautophagy is a well investigated process, as are the roles for members of the core autophagic machinery within it. Signaling and biogenesis of the double membraned autophagosome (the vesicle that transports autophagic cargo to the vacuole) require the core autophagic machinery.

While many studies generally focus on the mechanism of (selective) macroautophagy, micronucleophagy in yeast, termed piecemeal microautophagy of the nucleus (PMN), presents a model system for selective microautophagy. Microautophagy is representative of the umbrella terminology that covers processes which involve the direct abstraction of cellular materials into the vacuole/lysosome. The autophagic machinery is required only for certain subtypes [11,12,13]. PMN depends on the autophagic machinery [13,14]. Localisation of the autophagic machinery in PMN is reminiscent of the micropexophagic membrane apparatus (MIPA), required for micropexophagy in *Komagataella phaffii* (former *Pichia pastoris*) [10,15,16]. Its localisation is mirrored by the ESCRT-III complex in micro-ER-phagy [9]. The distinct roles of the autophagic machinery show that while macro- and micronucleophagic processes depend on the core autophagic machinery and can occur under the same conditions, there is an increasing number of features that set them apart.

Many cargos are shared between the nucleophagic modes, implicating nucleophagy in regulation of the nuclear envelope and its domains, as well as nucleoplasmic content, including the nucleolus (Figure 1) [5,14]. The nucleus vacuole junction (NVJ), a large compositionally unique membrane contact site between vacuolar membrane and nuclear envelope is an exclusive target of PMN [4,10]. Regulation of the NVJ by PMN has many implications towards its influence on lipid metabolism and both nuclear and vacuolar membrane composition. In contrast, NPCs are excluded from the NVJ and PMN does not result in their degradation [4,17,18,19,20]. Instead, a specialised form of macroautophagy targets NPCs. Nup159 was shown to specifically act as a cargo receptor for macroautophagic degradation of NPCs, which occurs largely independent of Atg39. Similar to Atg39 dependent macronucleophagy, autophagy of NPCs also includes nuclear envelope and a portion of the nucleoplasm [19,20].

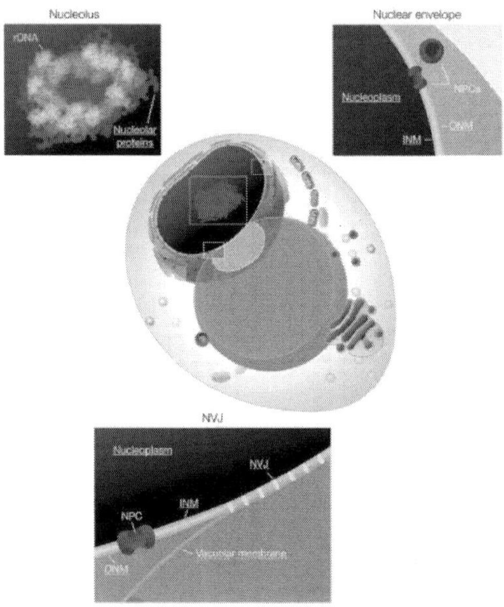

Figure 1: Nucleophagic cargo in context of an *S. cerevisiae* cell. Nucleophagy targets a variety of cargos in yeast. Nuclear compartments relevant for nucleophagy are additionally highlighted in a detailed depiction. Known cargos of nucleophagy are underlined, and cargos exclusive to micronucleophagy show a dashed underline. Nucleophagy results in formation of micronuclei, small portions of nuclear material that bud off of the main nucleus. Inner- and outer nuclear membrane (INM and ONM) are targeted by nucleophagy, excluding nuclear pore complexes (NPCs), which are degraded by a specialised form of selective autophagy. Nucleoplasm, as well as nucleolar proteins are degraded, while excluding ribosomal DNA (rDNA). Piecemeal microautophagy of the nucleus targets these components as well and additionally degrades the tightly clamped nucleus vacuole junction (NVJ), including a portion of vacuolar membrane.

Many aspects of the nucleophagic processes are still unknown. The budding process that results in the formation of micronuclei and the nature and composition of a potential microautophagic membrane have yet to be characterised in detail. As the role of the ESCRT machinery in PMN is still unknown, future studies will have to determine its involvement to further understand the details of the microautophagic classes. For many nucleophagic cargos such as the nucleoplasm, it is difficult to determine whether they are actual targets of the autophagic process, or degraded simply by circumstance. To understand the role of nucleophagy, detailed analyses of nucleophagic cargo compositions might distinguish designated from collateral cargos. Regulation of the NVJ by PMN and its implications on lipid metabolism have been hinted at, but direct evidence could substantiate a nucleophagic role in lipid regulation. Ultimately, nucleophagic influence on disease and its utilisation towards uncovering novel approaches to treatment is implied by present data, but still requires further research for meaningful assessments.

SIGNALING FOR NUCLEOPHAGY

Nucleophagy was initially shown to occur during nutrient limitation in yeast. TORC1 inactivation by nitrogen starvation as well as rapamycin induces nucleophagy independent of its mode [4,5,10]. Nucleophagic processes begin by formation of micronuclei and TORC1 inactivation leads to the recruitment of the core autophagic machinery to these micronuclei [10,13]. In addition to the core autophagic machinery, nucleophagy specifically requires cargo receptor Atg39 of the nuclear envelope [5,10].

In line with nucleophagic activity upon starvation, the upstream region of *ATG39* contains two stress-response element (STRE) repeats, which upregulate gene expression in stationary phase and nutrient deprivation conditions, supporting a link to nucleophagy [21]. Outer nuclear membrane (ONM) resident Nvj1 is essential for the formation of the NVJ and thereby micronucleophagy. The upstream region of *NVJ1* also contains two STRE repeat regions [22]. Noticeably, overexpression of Nvj1 leads to an enlargement of the NVJ and subsequently higher rates of micronucleophagy [6].

Besides similarities in expressional regulation, localisation of both Atg39 and Nvj1 depends on the Nem1/Spo7–Pah1 axis [8]. The Nem1/Spo7–Pah1 axis is activated as a consequence of TORC1 inactivation and its absence leads to impairment of both macro- and micronucleophagy [8,23]. Previously shown to be required for maintenance of nuclear envelope morphology, the Nem1/Spo7 complex is located in the nuclear envelope [24]. While its influence on nucleophagy is particularly extensive, the complex is also relevant for other modes of autophagy yet does not influence the Cvt-pathway [8]. Pah1 yields diacylglycerol by dephosphorylation of phosphatidic acids. Its capacity for maintenance of the nuclear envelope might be a factor for the Nem1/Spo7–Pah1 axis in modulating the nuclear membrane for micronuclei formation.

Whether a target or a side effect of microautophagic processes, vacuolar membrane degradation is a distinguishing feature of the process. Pah1 mediated diacylglycerol synthesis is required for functionality of overall macroautophagy and micronucleophagy as well as maintenance of nuclear morphology [8]. The Nem1/Spo7–Pah1 axis appears to be relevant beyond nucleophagy, in overall microautophagic context, as core autophagy independent and ESCRT dependent micro-ER-phagy also depends on it [9]. Lipid provision and/or metabolism after microautophagic membrane consumption and involvement in vacuolar domain formation were put forward as potential reasons for its conserved relevance in overall microautophagy [8,9,25,26].

Nucleophagic activity in mammalian cells is linked to oncogenic and genotoxic stress [3,27,28]. While nucleophagy in mammalian cells is initiated by pathological conditions, the Nem1/Spo7–Pah1 axis is conserved from yeast to mammalian cells, as the orthologous CTDNEP1/NEP1R1-lipin complex. The CTDNEP1/NEP1R1-lipin complex is, similar to its counterpart, located in the nuclear envelope [29]. Human CTDNEP1 can functionally complement yeast Nem1 and mutations in CTDNEP1 were discovered in tumor cells [29,30]. Whether nucleophagy is also observed in mammalian cells under physiological conditions will have to be clarified in future studies.

MECHANISM

After signaling, nucleophagic processes begin with the formation of micronuclei. These small portions of nuclear material are then recognised by the autophagic machinery. Their formation has not yet been described in detail. Recruitment of the autophagic machinery to micronuclei most likely follows cargo receptor accumulation [10].

Whereas the autophagosome envelops micronuclei in macronucleophagy the autophagic machinery accumulates at a MIPA-like structure in PMN (Figure 2). Presence of the autophagic machinery is not only highly reminiscent of micropexophagy in *K. phaffii*, but also of ESCRT involvement in micro-ER-phagy [9,31].

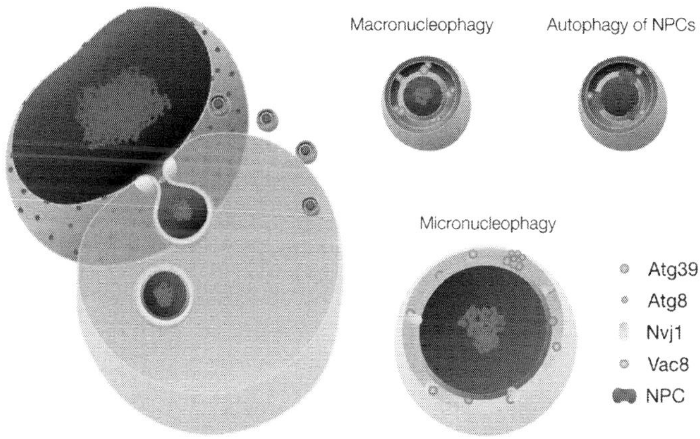

Figure 2: The two principle modes of nucleophagy in *S. cerevisiae*. Macronucleophagy is a selective macroautophagic mode dependent on cargo receptor Atg39. Atg39 interacts with Atg8 in the autophagic membrane to form autophagosomes around micronuclei, which are ultimately transferred to the vacuolar lumen. Autophagy of nuclear pore complexes (NPCs) was shown to be a process separate from macronucleophagy, occurring independent of Atg39, instead recognizing cargo by interaction of Atg8 with Nup159 within the NPC. Micronucleophagy (piecemeal microautophagy of the nucleus) also depends on Atg39. Its larger cargo diameter of ~700 nm for micro- when compared to ~380 nm of macronucleophagic vesicles, restriction to the nuclear vacuolar junction (NVJ) and a punctate Atg8 positive structure between the tips of the vacuolar invagination set it apart from macronucleophagy.

Process of Nucleophagy

Nucleophagic processes involve bulging of the nuclear envelope as a first step, followed by budding of the micronucleus. In the ER, membrane bulging is mediated by ER cargo receptor RETREG1. RETREG1 (similar to its *S. cerevisiae* counterpart, Atg40) contains a reticulon homology domain (RHD), which conveys curvature induction required for ER-phagy [32]. In *S. cerevisiae* nucleophagy depends on ONM resident autophagic cargo receptor Atg39 [5,10]. While Atg39 does not contain an RHD, initial bulging of the nuclear envelope for formation of micronuclei could be facilitated by curvature inducing factors that localise to the ONM in tandem with Atg39.

Atg39 interacts with Atg8 via an Atg8-interacting motif (AIM) within its cytosolic N-terminal region. Atg39 additionally interacts with cargo receptor adaptor Atg11 via an Atg11 binding region (11-BR), a feature commonly observed in selective modes of autophagy. Both interactions with Atg8 and with Atg11 are required for macronucleophagy [5]. Deletion of Atg39, its AIM, or 11-BR leads to disruptions of nuclear envelope and the nucleus and entails a decrease in cellular viability [5,33].

Atg8 recruitment follows heightened Atg39 accumulation. Atg39 was shown to accumulate around macronucleophagic cargo prior to Atg8 recruitment. Autophagosome biogenesis follows and the process is finalised by transfer of the vesicle to the vacuolar lumen. Autophagosome transfer of micronuclei coincides with a change in vacuolar membrane composition, presumably a shared feature of overall macroautophagy. In contrast, Atg39 is present surrounding PMN cargo, comparable to its presence in the ONM, and accumulates at a punctate structure between the neck of the vacuolar invagination. The accumulation also coincides with Atg8 and an extended Atg11 presence, displaying a unique role for the autophagic machinery in a microautophagic system [10].

Role of the Autophagic Machinery in Micronucleophagy

The punctate Atg8 positive structure between the tips of the vacuolar invagination in PMN is reminiscent of mechanisms observed in other microautophagic processes. A membrane structure involved in microautophagic subtypes that depend on the core autophagic machinery explains at least in part their requirement, as the components are essential for autophagic membrane biogenesis.

The MIPA is an extended membrane structure that closes the microautophagic structure around aggregated peroxisomes, linking vacuolar sequestering arms [15,16]. Its functionality as a component for closure of the vacuolar membrane surrounding microautophagic cargo, as well as Atg8 presence within the structure are shared between micropexophagy and PMN. Despite these similarities, comparison of the MIPA with the micronucleophagic structure also reveals compositional differences. The MIPA contains *Kp*Atg26, which is required for micropexophagy and formation of the MIPA [31]. In contrast, Atg26 is redundant for the functionality of PMN [13]. Phosphatidylinositol 4-phosphate is a signaling lipid for the MIPA and accumulates within it [31]. Whether the micronucleophagic structure observed in PMN has a distinct lipid composition from autophagosomal membrane has yet to be determined. Whereas PMN cargo bulges into the vacuole, the MIPA links vacuolar sequestering membranes that protrude from the vacuole and envelop cargo.

The MIPA presumably seals these sequestering membranes, while PMN is carried out without their formation. The MIPA is also larger than the autophagic structure in PMN. While this could be another distinguishing factor, the MIPA is observed in response to overabundance of peroxisomes, resulting in particularly large microautophagic structures. PMN vesicles on the other hand are formed without overexpression, which is a potential explanation for the smaller, punctate autophagic structure [10,31]. The Atg8 positive, micronucleophagic structure is transferred to the vacuolar lumen alongside its associated cargo vesicle. Its presence is speculated to allow access for vacuolar hydrolases to the cargo vesicle since it presents a site that is not shielded by vacuolar membrane [10].

Core autophagy independent microautophagy typically involves the ESCRT machinery [9]. ESCRT activity resembles microautophagy and is involved in membrane remodeling and severance in various processes including multivesicular endosome formation, NPC quality control, as well as nuclear envelope assembly and repair [34,35,36,37]. Whether the ESCRT machinery is also involved in micronucleophagy has yet to be determined. ESCRT involvement is not exclusive to microautophagy, but is also required for the functionality of macroautophagic processes [38,39,40].

Morphologically and functionally reminiscent of the micronucleophagic process, micro-ER-phagy shares its large vesicle diameter and also displays mechanistic activity at the neck of vacuolar membrane invagination. ESCRT-III component Snf7 is found here as a punctate structure, resembling the situation reported for Atg8 in PMN [9,10]. The ESCRT-III complex is required for membrane scission in the micro-ER-phagy process, rather than for the formation of a microautophagic membrane. Micro-ER-phagy is also a downstream process of the Nem1/Spo7–Pah1 axis [9]. A shared reliance on the Nem1/Spo7–Pah1 suggests requirement for lipid metabolism that preludes microautophagy as a consequence of its membrane consumption, or its potential for vacuolar domain formation.

The ESCRT machinery is also required for both microautophagy and macroautophagy in mammalian cells [39,40,41]. Here, micro-ER-phagy was shown to occur independent of core autophagic proteins required for autophagosome biogenesis (ULK1, ULK2, ATG13, ATG14), but dependent on those required for LC3 lipidation (ATG4B, ATG7, or ATG16L1). This select subset suggests, that micro-ER-phagy in mammalian cells does not require autophagic membrane formation [41].

NUCLEOPHAGIC CARGO

The distinction of nucleophagic processes with regard to cargo composition remains difficult, since many cargos are shared between the nucleophagic modes (Table 1). Presumably an effect of substrate composition, macro- and micronucleophagy form vesicles with highly distinct diameters, with micronucleophagy producing significantly larger vesicles than those observed for macroautophagy (Figure 2) [10].

Table 1: Known nuclear cargos and their associated autophagic modes described in *S. cerevisiae.*

Cargo	Autophagic Mode	Specifications	Reported Cargo Components	Reference
Nuclear envelope	Macronucleophagy, micronucleophagy, autophagy of NPCs	Macro- and micro-nucleophagy harbor Atg39	Sec63, Hmg1, Src1, Atg39	[4,5]
NPCs	Autophagy of NPCs, macronucleophagy (partially)	Harbor Nup159, the cargo receptor required for autophagy of NPCs	Many NPC subcomponents including Nup159	[19,20]
NVJ	Micronucleophagy	Requires Nvj1 and Vac8 for formation	Nvj1, Vac8, Osh1	[4,6,13]
Nucleoplasm	Macronucleophagy, micronucleophagy, autophagy of NPCs	Fusion of the Nab2 localisation signal with fluorescent proteins	Nuclear localisation signal of Nab2	[10,13,19]
Nucleolus	Macronucleophagy, micronucleophagy	rDNA and nucleolar proteins are separated prior to nucleophagy	Nop1	[5,14]

Overall, mammalian cells share nucleophagic cargo types observed in yeast, but additionally include the nuclear lamina, a feature absent from yeast cells. DNA is generally excluded from nucleophagic processes in yeast, but found in micronuclei in mammalian cells [42].

Nuclear Envelope

The nuclear envelope is the confining structure of the nucleus. It is a double membrane, representative of the perinuclear portion of the ER. Required for maintenance of structural integrity and segregation of nucleoplasm and cytoplasm, the nuclear envelope is also required for the spatial organisation of the nucleus. The inner nuclear membrane (INM) forms domains that interact with nuclear components and is associated with the nuclear lamina, a conserved feature of metazoan cells. Micronuclei, precursors to nucleophagic vesicle formation and the vesicles themselves, were shown to include nuclear envelope, independent of the mode. Degradation of nuclear envelope via nucleophagy is linked to regulation of nuclear extent, counteracting disruption of its components and morphology, as well as targeted removal of its subdomains.

As the nuclear envelope harbors cargo receptor Atg39, it is degraded by macro- and micronucleophagy [10]. Nucleophagy thereby targets both ONM as well as INM, resulting in double ringed structures detected within corresponding autophagosomes [4,5]. Both nuclear membranes are also core aspects of the NVJ, clamped together by Nvj1, and are degraded alongside vacuolar membrane by micronucleophagy [43].

Present in mammalian cells but not in yeast, the nuclear lamina is associated with the nuclear envelope. It is a cytoskeleton meshwork providing mechanical stability and organisation of chromatin, located at the INM. The lamina contains multiple lamins and associated proteins. Downregulation of nuclear lamina proteins occurs upon senescence within days, but their long-lived structure suggests active removal of present proteins by a targeted turnover process [3]. LC3 interacts with lamin A/C, emerin, as well as lamin B1 of the nuclear lamina [27,44]. As opposed to Atg8, which interacts with cargo receptors on the cytosolic face of the nuclear envelope for nucleophagy, LC3 is conditionally localised in the nucleus and directly interacts with lamins of the nuclear lamina. Nuclear localisation of LC3 was shown previously and is regulated by de-acetylation upon starvation, which results in its redistribution to the cytosol for the initiation of autophagy [45]. Interaction of LC3 with the nuclear lamina and nucleophagy of lamina components were shown to affect nuclear morphology and cellular viability [7,44].

LC3 interacts with lamin B1 at lamin-associated chromatin domains (LADs) [44]. LADs are regions of chromatin contact with the lamina. Autophagy was shown to degrade chromatin fragments that bud off of the main nucleus [46]. Nuclear envelope sites targeted by nucleophagy lack nuclear envelope protein SUN1, which mediates nuclear anchorage, suggesting a distinct composition of designated cargo sites [47]. Lamin B1 degradation does not occur under nutritional stress, but instead upon cellular senescence, induced by oncogenic and genotoxic conditions [39]. Lamin A/C and emerin deficient cells show LC3 positive autophagosomes tethered to the nuclear envelope. Inhibition of autophagy was shown to result in disturbance of nuclear morphology and compromises cellular vitality, a potential result of lacking means towards removal of damaged nuclear content [7].

NPCs

NPCs allow bidirectional transport of macromolecules from- and to the nucleus and are distributed over the nuclear envelope. Autophagic degradation of NPCs allows bulk degradation of these macromolecular structures and is a process separate from macronucleophagy, as it does not require Atg39 [19,20]. As a consequence of their absence from the NVJ, NPCs are not degraded by PMN [14]. Autophagy of NPCs is a countermeasure for disrupted complexes and regulates the overall quantity of NPCs within the nuclear envelope.

Nuclear protrusions were reported to contain NPCs and macroautophagy of NPCs was shown to enable NPC turnover either upon induction by starvation or NPC disruption [19,20,48]. Analogous to Atg39 located in the ONM, NPC component and Nup159 also contains a functional cytoplasmic AIM. As with macro- and micronucleophagy, and in line with its selective modality, macroautophagy of NPCs also additionally requires presence of- and interaction with Atg11. Autophagy of NPCs degrades multiple NPCs alongside nuclear envelope and a portion of the nucleoplasm [19]. Autophagic degradation of NPCs was observed under nitrogen starvation conditions and also upon introducing mutations that cause NPC clustering, suggesting regulatory in addition to bulk protein degradation observed as a result of starvation [19,20].

In mammalian cells, nuclear envelope protrusions associated with autophagic degradation were shown to contain NPCs [42]. Impairment of NPC assembly, as well as complex disruption, is linked to ageing, neurodegeneration and cancer [49,50,51].

Nucleolus

The nucleolus is the largest sub-nuclear structure. It is a site of ribosome biogenesis and assembles around ribosomal gene (rDNA) arrays. Beyond its relevance for ribosome biogenesis, the nucleolus continues to emerge as a hub for regulation of many cellular processes, including the regulation of gene expression, nuclear organisation and maintaining cellular homeostasis. Nucleolar components are degraded by nucleophagy and rDNA is retained by a coordinated separation mechanism. Pathological increase of nucleolar size has been observed in various diseases and its continuous expansion has been linked to age. Regulation of the nucleolar components is a requirement for maintenance of cellular viability and size reduction was correlated with a significant expansion of cell lifespan [1,52,53,54].

Nucleophagy results in the selective degradation of parts of the nucleolus. Fibrillarin ortholog and Ribi (ribosome biogenesis/maturation) protein Nop1 was shown to be a prominent cargo of both macro- and micronucleophagy [5,14]. TORC1 inactivation results in reduction of nucleolar size and condensation of rDNA in yeast and mammalian cells [55,56]. Of note, overabundance of Nop1 is toxic and while it is degraded under native conditions by nucleophagy, actively overexpressed Nop1 is additionally targeted by proteasomal degradation via the CUE domain-containing Def1 [57].

Nucleolar size has been shown to increase relative to cellular age in *Caenorhabtitis elegans* and mammalian culture of Hutchinson–Gilford progeria syndrome (HGPS) derived cells [1,52]. HGPS cells express progerin, a mutant form of lamin A. Depletion of lamin A and expression of progerin both result in nucleolar expansion [1]. Rapamycin induces mutant protein clearance in HGPS cells and counteracts its cellular phenotype [58]. Lifespan expanding *C. elegans* mutants generally maintain smaller nucleoli and in analogy to Nop1, fibrillarin expression expands nucleolar size and results in progeria. NCL-1, homologue of tumour suppressor TRIM2/Brat inhibits production of fibrillarin and promotes lifespan, whereas its deletion causes nucleolar expansion [53,54].

Nucleic Acids

DNA, specifically rDNA, is separated from the nucleolus and condensed in the initial stages of nucleophagy, allowing for degradation of nucleolar proteins while retaining rDNA. rDNA exclusion from nucleophagic cargo is mediated by two mechanisms, the CLIP cohibin system that retracts rDNA to the INM and condensation of rDNA [55,59,60,61]. DNA damage induces autophagy in yeast by a process termed genotoxin-induced targeted autophagy (GTA) and defective autophagy inhibits DNA repair. GTA is a selective autophagic mode that requires Atg11, but does not require any of the known cargo receptors including Atg39. Genotoxicity and double-strand break initiate GTA independent from TORC1 inactivation, instead requiring an alternate pathway that includes Mec1/ATR, Tel1/ATM, and Rad53/

CHEK2, which are largely dispensable for TORC1 mediated autophagy [62].

In contrast, DNA degradation and nucleophagy are interlinked in mammalian cells. LC3 positive micronuclei contain DNA alongside DNA damage marker γH2AX [63]. DNA damage prompts nucleophagy, resulting in degradation of lamin A/C alongside leaked DNA [28]. DNA degradation alongside chromatin was shown to be carried out by autophagy in cancer cells. Here, arginine starvation was used to induce DNA as well as chromatin leakage which leads to subsequent degradation by autophagy, specifically killing tumor cells [42].

NVJ

The NVJ is a uniquely composed membrane contact site between nucleus and vacuole [64,65]. A site for lipid metabolic activity and membrane lipid exchange, it is a prerequisite for and at the same time exclusively degraded by PMN [4]. Linking nucleophagic processes and lipid metabolism also explains identification of Atg39 as a factor for both [5,33].

Formation of the NVJ facilitated by interaction of Nvj1 in the nuclear envelope and Vac8 in the vacuolar membrane [18]. The competitive interaction with Atg13 and Nvj1 differentially regulates the Cvt and PMN pathways [66,67]. Vac8 is required for many micro- and macroautophagic modes [68,69,70,71,72]. The unique composition of the NVJ is at least in part due to Nvj1, which recruits various other proteins. Osh1, a functional homologue to the oxysterol-binding proteins (OSBPs) within the mammalian system, was shown to transport sterols via its OSBP-related domain (ORD) presumably by exchange with PI4P [73]. The NVJ was shown to be a sterol enriched membrane domain, which is typically liquid ordered [14,25]. The essential enoyl-CoA reductase Tsc13 is also recruited to the NVJ in a Nvj1 dependent manner and is presumed to catalyse synthesis of very long chain fatty acids. Attenuation of its fatty acid elongation activity was shown to result in a significant decrease of PMN vesicle diameter, implying a compositional dependence on vesicle size [74].

Mdm1 additionally tethers the NVJ and its overexpression results in hypertethering of ONM to the vacuole, creating contact sites independent of Nvj1. Mdm1 is a substrate of PMN, yet dispensable for its functionality [75,76]. It is highly conserved in mammals as four sorting nexins (Snx13, Snx14, Snx19, Snx25). Snx14 mutations are implicated in neurological disease (autosomal-recessive cerebellar ataxia) and introducing disease analogous mutations in Mdm1 impairs its tethering functionality [75,77]. Lipid droplets accumulate at the edges of the NVJ and Mdm1 is involved in their recruitment and synthesis [76,78]. Recruitment of lipid droplets to the NVJ is speculated to be a mechanism for the removal of toxic lipids, such as free fatty acids, from the ER, which could subsequently be degraded by microlipophagy [79].

Noticeably Atg39 was also identified as ER-stress-induced microlipophagy gene 1 (Esm1), linked to microlipophagy and ESCRT machinery, required for adaption to lipid imbalance after lipid- and ER-stress [33]. Lipid homeostasis is influenced by the NVJ and its autophagy dependent regulation via Atg39-dependent micronucleophagy. Due to association of the aminoterminus of Nvj1 with the INM within the NVJ, the nuclear envelope shows a distance of approximately 8.7 nm from ONM to INM, whereas outside it shows a distance of approximately 18.7 nm [43]. The tight clamping might prevent simple dissociation of the NVJ. Instead, micronucleophagy as a regulatory process of the NVJ provides a fast and comprehensive degradation mechanism [10,74].

OUTLOOK

Nucleophagy has recently emerged as a highly investigated topic, yet many of its aspects remain to be elucidated. While it is a selective mode of autophagy, and reference with other selective autophagic processes furthers the understanding of nucleophagy, many comparisons have limited relevance for the process due to the fact that nucleophagy targets only a select portion of the organelle, adding an additional level of complexity to its selective modality.

Involvement of reticulons, as observed for autophagic turnover of cortical ER via Atg40 for initial membrane bulging, could also be an intriguing mechanism for labelling sub-organellar sites for nucleophagic turnover and requires further investigation.

Identification of a structure resembling the MIPA needs to be addressed in compositional reference with the situation observed in *K. phaffii*. Analysis of the microautophagic structure will likely allow for a better understanding of the role the core autophagic machinery assumes in microautophagy.

Parallels regarding ESCRT involvement in micro-ER-phagy should be addressed by investigation of ESCRT components in relation to nucleophagy and whether their absence has an impact on nucleophagic cargo turnover.

A more comprehensive understanding of nucleophagic cargo with regard to underlying coordination of its components such as DNA, but also protein content and thereby a possible differentiation of the nucleoplasmic portion would greatly benefit the understanding of nucleophagy. Compositional analysis of micronuclei and a connection of NVJ, micronucleophagy and lipid metabolism will enable a more comprehensive insight into the relevance of nucleophagic processes regarding cellular homeostasis and its associated disease conditions.

REFERENCES

1. Buchwalter, A.; Hetzer, M.W. Nucleolar expansion and elevated protein translation in premature aging. Nat. Commun. 2017, 8, 328.
2. Baron, O.; Boudi, A.; Dias, C.; Schilling, M.; Nolle, A.; Vizcay-Barrena, G.; Rattray, I.; Jungbluth, H.; Scheper, W.; Fleck, R.A.; et al. Stall in Canonical Autophagy-Lysosome Pathways Prompts Nucleophagy-Based Nuclear Breakdown in Neurodegeneration. Curr. Biol. 2017, 27, 3626–3642.e6.
3. Dou, Z.; Ivanov, A.; Adams, P.D.; Berger, S.L. Mammalian autophagy degrades nuclear constituents in response to tumorigenic stress. Autophagy 2016, 12, 1416–1417.
4. Roberts, P.; Moshitch-Moshkovitz, S.; Kvam, E.; O'Toole, E.; Winey, M.; Goldfarb, D.S. Piecemeal microautophagy of nucleus in Saccharomyces cerevisiae. Mol. Biol. Cell 2003, 14, 129–141.
5. Mochida, K.; Oikawa, Y.; Kimura, Y.; Kirisako, H.; Hirano, H.; Ohsumi, Y.; Nakatogawa, H. Receptor-mediated selective autophagy degrades the endoplasmic reticulum and the nucleus. Nature 2015, 522, 359–362.
6. Millen, J.I.; Krick, R.; Prick, T.; Thumm, M.; Goldfarb, D.S. Measuring piecemeal microautophagy of the nucleus in Saccharomyces cerevisiae. Autophagy 2009, 5, 75–81.
7. Park, Y.E.; Hayashi, Y.K.; Bonne, G.; Arimura, T.; Noguchi, S.; Nonaka, I.; Nishino, I. Autophagic degradation of nuclear components in mammalian cells. Autophagy 2009, 5, 795–804.
8. Rahman, M.A.; Mostofa, M.G.; Ushimaru, T. The Nem1/Spo7-Pah1/lipin axis is required for autophagy induction after TORC1 inactivation. FEBS J. 2018, 285, 1840–1860.
9. Schafer, J.A.; Schessner, J.P.; Bircham, P.W.; Tsuji, T.; Funaya, C.; Pajonk, O.; Schaeff, K.; Ruffini, G.; Papagiannidis, D.; Knop, M.; et al. ESCRT machinery mediates selective microautophagy of endoplasmic reticulum in yeast. EMBO J. 2020, 39, e102586.
10. Otto, F.B.; Thumm, M. Mechanistic dissection of macro- and micronucleophagy. Autophagy 2020, 1–14.
11. Schuck, S.; Gallagher, C.M.; Walter, P. ER-phagy mediates selective degradation of endoplasmic reticulum independently of the core autophagy machinery. J. Cell Sci. 2014, 127, 4078–4088.

12. Oku, M.; Sakai, Y. Three Distinct Types of Microautophagy Based on Membrane Dynamics and Molecular Machineries. Bioessays 2018, 40, e1800008.
13. Krick, R.; Muehe, Y.; Prick, T.; Bremer, S.; Schlotterhose, P.; Eskelinen, E.L.; Millen, J.; Goldfarb, D.S.; Thumm, M. Piecemeal microautophagy of the nucleus requires the core macroautophagy genes. Mol. Biol. Cell 2008, 19, 4492–4505.
14. Dawaliby, R.; Mayer, A. Microautophagy of the nucleus coincides with a vacuolar diffusion barrier at nuclear-vacuolar junctions. Mol. Biol. Cell 2010, 21, 4173–4183.
15. Mukaiyama, H.; Baba, M.; Osumi, M.; Aoyagi, S.; Kato, N.; Ohsumi, Y.; Sakai, Y. Modification of a ubiquitin-like protein Paz2 conducted micropexophagy through formation of a novel membrane structure. Mol. Biol. Cell 2004, 15, 58–70.
16. Oku, M.; Warnecke, D.; Noda, T.; Muller, F.; Heinz, E.; Mukaiyama, H.; Kato, N.; Sakai, Y. Peroxisome degradation requires catalytically active sterol glucosyltransferase with a GRAM domain. EMBO J. 2003, 22, 3231–3241.
17. Severs, N.J.; Jordan, E.G.; Williamson, D.H. Nuclear pore absence from areas of close association between nucleus and vacuole in synchronous yeast cultures. J. Ultrastruct. Res. 1976, 54, 374–387.
18. Pan, X.; Roberts, P.; Chen, Y.; Kvam, E.; Shulga, N.; Huang, K.; Lemmon, S.; Goldfarb, D.S. Nucleus-vacuole junctions in Saccharomyces cerevisiae are formed through the direct interaction of Vac8p with Nvj1p. Mol. Biol. Cell 2000, 11, 2445–2457.
19. Lee, C.W.; Wilfling, F.; Ronchi, P.; Allegretti, M.; Mosalaganti, S.; Jentsch, S.; Beck, M.; Pfander, B. Selective autophagy degrades nuclear pore complexes. Nat. Cell Biol. 2020, 22, 159–166.
20. Tomioka, Y.; Kotani, T.; Kirisako, H.; Oikawa, Y.; Kimura, Y.; Hirano, H.; Ohsumi, Y.; Nakatogawa, H. TORC1 inactivation stimulates autophagy of nucleoporin and nuclear pore complexes. J. Cell Biol. 2020, 219, e201910063.
21. Gasch, A.P.; Spellman, P.T.; Kao, C.M.; Carmel-Harel, O.; Eisen, M.B.; Storz, G.; Botstein, D.; Brown, P.O. Genomic expression programs in the response of yeast cells to environmental changes. Mol. Biol. Cell 2000, 11, 4241–4257.

22. Moskvina, E.; Schuller, C.; Maurer, C.T.; Mager, W.H.; Ruis, H. A search in the genome of Saccharomyces cerevisiae for genes regulated via stress response elements. Yeast 1998, 14, 1041–1050.
23. Dubots, E.; Cottier, S.; Peli-Gulli, M.P.; Jaquenoud, M.; Bontron, S.; Schneiter, R.; De Virgilio, C. TORC1 regulates Pah1 phosphatidate phosphatase activity via the Nem1/Spo7 protein phosphatase complex. PLoS ONE 2014, 9, e104194.
24. Siniossoglou, S.; Santos-Rosa, H.; Rappsilber, J.; Mann, M.; Hurt, E. A novel complex of membrane proteins required for formation of a spherical nucleus. EMBO J. 1998, 17, 6449–6464.
25. Toulmay, A.; Prinz, W.A. Direct imaging reveals stable, micrometer-scale lipid domains that segregate proteins in live cells. J. Cell Biol. 2013, 202, 35–44.
26. Tsuji, T.; Fujimoto, M.; Tatematsu, T.; Cheng, J.; Orii, M.; Takatori, S.; Fujimoto, T. Niemann-Pick type C proteins promote microautophagy by expanding raft-like membrane domains in the yeast vacuole. eLife 2017, 6.
27. Lan, Y.Y.; Londono, D.; Bouley, R.; Rooney, M.S.; Hacohen, N. Dnase2a deficiency uncovers lysosomal clearance of damaged nuclear DNA via autophagy. Cell Rep. 2014, 9, 180–192.
28. Li, Y.; Jiang, X.; Zhang, Y.; Gao, Z.; Liu, Y.; Hu, J.; Hu, X.; Li, L.; Shi, J.; Gao, N. Nuclear accumulation of UBC9 contributes to SUMOylation of lamin A/C and nucleophagy in response to DNA damage. J. Exp. Clin. Cancer Res. 2019, 38, 67.
29. Kim, Y.; Gentry, M.S.; Harris, T.E.; Wiley, S.E.; Lawrence, J.C., Jr.; Dixon, J.E. A conserved phosphatase cascade that regulates nuclear membrane biogenesis. Proc. Natl. Acad. Sci. USA 2007, 104, 6596–6601.
30. Pugh, T.J.; Weeraratne, S.D.; Archer, T.C.; Pomeranz Krummel, D.A.; Auclair, D.; Bochicchio, J.; Carneiro, M.O.; Carter, S.L.; Cibulskis, K.; Erlich, R.L.; et al. Medulloblastoma exome sequencing uncovers subtype-specific somatic mutations. Nature 2012, 488, 106–110.
31. Yamashita, S.; Oku, M.; Wasada, Y.; Ano, Y.; Sakai, Y. PI4P-signaling pathway for the synthesis of a nascent membrane structure in selective autophagy. J. Cell Biol. 2006, 173, 709–717.
32. Bhaskara, R.M.; Grumati, P.; Garcia-Pardo, J.; Kalayil, S.; Covarrubias-Pinto, A.; Chen, W.; Kudryashev, M.; Dikic, I.; Hummer, G. Curvature

induction and membrane remodeling by FAM134B reticulon homology domain assist selective ER-phagy. Nat. Commun. 2019, 10, 2370.

33. Vevea, J.D.; Garcia, E.J.; Chan, R.B.; Zhou, B.; Schultz, M.; Di Paolo, G.; McCaffery, J.M.; Pon, L.A. Role for Lipid Droplet Biogenesis and Microlipophagy in Adaptation to Lipid Imbalance in Yeast. Dev. Cell 2015, 35, 584–599.
34. Babst, M.; Katzmann, D.J.; Snyder, W.B.; Wendland, B.; Emr, S.D. Endosome-associated complex, ESCRT-II, recruits transport machinery for protein sorting at the multivesicular body. Dev. Cell 2002, 3, 283–289.
35. Webster, B.M.; Colombi, P.; Jager, J.; Lusk, C.P. Surveillance of nuclear pore complex assembly by ESCRT-III/Vps4. Cell 2014, 159, 388–401.
36. Olmos, Y.; Hodgson, L.; Mantell, J.; Verkade, P.; Carlton, J.G. ESCRT-III controls nuclear envelope reformation. Nature 2015, 522, 236–239.
37. Raab, M.; Gentili, M.; de Belly, H.; Thiam, H.R.; Vargas, P.; Jimenez, A.J.; Lautenschlaeger, F.; Voituriez, R.; Lennon-Dumenil, A.M.; Manel, N.; et al. ESCRT III repairs nuclear envelope ruptures during cell migration to limit DNA damage and cell death. Science 2016, 352, 359–362.
38. Filimonenko, M.; Stuffers, S.; Raiborg, C.; Yamamoto, A.; Malerod, L.; Fisher, E.M.; Isaacs, A.; Brech, A.; Stenmark, H.; Simonsen, A. Functional multivesicular bodies are required for autophagic clearance of protein aggregates associated with neurodegenerative disease. J. Cell Biol. 2007, 179, 485–500.
39. Zhou, F.; Wu, Z.; Zhao, M.; Murtazina, R.; Cai, J.; Zhang, A.; Li, R.; Sun, D.; Li, W.; Zhao, L.; et al. Rab5-dependent autophagosome closure by ESCRT. J. Cell Biol. 2019, 218, 1908–1927.
40. Takahashi, Y.; Liang, X.; Hattori, T.; Tang, Z.; He, H.; Chen, H.; Liu, X.; Abraham, T.; Imamura-Kawasawa, Y.; Buchkovich, N.J.; et al. VPS37A directs ESCRT recruitment for phagophore closure. J. Cell Biol. 2019, 218, 3336–3354.
41. Loi, M.; Raimondi, A.; Morone, D.; Molinari, M. ESCRT-III-driven piecemeal micro-ER-phagy remodels the ER during recovery from ER stress. Nat. Commun. 2019, 10, 5058.
42. Changou, C.A.; Chen, Y.R.; Xing, L.; Yen, Y.; Chuang, F.Y.; Cheng, R.H.; Bold, R.J.; Ann, D.K.; Kung, H.J. Arginine starvation-associated

atypical cellular death involves mitochondrial dysfunction, nuclear DNA leakage, and chromatin autophagy. Proc. Natl. Acad. Sci. USA 2014, 111, 14147–14152.

43. Millen, J.I.; Pierson, J.; Kvam, E.; Olsen, L.J.; Goldfarb, D.S. The luminal N-terminus of yeast Nvj1 is an inner nuclear membrane anchor. Traffic 2008, 9, 1653–1664.
44. Dou, Z.; Xu, C.; Donahue, G.; Shimi, T.; Pan, J.A.; Zhu, J.; Ivanov, A.; Capell, B.C.; Drake, A.M.; Shah, P.P.; et al. Autophagy mediates degradation of nuclear lamina. Nature 2015, 527, 105–109.
45. Huang, R.; Xu, Y.; Wan, W.; Shou, X.; Qian, J.; You, Z.; Liu, B.; Chang, C.; Zhou, T.; Lippincott-Schwartz, J.; et al. Deacetylation of nuclear LC3 drives autophagy initiation under starvation. Mol. Cell 2015, 57, 456–466.
46. Ivanov, A.; Pawlikowski, J.; Manoharan, I.; van Tuyn, J.; Nelson, D.M.; Rai, T.S.; Shah, P.P.; Hewitt, G.; Korolchuk, V.I.; Passos, J.F.; et al. Lysosome-mediated processing of chromatin in senescence. J. Cell Biol. 2013, 202, 129–143.
47. Lu, X.; Djabali, K. Autophagic Removal of Farnesylated Carboxy-Terminal Lamin Peptides. Cells 2018, 7, 33.
48. Meszaros, N.; Cibulka, J.; Mendiburo, M.J.; Romanauska, A.; Schneider, M.; Kohler, A. Nuclear pore basket proteins are tethered to the nuclear envelope and can regulate membrane curvature. Dev. Cell 2015, 33, 285–298.
49. Rempel, I.L.; Crane, M.M.; Thaller, D.J.; Mishra, A.; Jansen, D.P.; Janssens, G.; Popken, P.; Aksit, A.; Kaeberlein, M.; van der Giessen, E.; et al. Age-dependent deterioration of nuclear pore assembly in mitotic cells decreases transport dynamics. eLife 2019, 8.
50. Chou, C.C.; Zhang, Y.; Umoh, M.E.; Vaughan, S.W.; Lorenzini, I.; Liu, F.; Sayegh, M.; Donlin-Asp, P.G.; Chen, Y.H.; Duong, D.M.; et al. TDP-43 pathology disrupts nuclear pore complexes and nucleocytoplasmic transport in ALS/FTD. Nat. Neurosci. 2018, 21, 228–239.
51. Rodriguez-Bravo, V.; Pippa, R.; Song, W.M.; Carceles-Cordon, M.; Dominguez-Andres, A.; Fujiwara, N.; Woo, J.; Koh, A.P.; Ertel, A.; Lokareddy, R.K.; et al. Nuclear Pores Promote Lethal Prostate Cancer by Increasing POM121-Driven E2F1, MYC, and AR Nuclear Import. Cell 2018, 174, 1200–1215.e20.

52. Tiku, V.; Jain, C.; Raz, Y.; Nakamura, S.; Heestand, B.; Liu, W.; Spath, M.; Suchiman, H.E.D.; Muller, R.U.; Slagboom, P.E.; et al. Small nucleoli are a cellular hallmark of longevity. Nat. Commun. 2017, 8, 16083.
53. Lee, L.W.; Lee, C.C.; Huang, C.R.; Lo, S.J. The nucleolus of Caenorhabditis elegans. J. Biomed. Biotechnol. 2012, 2012, 601274.
54. Yi, Y.H.; Ma, T.H.; Lee, L.W.; Chiou, P.T.; Chen, P.H.; Lee, C.M.; Chu, Y.D.; Yu, H.; Hsiung, K.C.; Tsai, Y.T.; et al. A Genetic Cascade of let-7-ncl-1-fib-1 Modulates Nucleolar Size and rRNA Pool in Caenorhabditis elegans. PLoS Genet. 2015, 11, e1005580.
55. Tsang, C.K.; Bertram, P.G.; Ai, W.; Drenan, R.; Zheng, X.F. Chromatin-mediated regulation of nucleolar structure and RNA Pol I localization by TOR. EMBO J. 2003, 22, 6045–6056.
56. Tsang, C.K.; Li, H.; Zheng, X.S. Nutrient starvation promotes condensin loading to maintain rDNA stability. EMBO J. 2007, 26, 448–458.
57. Morshed, S.; Mochida, T.; Shibata, R.; Ito, K.; Mostofa, M.G.; Rahman, M.A.; Ushimaru, T. Def1 mediates the degradation of excess nucleolar protein Nop1 in budding yeast. Biochem. Biophys. Res. Commun. 2019, 519, 302–308.
58. Cao, K.; Graziotto, J.J.; Blair, C.D.; Mazzulli, J.R.; Erdos, M.R.; Krainc, D.; Collins, F.S. Rapamycin reverses cellular phenotypes and enhances mutant protein clearance in Hutchinson-Gilford progeria syndrome cells. Sci. Transl. Med. 2011, 3, 89ra58.
59. Mostofa, M.G.; Rahman, M.A.; Koike, N.; Yeasmin, A.M.; Islam, N.; Waliullah, T.M.; Hosoyamada, S.; Shimobayashi, M.; Kobayashi, T.; Hall, M.N.; et al. CLIP and cohibin separate rDNA from nucleolar proteins destined for degradation by nucleophagy. J. Cell Biol. 2018, 217, 2675–2690.
60. Mostofa, M.G.; Morshed, S.; Shibata, R.; Takeichi, Y.; Rahman, M.A.; Hosoyamada, S.; Kobayashi, T.; Ushimaru, T. rDNA Condensation Promotes rDNA Separation from Nucleolar Proteins Degraded for Nucleophagy after TORC1 Inactivation. Cell Rep. 2019, 28, 3423–3434.e2.
61. Mekhail, K.; Seebacher, J.; Gygi, S.P.; Moazed, D. Role for perinuclear chromosome tethering in maintenance of genome stability. Nature 2008, 456, 667–670.

62. Eapen, V.V.; Waterman, D.P.; Bernard, A.; Schiffmann, N.; Sayas, E.; Kamber, R.; Lemos, B.; Memisoglu, G.; Ang, J.; Mazella, A.; et al. A pathway of targeted autophagy is induced by DNA damage in budding yeast. Proc. Natl. Acad. Sci. USA 2017, 114, E1158–E1167.
63. Rello-Varona, S.; Lissa, D.; Shen, S.; Niso-Santano, M.; Senovilla, L.; Marino, G.; Vitale, I.; Jemaa, M.; Harper, F.; Pierron, G.; et al. Autophagic removal of micronuclei. Cell Cycle 2012, 11, 170–176.
64. Jeong, H.; Park, J.; Kim, H.I.; Lee, M.; Ko, Y.J.; Lee, S.; Jun, Y.; Lee, C. Mechanistic insight into the nucleus-vacuole junction based on the Vac8p-Nvj1p crystal structure. Proc. Natl. Acad. Sci. USA 2017, 114, E4539–E4548.
65. Honscher, C.; Ungermann, C. A close-up view of membrane contact sites between the endoplasmic reticulum and the endolysosomal system: From yeast to man. Crit. Rev. Biochem. Mol. Biol. 2014, 49, 262–268.
66. Eisenberg-Bord, M.; Shai, N.; Schuldiner, M.; Bohnert, M. A Tether Is a Tether Is a Tether: Tethering at Membrane Contact Sites. Dev. Cell 2016, 39, 395–409.
67. Park, J.; Kim, H.I.; Jeong, H.; Lee, M.; Jang, S.H.; Yoon, S.Y.; Kim, H.; Park, Z.Y.; Jun, Y.; Lee, C. Quaternary structures of Vac8 differentially regulate the Cvt and PMN pathways. Autophagy 2019, 6, 991–1006.
68. Scott, S.V.; Nice, D.C., 3rd; Nau, J.J.; Weisman, L.S.; Kamada, Y.; Keizer-Gunnink, I.; Funakoshi, T.; Veenhuis, M.; Ohsumi, Y.; Klionsky, D.J. Apg13p and Vac8p are part of a complex of phosphoproteins that are required for cytoplasm to vacuole targeting. J. Biol. Chem. 2000, 275, 25840–25849.
69. Hollenstein, D.M.; Gomez-Sanchez, R.; Ciftci, A.; Kriegenburg, F.; Mari, M.; Torggler, R.; Licheva, M.; Reggiori, F.; Kraft, C. Vac8 spatially confines autophagosome formation at the vacuole in S. cerevisiae. J. Cell Sci. 2019, 132.
70. Boutouja, F.; Stiehm, C.M.; Reidick, C.; Mastalski, T.; Brinkmeier, R.; Magraoui, F.E.; Platta, H.W. Vac8 Controls Vacuolar Membrane Dynamics during Different Autophagy Pathways in Saccharomyces cerevisiae. Cells 2019, 8, 661.
71. Gatica, D.; Damasio, A.; Pascual, C.; Klionsky, D.J.; Ragusa, M.J.; Popelka, H. The carboxy terminus of yeast Atg13 binds phospholipid

membrane via motifs that overlap with the Vac8-interacting domain. Autophagy 2020, 16, 1007–1020.

72. Munzel, L.; Neumann, P.; Otto, F.B.; Krick, R.; Metje-Sprink, J.; Kroppen, B.; Karedla, N.; Enderlein, J.; Meinecke, M.; Ficner, R.; et al. Atg21 organizes Atg8 lipidation at the contact of the vacuole with the phagophore. Autophagy 2020, 1–21.
73. Levine, T.P.; Munro, S. Dual targeting of Osh1p, a yeast homologue of oxysterol-binding protein, to both the Golgi and the nucleus-vacuole junction. Mol. Biol. Cell 2001, 12, 1633–1644.
74. Kvam, E.; Gable, K.; Dunn, T.M.; Goldfarb, D.S. Targeting of Tsc13p to nucleus-vacuole junctions: A role for very-long-chain fatty acids in the biogenesis of microautophagic vesicles. Mol. Biol. Cell 2005, 16, 3987–3998.
75. Henne, W.M.; Zhu, L.; Balogi, Z.; Stefan, C.; Pleiss, J.A.; Emr, S.D. Mdm1/Snx13 is a novel ER-endolysosomal interorganelle tethering protein. J. Cell Biol. 2015, 210, 541–551.
76. Hariri, H.; Rogers, S.; Ugrankar, R.; Liu, Y.L.; Feathers, J.R.; Henne, W.M. Lipid droplet biogenesis is spatially coordinated at ER-vacuole contacts under nutritional stress. EMBO Rep. 2018, 19, 57–72.
77. Thomas, A.C.; Williams, H.; Seto-Salvia, N.; Bacchelli, C.; Jenkins, D.; O'Sullivan, M.; Mengrelis, K.; Ishida, M.; Ocaka, L.; Chanudet, E.; et al. Mutations in SNX14 cause a distinctive autosomal-recessive cerebellar ataxia and intellectual disability syndrome. Am. J. Hum. Genet. 2014, 95, 611–621.
78. Eisenberg-Bord, M.; Mari, M.; Weill, U.; Rosenfeld-Gur, E.; Moldavski, O.; Castro, I.G.; Soni, K.G.; Harpaz, N.; Levine, T.P.; Futerman, A.H.; et al. Identification of seipin-linked factors that act as determinants of a lipid droplet subpopulation. J. Cell Biol. 2018, 217, 269–282.
79. Hariri, H.; Speer, N.; Bowerman, J.; Rogers, S.; Fu, G.; Reetz, E.; Datta, S.; Feathers, J.R.; Ugrankar, R.; Nicastro, D.; et al. Mdm1 maintains endoplasmic reticulum homeostasis by spatially regulating lipid droplet biogenesis. J. Cell Biol. 2019, 218, 1319–1334.

Index

D

E

F

G

M

N

O

P

R

S